81 6503895 7
TELEPEN
DILEWYD O STOC
WITHDRAWN FROM STOCK
LLYFRGELL
LIBRARY

AF598698

LIQUID SCINTILLATION COUNTING

Volume 4

LIQUID SCINTILLATION COUNTING

Volume 4

Proceedings of a Symposium on
Liquid Scintillation Counting
organised by
Radiochemical Methods Group
(Analytical Division, The Chemical Society)
Bath, England
September 16–19 1975

Editors:

M. A. Crook (Polytechnic of the South Bank)
P. Johnson (Hoechst Pharmaceutical Research Laboratories)

London · New York · Rheine

Heyden & Son Ltd., Spectrum House, Alderton Crescent, London NW4 3XX
Heyden & Son Inc., 225 Park Avenue, New York, N.Y. 10017, U.S.A.
Heyden & Son GmbH, Münsterstrasse 22, 4440 Rheine/Westf., Germany

Library of Congress Catalog Card No. 70-156826

ISBN 0 85501 209 9

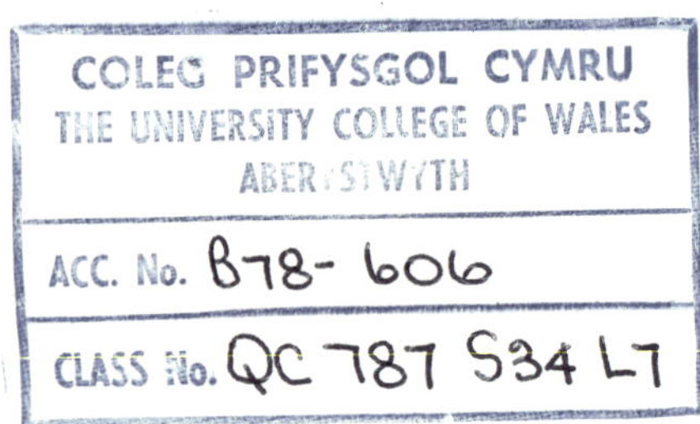

Printed in Great Britain by J.W. Arrowsmith Ltd, Bristol BS3 2NT.

Contents

Preface

This volume is the fourth in the series on Liquid Scintillation Counting and records the proceedings of the International Symposium held in Bath, England, on September 16th–19th, 1975, organised by the Radiochemical Methods Group (Analytical Division, The Chemical Society). Invited plenary Lectures were given by Dr. J.B. Birks, Dr. B.W. Fox, Dr. P. Tothill, Dr. J.E. Noakes and Dr. L.A. Currie, with the Chairmen of the five sessions being Dr. G.W.A. Newton, Dr. D. Bowyer, Dr. R. Parker, Mr. R. Burleigh and Dr. A. Dyer. In addition to their support, the undoubted success of the meeting was assured by the provision of sixteen submitted papers, and the attendance of 250 delegates from eighteen countries.

The tendency to begin to refer to these Symposia as the 'Brighton' meetings was broken by the choice of Bath as the new venue, a decision which was not regretted scientifically, culturally or socially, and if delegates are to be congratulated on the first of these, the Mayor and Council of the City of Bath equally deserve our thanks in respect of the second and third.

As is now customary an excellent exhibition was mounted by various manufacturers:

Anachem
Beckman R.I.I.C.
Berthold Frieseke
E.S.I. Nuclear
Fisons Scientific Apparatus
G.E.C. Medical
I.C.N. Tracerlab
Intertechnique
Koch Light
L.K.B. Instruments
G.D. Searle & Co. (Nuclear Chicago)
N.E.N.
Nuclear Enterprises
Packard Instruments
Pye Unicam
Radiochemical Centre
Hughes Whitlock

whose representatives on the organising committee, Messrs. D.A. Ginger, B.R. Lumb and M. Moseley, lent strong support.

The trend noted in the Preface to Volume 3, of diminishing scope for progress in the field of Liquid Scintillation Counting, has continued. Future Symposia will undoubtedly be held, but it is suggested that the time has come for discussing wider aspects and applications of the Scintillation Counting process and to narrow the discussion by the use of the term 'Liquid' may no longer be possible.

September 1976

M.A. Crook
P. Johnson

SECTION I
SCINTILLATION PROCESSES

Chapter 1

Impurity Quenching of Organic Liquid Scintillators

J. B. Birks

The Schuster Laboratory, University of Manchester, Manchester, England

INTRODUCTION

For the English-speaking world 1975 is the silver jubilee of liquid scintillation counting, although the Italians could have celebrated the event last year. The 1950 issues of *Physical Review* included three papers, by Reynolds *et al.*[1] from Princeton, Kallmann[2] from New York and Ageno *et al.*[3] from Italy, each claiming the first discovery of scintillations from organic liquid solutions. These were the first reports of liquid scintillators published in English, but priority belongs to Ageno *et al.*[4] who also published their results in a small Italian journal in 1949.

Organic scintillation counting originated in 1947 when Kallmann[5] discovered that crystalline naphthalene, grown from moth-balls garnered from the chemists' shops of war-ravaged Berlin, and irradiated with β-particles or γ-rays, emitted scintillations which could be detected and converted into electrical pulses with a photomultiplier. In 1948, Bell[6] found crystalline anthracene to have a much higher scintillation efficiency than crystalline naphthalene, and Collins[7] measured its scintillation lifetime to be about 10 ns. This is much less than that of impurity-activated zinc sulphide, the visual scintillator which dominated the first 25 years of experimental nuclear physics, or of thallium-activated sodium iodide, introduced by Hofstadter[8] in 1948 and since used extensively for γ-ray detection and spectroscopy. It was thus shown that organic scintillators offer major advantages in time resolution and counting rates compared with inorganic scintillators.

As mentioned previously, organic liquid solution scintillators were discovered in Italy in 1949[3,4] and in the United States in 1950.[1,2] Two other types of organic solution scintillator were also discovered in 1950. Schorr and Torney[9] introduced plastic solution scintillators, destined to have a major impact in high-energy physics. Birks[10] discovered organic crystal solution scintillators (mixed crystals), but these have found little practical application compared with their liquid and plastic solution counterparts. The first paper on the organic scintillation process was also published in 1950, in which Birks[11] proposed a relation between the specific scintillation efficiency of an anthracene crystal and the specific energy loss of an ionising particle, a relation which later proved applicable to all types of organic scintillator. Nineteen seventy-five is thus a personal silver jubilee. My interest in the subject started in 1948, however, so that the Sydney symposium in 1973 was chosen as the occasion for personal argento-jubilation.[12] Participation in the early stages of any new branch of science offers several subsequent advantages. One is strategically placed to write books on the subject to inform newcomers to the field.[13,14] One is invited to participate in meetings in many interesting places.

At the 1973 Sydney symposium[15] I drew an analogy between a liquid scintillation counting symposium and a theological assembly: so the choice of topic here is between two opposing themes: good or evil, virtue or vice, purity or impurity. Those interested in the fundamental physical processes in liquid scintillators usually study them under conditions of extreme purity, even excluding the oxygen dissolved in the solvent.

Those concerned with the utilisation of internal liquid scintillation counting for radioassay do not usually operate under such high-purity conditions. The addition of the radioactive specimen and any associated incorporating agents to the scintillator introduces impurity quenching which reduces the scintillation efficiency of the system. Since most of the present audience are concerned with the practical world of the impure scintillator, and not with the monastic world of the pure scintillator, the present talk will deal with the physics of impurity quenching, commonly but incorrectly known as chemical quenching.

The theme will thus be impurity rather than purity, pollution rather than perfection. The theological overtone is thus to consider how one can reduce the harmful effects of impurity in order to see more clearly the inner light.

THE SCINTILLATION PROCESS

The following sequence of processes in a liquid scintillator solution leads to the prompt scintillation emission.[14] Processes involving thermal dissipation of the excitation energy by internal quenching, or the formation and internal conversion of excited triplet states leading to delayed scintillation emission via triplet–triplet interaction, are omitted to simplify the discussion.

(i) The ionising particle excites solvent molecules (^1X) into higher excited singlet ($^1X^{**}$) states.

(ii) The particle also ionises solvent molecules yielding molecular ions ($^2X^+$) and electrons ($^2e^-$).

(iii) $^2X^+$ and $^2e^-$ recombine to yield $^1X^{**}$.

(iv) $^1X^{**}$ undergoes internal conversion to the lowest excited singlet ($^1X^*$) state of the solvent molecule.

(v) $^1X^*$ transfers its energy radiationlessly, by a process involving molecular diffusion, excimer ($^1D^*$) formation and dissociation, and excitation migration, to a primary solute molecule (^1Y), thereby exciting it into its lowest excited singlet ($^1Y^*$) state.

(vi) The $^1Y^*$ excitation energy is either transferred radiationlessly to a secondary solute molecule (^1Z), thereby exciting it into its lowest excited singlet ($^1Z^*$) state, or it is emitted as a fluorescence photon, which may be absorbed by ^1Z, so that the $^1Y^*$ energy is transferred radiatively to ^1Z yielding $^1Z^*$.

(vii) The $^1Z^*$ fluorescence (and any unabsorbed fluorescence of $^1Y^*$) constitutes the scintillation emission of the solution.

The introduction of an impurity molecular species Q into the pure scintillator solution (^1X, ^1Y, ^1Z) reduces the scintillation efficiency by interfering, through competitive processes, at various stages in the above sequence. Very fast processes are unaffected by Q. Thus the excitation and ionisation processes (i) and (ii) which occur in about 10^{-15} s, and the internal conversion process (iv) which occurs in about 10^{-12} s, are immune to impurity quenching. The processes which are susceptible to impurity quenching are those that occur on the 10^{-10}–10^{-9} s time scale, namely (iii) solvent ion recombination, (v) solvent ($^1X^*$) excitation, (vi) primary solute ($^1Y^*$) excitation, and (vii) secondary solute ($^1Z^*$) excitation.

ELECTRON CAPTURE QUENCHING

Because of its high mobility, an electron ($^2e^-$) produced by the primary ionisation process (ii), is susceptible to capture by an impurity species Q, particularly if it has sufficient energy to be removed from the Coulombic field of the associated molecular ion ($^2X^+$). Such electron capture competes with ion recombination (iii) and thus reduces the yield of $^1X^{**}$ (and hence the subsequent yields of $^1X^*$, $^1Y^*$ and $^1Z^*$).

The electron capture cross-section of Q increases with its electron affinity A. Molecules with relatively high values of A include chlorine, bromine, iodine, iodine chloride, *p*-chloranil, *p*-bromanil, *p*-iodanil, *p*-benzoquinone, trinitrobenzene and tetracyanoethylene.[16] Oxygen, which is a very efficient excitation quencher, has a relatively low value of A. Blaunstein and Christophorou[17] have compiled a list of the electron affinities of 198 molecules which may be used to assess the probability of electron capture quenching by a particular molecular species Q.

In most liquid scintillator solutions, excitation ($^1X^*$, $^1Y^*$, $^1Z^*$) quenching appears to be more important than electron capture quenching.

EXCITATION QUENCHING

The fluorescence quantum yield q_{FM} of an excited molecule $^1M^*$ in a 'pure' solution, i.e. in the absence of Q, is

$$q_{FM} = \frac{k_{FM}}{k_{FM} + k_{NM}} = \frac{k_{FM}}{k_M} \qquad (1)$$

where k_{FM} is the radiative (fluorescence) rate parameter and k_{NM} is the total radiationless rate parameter. k_{NM} is the sum of the rates of all radiationless processes operating on $^1M^*$, including internal quenching (rate k_{IM}) due to intersystem crossing and internal conversion,[16] oxygen quenching (rate k_{OM} $[O_2]$, where $[O_2]$ is the molar concentration of dissolved oxygen), and radiationless energy transfer (rate k_{BM} $[^1B]$) to another molecular species, molar concentration $[^1B]$, in the solution. k_M is the total $^1M^*$ decay rate and $\tau_M (= 1/k_M)$ is the $^1M^*$ fluorescence (excitation) lifetime.

The introduction of a molar concentration [Q] of a quenching species Q reduces the fluorescence quantum yield to

$$\phi_{FM} = \frac{k_{FM}}{k_M + k_{QM}[Q]}$$

$$= \frac{q_{FM}}{1 + \tau_M k_{QM}[Q]} = \frac{q_{FM}}{1 + K_{QM}[Q]} \qquad (2)$$

where k_{QM} (units M^{-1} s^{-1}) is the bimolecular rate parameter of impurity quenching of $^1M^*$ by Q. The parameter K_{QM} (units M^{-1}) is the Stern–Volmer coefficient of impurity quenching. Its reciprocal $[Q]_{0.5}$ ($= 1/K_{QM}$) is the half-value quencher concentration at which $\phi_{FM} = 0.5q_{FM}$.

The impurity quenching rate parameter k_{QM} is given by the relation

$$k_{QM} = \frac{4\pi NpDR'}{10^3}\left(1 + \frac{pR'}{(\pi Dt)^{1/2}}\right) \qquad (3)$$

obtained from Einstein-Smolochowski diffusion theory,[16] where N is Avogadro's number, D ($= D_M + D_Q$) is the sum of the diffusion coefficients of the reactant species, R' ($= R'_M + R'_Q$) is the sum of their interaction radii, t is time and p ($\leqslant 1$) is the quenching probability per molecular collision. The term in parentheses is the so-called transient term. Except at very short times, it is of the order of unity and it can usually be disregarded, so that Eqn. (3) simplifies to

$$k_{QM} = \frac{4\pi NpDR'}{10^3} \qquad (4)$$

The Stern–Volmer coefficient of impurity quenching of $^1M^*$ by Q may be written as

$$K_{QM} = \tau_M k_{QM} = a\tau_M DR'p \qquad (5)$$

where a ($= 4\pi N \times 10^{-3}$) is constant.

The factors determining τ_M, D and R' are known. We shall now discuss the factors determining p, the probability of quenching during a molecular collision between $^1M^*$ and Q. Such a collision usually results in the formation (rate $k_{EM}[Q]$) of an excited complex or *exciplex* $^1(M.Q)^*$,[16] which undergoes internal quenching (rate k_E) or dissociation (rate k_{ME}) into $^1M^*$ and Q as follows:

$$^1M^* + Q \underset{k_{ME}}{\overset{k_{EM}[Q]}{\rightleftharpoons}} {}^1(M.Q)^* \xrightarrow{k_E} \qquad (6)$$

The quenching rate parameter k_{QM} is given by

$$k_{QM} = k_{EM} \frac{k_E}{k_E + k_{ME}} = pk_{EM} \tag{7}$$

where k_{EM} is the diffusion-controlled rate parameter, given by Eqn. (4) with $p = 1$,

$$k_{EM} = \frac{4\pi \mathrm{N} D R'}{10^3} \tag{4a}$$

Thus p is given by

$$p = \frac{k_E}{k_E + k_{ME}} \tag{8}$$

p is large when k_E is large, i.e. strong internal quenching of $^1(M.Q)^*$, or when k_{ME} is small, i.e. low $^1(M.Q)^*$ dissociation rate.

k_E is large when Q contains heavy atoms of high atomic number. These increase the spin-orbit coupling in the complex (the heavy-atom effect) and thus enhance the rate of intersystem crossing from the singlet exciplex $^1(M.Q)^*$ to the triplet exciplex $^3(M.Q)^*$.[16] This explains the strong fluorescence quenching ($p = 1$) of scintillator solvents and solutes by carbon tetrabromide.[18]

The magnitude of k_{ME} depends on the exciplex thermodynamics.[16] The molar equilibrium constant of Eqn. (6) is given by

$$\frac{k_{EM}}{k_{ME}} = \exp(\Delta S/\mathbf{R}) \exp(-\Delta H/\mathbf{R}T) \tag{9}$$

so that

$$k_{ME} = k_{EM} \exp(-\Delta S/\mathbf{R}) \exp(\Delta H/\mathbf{R}T) \tag{9a}$$

where ΔH and ΔS are the enthalpy and entropy of the exciplex, **R** is the gas constant and T is the absolute temperature. The parameter B $(= -\Delta H/\mathbf{N})$ is the exciplex binding energy. k_{ME} is proportional to k_{EM} (Eqn. (4a)), so that it increases with the diffusion coefficient D. k_{ME} also increases with increase in temperature T, and with decrease in the exciplex binding energy B. k_{ME} is large, and hence p tends to be small (Eqn. (8)), in a system of high diffusion coefficient D (i.e. low viscosity), high temperature T and low exciplex binding energy B. The magnitude of B is determined mainly by the ionisation potential I and electron affinity A of the exciplex constituents. The theory of exciplexes has been developed to an advanced state,[19] but no simple rules have been formulated for the evaluation of B or p for a given pair of molecules. Two empirical approaches[21,22] to the problem will be considered later.

PRACTICAL ASPECTS OF QUENCHING

The magnitude of the quenching of a singlet-excited molecule $^1M^*$ by Q is, from Eqn. (2),

$$\frac{q_{FM}}{\phi_{FM}} = 1 + K_{QM}[Q] \tag{10}$$

where

$$K_{QM} = a\tau_M DR'p \tag{5}$$

Thus the quenching depends on [Q], τ_M, D, R' and p.

There are several alternative methods by which the quenching of a liquid scintillator can be reduced.

Oxygen removal

The dissolved oxygen in a liquid scintillator, which introduces up to 20% quenching, can be eliminated by nitrogen bubbling or otherwise. Operation at a lower temperature, where the oxygen solubility in the solvent is reduced, is an alternative method of reducing oxygen quenching.

Decreased specimen concentration

The specimen concentration which determines [Q] can sometimes be reduced. The disadvantage of an increased counting time is offset by the smaller quench correction required.

Reduced diffusion coefficient

The diffusion coefficient D can be reduced by operating at low temperatures or by using a viscous solvent (e.g. methylnaphthalene) or a viscous emulsion. The associated decrease in k_{QM} may be offset by an increase in p.

Increased primary solute concentration

The solvent ($^1X^*$) lifetime is given by

$$\tau_X = \left(k_{FX} + k_{IX} + k_{YX}\,[^1Y]\right)^{-1} \tag{11}$$

$$= \left(k_{FX} + k_{IX}\right) \text{ for } [^1Y] = 0 \tag{12}$$

$$\cong 1/k_{YX}\,[^1Y] \text{ for } [^1Y] \text{ large} \tag{13}$$

In the absence of the primary solute (Eqn. (12)) $\tau_X \cong 30$ ns, but in the scintillator solution this is reduced considerably due to radiationless energy transfer (rate parameter k_{YX}) to the primary solute, molar concentration $[^1Y]$. For toluene solutions at room temperature, $k_{YX} = 5.6 \times 10^{10}\ M^{-1}\ s^{-1}$,[16] so that for $[^1Y] = 5 \times 10^{-2}$ M, $\tau_X = 0.36$ ns. An increase in the primary solute concentration $[^1Y]$ thus provides a simple method of reducing τ_X (Eqn. (13)) and hence reducing the $^1X^*$ Stern-Volmėr quenching coefficient K_{QX} (cf. Eqn. (5)).

Choice of primary solute

The primary solute ($^1Y^*$) lifetime is

$$\tau_Y = \left(k_{FY} + k_{IY} + k_{ZY} \, [^1Z] \right)^{-1} \tag{14}$$

where k_{ZY} is the rate parameter of radiationless energy transfer to the secondary solute, molar concentration $[^1Z]$. In the absence of $[^1Z]$, the experimental values of τ_Y for various primary solutes, mainly in toluene solution, are as follows:[16]

diphenylstilbene	1.1 ns
BBOT	1.1 ns
p-terphenyl	1.2 ns
BBO	1.2 ns
PBD	1.3 ns
butyl PBD	1.3 ns
PPO	1.8 ns
a-NPO	2.0 ns
a-NPD	2.0 ns

The $^1Y^*$ quenching coefficient K_{QY} (Eqn. (5)) can be minimised by the choice of a primary solute with a low fluorescence lifetime τ_Y.

Increased secondary solute concentration

In a typical scintillator, τ_Y is reduced to about 0.5–0.75 of its initial value $(k_{FY} + k_{IY})^{-1}$ by radiationless energy transfer to ^1Z. τ_Y and the $^1Y^*$ quenching coefficient K_{QY} can be further reduced by increasing the secondary solute molar concentration $[^1Z]$.

Omission of secondary solute

The secondary solute ($^1Z^*$) lifetime is

$$\tau_Z = \left(k_{FZ} + k_{IZ} \right)^{-1} \tag{15}$$

and it is unaffected by energy transfer. τ_Z (POPOP 1.3 ns, dimethyl POPOP 1.5 ns, *a*-NPO 2.0 ns[16]) is longer than τ_X (~ 0.4 ns) or τ_Y (~ 0.7 ns) in a typical liquid scintillator solution. This makes $^1Z^*$ particularly susceptible to impurity quenching. The omission of the secondary solute is the simplest method to eliminate this quenching. This can usually be done without loss in scintillation efficiency, provided there is no appreciable colour quenching present in the solution. Although the omission of ^1Z makes the primary solute excitation $^1Y^*$ more vulnerable to quenching because of its increased lifetime in the absence of $[^1Z]$ (Eqn. (14)), the elimination of one step in the scintillation process reduces the overall quenching. As noted previously,[20] in the absence of colour quenching, secondary solutes are redundant with modern bialkali cathode photomultiplier tubes with maximum spectral response at 380 nm wavelength, provided primary solutes like PBD and butyl PBD whose fluorescence emission matches this response are used in association with vials which do not attenuate the scintillation emission.

Chemical treatment of specimen

Kerr *et al*[21] investigated the quenching of three standard solutions (4 g l^{-1} PPO and 0.1 g l^{-1} POPOP in toluene; 7 g l^{-1} PPO, 0.05 g l^{-1} POPOP and 50 g l^{-1} naphthalene in *p*-dioxan; and 8 g l^{-1} *p*-terphenyl in toluene) by the addition of seventy different organic compounds. They found that aliphatic compounds, as a group, quench less than aromatic compounds. They classified the degree of quenching by aliphatic compounds in terms of functional groups as follows:

Diluters (weak quenchers): R – H, R – F, R – O – R, $(RO)_3$ PO, R – CN, R – OH, R – COO – R, R – Cl

Mild quenchers: R – COOH, R – NH_2, R – CH = CH – R, R – Br, R – S – R

Strong quenchers: R – SH, R – OCOCO – R, R – CO – R, R – COX, R – NH – R, R – CHO, R_2N – R, R – I, R – NO_2

In the light of subsequent experience,[18] the bromides (R – Br) should be reclassified as strong, rather than mild, quenchers.

Chemical treatment of the specimen prior to its incorporation into the scintillator, aimed at the elimination of functional groups which act as strong quenchers, represents a further approach to the reduction of impurity quenching.

Choice of scintillator solution

Extensive tables[14] and graphs[22] have been published comparing the relative scintillation efficiencies of numerous liquid scintillator solutions, under oxygen-free or air-equilibrated conditions. These data are commonly used in the selection of a liquid scintillator. This can be misleading, since liquid scintillators differ in their susceptibility to impurity quenching.

Birks and Poullis[22] have studied the quenching of a series of air-equilibrated binary liquid solution scintillators by carbon tetrachloride. Table 1 lists the values of the half-value quencher concentration $[Q]_{0.5}$, the reciprocal of the Stern–Volmer quenching coefficient. A low value of $[Q]_{0.5}$ corresponds to strong quenching and a high value to weak quenching. The solvents and solutes are arranged in horizontal and vertical columns in order of decreasing $[Q]_{0.5}$, i. e. of increasing quenching susceptibility, in Table 1 and there are only minor deviations from the general pattern. The quenching susceptibility of the alkyl benzene solvents increases with increasing molecular weight and with decreasing ionisation potential, a result to be expected theoretically. *p*-Dioxan containing 100 g l^{-1} naphthalene is more resistant to quenching than any of the solvents except benzene. In all solvents, PBD is the solute most resistant to quenching. Butyl PBD, PBO and BBOT have a reasonable quenching resistance, but PPO, BIBUQ and *p*-terphenyl are rather strongly quenched.

The data are presented in a different form in Table 2, which lists the relative scintillation pulse heights of the solutions in the absence (V_0) and presence (V_Q, V'_Q) of 0.1 M carbon tetrachloride. V_0 and V_Q are expressed relative to V_0= 100 for 10 g l^{-1} PBD in toluene. V'_Q is the value of V_Q normalised to V'_Q = 100 for 10 g l^{-1} PBD in toluene. The relative orders of merit of the values of V_0 and V'_Q are shown in parentheses in the V_0 and V'_Q columns, respectively, in Table 2.

The addition of the quencher produces some dramatic changes in the order of merit. The BIBUQ solutions in *p*-xylene and toluene drop from (1)= to (16) and (17), respectively. The BIBUQ solutions in *p*-dioxan and benzene slip from (9) and (12)=, respectively, to (19)=. Clearly BIBUQ is very susceptible to CCl_4 quenching. On

Table 1. Binary solutions. Quenching by carbon tetrachloride. Half-value quencher concentration $[Q]_{0.5}$ in units of 10^{-2} M.[22]

		Solvent					
Solute conc (g l^{-1})	Solute	Benzene	*p*-Dioxan[a]	Toluene	Xylene	*p*-Xylene	Mesitylene
10	PBD	8.17	6.44	5.79	5.06	4.83	4.50
10	butyl PBD	6.31	6.03	4.66	4.60	4.22	3.98
8	PBO	6.12	5.25	4.50	3.71	3.51	3.42
8	BBOT	5.64	7.23	3.82	3.56	3.17	3.09
6	PPO	3.96	3.25	2.78	2.64	2.72	2.44
15	BIBUQ	2.98	2.85	2.24	2.19	2.46	1.93
5	TP	2.74	–	2.22	2.12	1.91	1.97

[a]Plus 100 g l^{-1} naphthalene.

the other hand, the benzene solutions of PBD, butyl PBD and PBO climb from (15)=, (10)=, and (15)= to (2), (4)= and (7), respectively. Benzene is relatively immune to CCl_4 quenching. All the *p*-dioxan/100 g l^{-1} naphthalene solutions, except that containing BIBUQ, improve their relative positions, showing the quench resistance of this solvent. The toluene solution of PBD, which improves from (3) to (1), is the most efficient scintillator in the presence of 0.1 M carbon tetrachloride.

The results show that impurity quenching can be minimised by the proper choice of scintillator solution. Although the numerical data refer only to carbon tetrachloride (Table 1) and to a particular quencher concentration (Table 2), the results are of more general application. The quenching susceptibility of an excited solvent or solute molecule to *any quencher* depends on its inherent molecular properties (singlet excitation energy and lifetime, ionisation potential, electron affinity). It is therefore proposed that the *order* of decreasing quench resistance to CCl_4 quenching of the solvents and solutes listed in Table 1 applies also to other quenchers. Table 1, together with the data on V_0,[22] the relative scintillation efficiency of the unquenched solution, thus provides a general guide to the choice of quench-resistant scintillator solutions. Benzene, *p*-dioxan/100 g l^{-1} naphthalene, and toluene are the best quench-resistant solvents, and PBD, butyl PBD, and PBO are the best quench-resistant solutes, of the materials considered. Similar studies using other quenchers are required to test the general validity of the order of quench resistance shown in Table 1, and provide further numerical data on impurity quenching.

Micellar scintillators

A further method of reducing impurity quenching is to separate the radioactive specimen and its associated quenching component Q from the liquid scintillator solution (^1X, ^1Y, ^1Z) by the use of a micellar scintillator. Ewer and Harding[23] have developed such a micellar scintillator (available as NE 260 from Nuclear Enterprises, Ltd.) which segregates proteins, nucleotides, salts and sugars from the xylene-based liquid scintillator solution. If there is complete segregation of Q from the liquid scintillator, the impurity quenching should be eliminated, although the micelle walls absorb some of the β-particle energy and thus reduce the scintillation efficiency to some extent.

CONCLUSION

Impurity quenching in liquid scintillators can be minimised by (i) removal of dissolved oxygen, (ii) reduction of the specimen concentration, (iii) reduction of the diffusion coefficient by the use of a viscous solvent or emulsion, (iv) increase in the primary solute concentration, (v) choice of the primary solute, (vi) increase in the secondary solute concentration, (vii) omission of the secondary solute, (viii) chemical treatment of the specimen, (ix) choice of the scintillator solution (solvent and solute) and (x) the use of miscellar scintillators.

The purpose of internal liquid scintillation counting is to obtain an accurate radioassay of the specimen. The raw data available for this purpose are the photomultiplier pulses generated by the scintillator photons. The subsequent stages of data processing,

Table 2. Liquid scintillators. Relative pulse heights in the absence (V_0) and presence (V_Q, V'_Q) of 0.1 M carbon tetrachloride.[12,22]

Solute	Solvent	V_0	V_Q	V'_Q
10 g l^{-1} PBD	toluene	100 (3)	37	100 (1)
10 g l^{-1} PBD	benzene	77 (15)=	35	95 (2)
10 g l^{-1} butyl PBD	toluene	99 (4)=	31.5	86 (3)
10 g l^{-1} PBD	*p*-xylene	96 (8)	31	84 (4)=
10 g l^{-1} butyl PBD	benzene	80 (10)=	31	84 (4)=
8 g l^{-1} PBO	toluene	98 (6)	30	82 (6)
8 g l^{-1} PBO	benzene	77 (15)=	29.5	81 (7)
10 g l^{-1} butyl PBD	*p*-xylene	97 (7)	29	79 (8)
8 g l^{-1} BBOT	*p*-dioxan[a]	67 (21)	28	77 (9)
8 g l^{-1} PBO	*p*-dioxan[a]	80 (10)=	27.5	75 (10)
10 g l^{-1} butyl PBD	*p*-dioxan[a]	72 (18)=	27	74 (11)=
10 g l^{-1} PBD	*p*-dioxan[a]	70 (20)	27	74 (11)=
8 g l^{-1} PBO	*p*-xylene	99 (4)=	25.5	70 (13)
8 g l^{-1} BBOT	toluene	79 (12)=	22	60 (14)=
8 g l^{-1} BBOT	benzene	61 (24)	22	60 (14)=
15 g l^{-1} BIBUQ	*p*-xylene	103 (1)=	20.5	56 (16)
8 g l^{-1} BBOT	*p*-xylene	79 (12)=	18.5	51 (17)=
15 g l^{-1} BIBUQ	toluene	103 (1)=	18.5	51 (17)=
15 g l^{-1} BIBUQ	benzene	79 (12)=	18	49 (19)=
15 g l^{-1} BIBUQ	*p*-dioxan[a]	82 (9)	18	49 (19)=
6 g l^{-1} PPO	benzene	58 (26)	16	44 (21)=
6 g l^{-1} PPO	*p*-xylene	76 (17)	16	44 (21)=
6 g l^{-1} PPO	toluene	72 (18)=	16	44 (21)=
6 g l^{-1} PPO	*p*-dioxan[a]	59 (25)	13	35 (24)
5 g l^{-1} TP	toluene	65 (22)	11.5	32 (25)
5 g l^{-1} TP	benzene	52 (27)	11	30 (26)
5 g l^{-1} TP	*p*-xylene	64 (23)	10	28 (27)

[a]Plus 100 g l^{-1} naphthalene.

however sophisticated, cannot improve, they may even degrade, these data. The source of the data, the scintillator, is the most critical component of the system and it is the only one under the full control of the experimenter.

Quench reduction is always preferable to quench correction. Quench reduction improves the quality of the data. Quench correction accepts inferior data and attempts by various approximate methods to normalise them to unquenched conditions. The methods proposed in this paper should enable impurity quenching to be reduced to a sufficiently low level that quench correction methods can be applied more reliably to obtain an accurate radio-assay.

REFERENCES

1. G. T. Reynolds, F. B. Harrison and G. Salvani, *Phys. Rev.* **78**, 488 (1950).
2. H. Kallmann, *Phys. Rev.* **78**, 621 (1950).
3. M. Ageno, M. Chiozzotto and R. Querzoli, *Phys. Rev.* **79**, 720 (1950).
4. M. Ageno, M. Chiozzotto and R. Querzoli, *Acad. naz. Lincei* **6**, 629 (1949).
5. H. Kallmann, *Natur und Technik* (July 1947); I. Broser and H. Kallmann, *Z. Naturforsch.* **2a**, 439, 632 (1947).
6. P. R. Bell, *Phys. Rev.* **73**, 1405 (1948).
7. G. B. Collins, *Phys. Rev.* **74**, 1543 (1948).
8. R. Hofstadter, *Phys. Rev.* **74**, 100 (1948).
9. M. G. Schorr and F. L. Torney, *Phys. Rev.* **80**, 474 (1950).
10. J. B. Birks, *Proc. Phys. Soc.* **A63**, 1044 (1950).
11. J. B. Birks, *Proc. Phys. Soc.* **A63**, 1294 (1950); **A64**, 874 (1951).
12. J. B. Birks, in *Liquid Scintillation Counting. Recent Developments* (eds. P. E. Stanley and B. A. Scoggins), Academic Press, New York and London, 1974, p. 1.
13. J. B. Birks, *Scintillation Counters,* Pergamon Press, Oxford, 1953.
14. J. B. Birks, *The Theory and Practice of Scintillation Counting,* Pergamon Press, Oxford, 1964.
15. J. B. Birks, in *Liquid Scintillation Counting. Recent Developments* (eds. P. E. Stanley and B. A. Scoggins), Academic Press, New York and London, 1974, p. 477.
16. J. B. Birks, *Photophysics of Aromatic Molecules,* Wiley-Interscience, London and New York, 1970.
17. R. P. Blaunstein and L. G. Christophorou, *Radiation Res. Rev.* **3**, 69 (1971).
18. J. B. Birks, H. Y. Najjar and M. D. Lumb, *J. Phys. B (Atom. Molec. Phys.)* **4**, 1516 (1971).
19. H. Beens and A. Weller, in *Organic Molecular Photophysics,* Vol. 2 (ed. J. B. Birks), Wiley-Interscience, London and New York, 1975, p. 159.
20. J. B. Birks, *An Introduction to Liquid Scintillation Counting,* Philips, Eindhoven; Koch-Light Laboratories, Colnbrook, 1968.
21. V. N. Kerr, F. N. Hayes and D. G. Ott, *Intern. J. Appl. Radn. Isotopes* **1**, 284 (1957).
22. J. B. Birks and G. C. Poullis, in *Liquid Scintillation Counting,* Vol. 2 (eds. M. A. Crook P. Johnson and B. Scales), Heyden, London, 1972, p. 1; *Liquid Scintillators,* Philips, Eindhoven, 1972.
23. M. J. Ewer and N. G. L. Harding, in *Liquid Scintillation Counting,* Vol. 3 (eds. M. A. Crook and P. Johnson), Heyden, London, 1974, p. 220.

DISCUSSION

A. Dyer: You have drawn our attention to the effective choice of quenched and unquenched cocktails. The quenching agent you have used is carbon tetrachloride. Have you continued this work to determine the relative effect of quenchers of a different chemical type – such as acetone, for example?

J. B. Birks: No – but I am sure that a different quencher would produce the same order of ranking as that seen for carbon tetrachloride.

Chapter 2

Systematically Understanding the Liquid Scintillation Counting Process: A Stochastic Computer Model

Philip J. Malcolm

Biometry Section, Waite Agricultural Research Institute, Glen Osmond, South Australia 5064, Australia

Philip E. Stanley

Department of Clinical Pharmacology, The Queen Elizabeth Hospital, Woodville, South Australia 5011, Australia

INTRODUCTION

There exists a variety of methodological and computational techniques for estimating the true rate of nuclear disintegration from the rate observed using Liquid Scintillation Counting (LSC). There has, however, been no comprehensive approach treating the subject as a whole, and a very considerable knowledge and skill is therefore required to effectively utilise the full potential of the instrument.

We intended, in a project initiated in 1972, to provide just such a comprehensive approach. Especial attention was to be paid to setting up the instrument correctly in the first place, to gathering an appropriate amount of data and to accurately estimating the error distributions of the final estimates of the actual disintegration rate, especially for multi-labelled samples. It was felt that existing techniques could be blended into a single, comprehensive approach.

However, it soon became apparent that there was a reason why the many individually excellent but relatively limited approaches had been developed without the resultant emergence of a unified approach; this was because there were gaps in the scientific knowledge necessary for the development of a unified approach. While the LSC process is fundamentally based upon well-established laws of physics,[1] techniques for utilising the process tend to be based either upon a subset of the relevant laws (e.g. external standardisation[2]) or upon essentially empirical observations of the whole process (e.g. channels ratio[3]). There was at that time no published work dealing exhaustively and rigorously with the behaviour of the whole LSC process in terms of its fundamental components.

Thus, in adopting a unified approach, we continually found ourselves asking questions which could be answered definitively and rigorously only in special cases and not in the general case. The next step, therefore, was to develop an exhaustive and rigorous understanding of the whole LSC process in terms of its component processes. The obvious approach, that of computer simulation, was taken, with work beginning early in 1973. A very simple model, built to fix ideas, indicated that attention should first be paid to processes occurring within the vial itself, and further that it was necessary, initially at least, to deal individually with each photon, rather than deal as a whole with all the photons arising from a particular event. This last conclusion was based upon an appreciation of the fundamentally discontinuous nature of the sequence of processes taking place within the cylindrical vial.

A model has since been built which details, on a photon by photon basis, the processes taking place within the solution and vial walls. The behaviour of the two photocathodes and the overall response of the detector chamber have also been modelled. The asymmetric and diverse nature of the vial cap and elevator platform has forced us to restrict attention to vials with 'perfectly' absorbing tops and bottoms; counting efficiency is somewhat reduced in practice by these absorbers. Detector response has not been modelled in detail first because of design variations between instruments and second because it is doubtful that the large amount of effort required would be adequately repaid in improved model performance. 'Background' counts have similarly been ignored because of their diverse and complicated origins.

The aim of this modelling project, namely to relate existing scientific knowledge, has nonetheless been satisfactorily fulfilled in that:

(i) otherwise inaccessible empirical relationships, between primary- and secondary-emission photons for example, can now be readily ascertained;
(ii) areas of possible improvement in instrument design are indicated; and
(iii) a unified framework has been provided within which
  (a) the workings of the instrument can be more easily understood, and
  (b) its utilisation improved.

This chapter contains a general (non-mathematical) outline and discussion of the model, and also the main theoretical thrust of its results (items (i) and (iii a) above). The following chapter[4] deals with the more practical aspects of this modelling project and also discusses improvements in methodology resulting from it (item (iii) above). Item (ii) has been discussed elsewhere in a non-mathematical fashion,[5] wherein an optimised optical design is developed for an LSC system which is spherically – rather than cylindrically – symmetrical (as is the normal case). The logical design and mathematical details of the normal cylindrical model, the optimised design of the spherical system and a detailed model of the spherical system have been given in Refs. 6a and 6b.

As discussed elsewhere,[7] several other workers[8-12] have proposed fairly complete models of the LSC process, but none yet appear to have treated it rigorously in detail.

OUTLINE OF THE NORMAL LSC PROCESS AS MODELLED

We restrict attention to binary scintillation systems, PPO and toluene, first because such systems have been thoroughly investigated by other workers and found to be well behaved and second because these systems are simpler and therefore more easily described mathematically. It is assumed that the scintillation system is excited only by low-energy β-emitting radioisotopes, ^{14}C and ^{3}H for example, first because samples including such radionuclides produce relatively few photons and are therefore particularly sensitive to colour and chemical quenching and second because β-particles from such sources have a path length which is negligible compared to vial radius (less than 0.5% for β-mean ^{14}C, less than 0.2% for β-mean ^{3}H[13]).

We assume that the radionuclide's β-spectrum follows the theoretical Fermi distribution,[6a] and also that nuclear disintegrations or β-events occur randomly within the volume of the scintillation solution, henceforth called solution. We also assume that only one β-event is being processed by the LSC system at any one time, thereby restricting attention to samples whose activity is less than about 2×10^5 disintegrations min^{-1}.

Nuclear disintegrations are therefore characterised by position and β-energy, position being uniformly random within the solution volume. The β-energy of any individual disintegration is also random in the sense that while it is unpredictable and independent of other β-events, the β-spectrum built up from a very large number of disintegrations follows the Fermi distribution for that isotope.

Such behaviour can be described mathematically at two different levels; at the *event* level in terms of the details of individual events; or alternatively at the *spectral* level whereupon the basic mathematical entity is a whole spectrum.[12] The event level was chosen for this work, first because many of the important processes in the normal cylindrical LSC system are fundamentally discontinuous (e.g. total internal reflection and secondary emission) and second because the mathematical form of the equations at the spectral level would be very complicated and intractable, and their dynamical behaviour could therefore be incomprehensible to the average LSC user.

A stochastic, probabilistic or Monte Carlo description at the event level has therefore been developed, individual β-events for example being characterised by generating:

(i) an event position which is random three-dimensionally within the simulated solution; and

(ii) a random β-energy within the range of the emitter, this being generated so that β-energies from a very large number of simulated events cumulatively follow the Fermi distribution characterising the observed β-spectrum.

Each β-event is therefore simulated individually, and the fate of each pseudo-photon arising from it is individually traced until either it generates a pseudo-photoelectron or it is absorbed and lost from the counting system. Variation in behaviour is achieved by randomly generating characteristics such as event position, β-energies, energies and escape directions of photons, etc. Such random behaviour is, however, constrained, either geometrically (e.g. a β-event must occur within the solution), or physically since spectra determining population behaviour are followed. A computer program is used to generate the required pseudo-random numbers, and, while these numbers are not 'perfectly' random, they are sufficiently so for the overall behaviour of the model to be independent of the pseudo-random number generator used.

Such a detailed stochastic simulation project assumes that there is effectively unlimited access to a large-scale computer. A Control Data 6400 computer was used for this work, and since individual simulation runs typically consumed several hours of central processor time, programs were usually loaded in the evening and then allowed to run until the following morning (the program occasionally monitors the computer's real-time clock, and, when necessary, automatically terminates itself just before the engineers begin daily maintenance at 7 a.m.).

A non-standard version of FORTRAN was chosen as the programming language, first because it is maintained in a reasonably reliable fashion by Control Data and second because reasonably efficient object code could be produced using a partially optimising compiler. It should be noted that a useful characteristic of the program was that machine or system errors occurring during its long period of execution usually resulted in its catastrophic termination, this situation being far preferable to them passing unnoticed.

But, to return to our central discussion, most ($>$90%) of the β-energy released by a particular disintegration does not contribute to primary photon production by the scintillation system, but is lost from the counting process as low-grade heat energy.

Horrocks[14] has noted that the proportion of the β-energy contributing to primary photon production depends upon the β-energy E. This portion is known as the scintillation efficiency S_E and takes values of 0.024 for 0.5 keV up to 0.062 for energies in excess of 300 keV. Thus, if we let

$$E_p = S_E E$$

then E_p is the energy available for primary photon production. Impurity or chemical quenching lowers E_p still further, this being modelled by multiplying E_p by a fraction q where $(1-q)$ is the portion of energy lost by impurity quenching. $(1-q)$ is assumed to be independent of E and the energy conversion equation now becomes

$$E_p = qS_E E$$

Event position and E_p therefore characterise the disintegration in respect of light production and it must be noted that, even at this stage, several simplifying assumptions have been made in treating the system mathematically.

For example, Horrocks' data[14] relating scintillation efficiency and β-energy have been scaled to an S_E value of 0.052 for the average energy (≅ 50 keV) of the ^{14}C β-spectrum,[15] but we require primary scintillation efficiencies (i.e. efficiency of production of primary photons) for monoenergetic sources. We require, for example, the primary scintillation efficiency for a monoenergetic 50 keV source rather than overall scintillation efficiency for β-energies *averaged* to 50 keV over the broad range of 0–156 keV for ^{14}C. Monoenergetic data for primary photon production are unfortunately unavailable, so monoenergetic scintillation efficiences calculated from averaged data must therefore be used.[6a] It has not been possible to correct these data for secondary and higher order emissions; use of a spherically symmetric instrument would, however, enable data to be corrected for secondary emission.[6a, 6b] The model is at this stage limited by the availability and accuracy of data, such as scintillation efficiences, rather than a fundamental lack of theory (see the following chapter).

Other modelling constraints of a different flavour have also arisen, particularly with regard to the simulation of the detector apparatus and the simulation of the top and bottom of the vial. The optical properties of the vial shoulders and screw cap assembly, for example, are probably impossible to describe accurately mathematically because of their variable asymmetry and diffuse optical behaviour. We have therefore been forced to place a 'perfect' absorber above the scintillation solutions counted experimentally, this effect being quite easy to model.

Similarly, elevator platforms are in practice anything but perfect reflectors, and neither they nor vial bases are normally perfectly flat. Black photographic paper was therefore placed beneath the vial to act as a perfect absorber which was easily modelled. The paper was assumed not to contact the bottom of the vial, this last being taken to be perfectly flat. It was also assumed that the vials were perfectly cylindrical[4] and transparent, and that the refractive indices of the scintillation solution (1.502) and the vial glass (1.474 for Packard low background vials[16]) were identical, thus giving an optically continuous system; it should be noted that such assumptions, while invalid, give rise to relatively small errors.

The vials used in our counting experiments are compared with the simulated vial in Fig. 1. The height H of the solution was given a value of 4.0 cm in both the experiments and the model. (Note that it can therefore be assumed that photons can only escape to the detection chamber through the curved walls of the vial.)

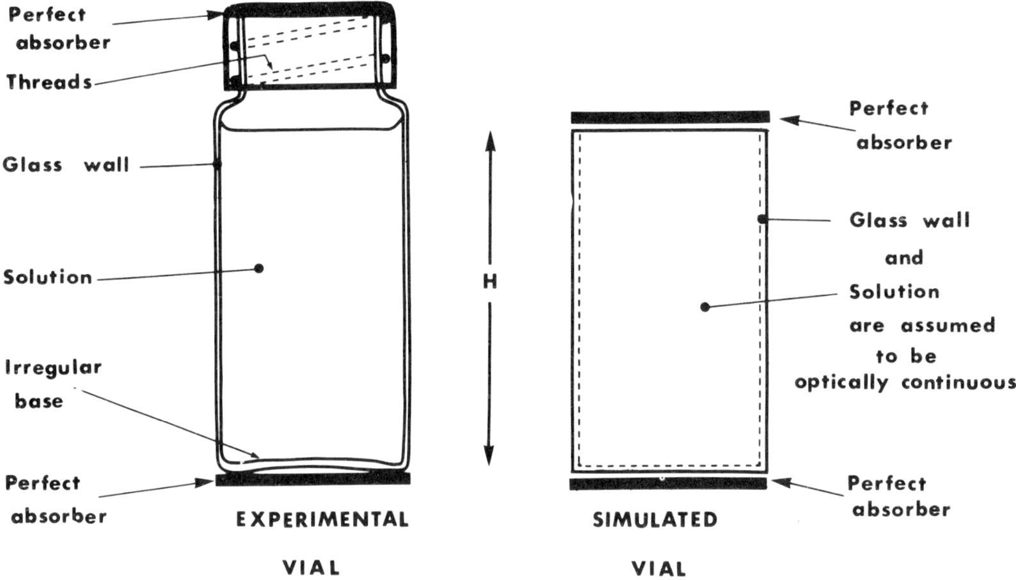

Fig. 1. Diagram comparing experimental and simulated vials.

Let us now consider photon production within the solution. We assume that excited fluor molecules emit photons in random directions three-dimensionally, and that these photons have random energies following the probability distribution given by the corrected fluorescence spectrum[4] of the fluor.

The energy E_p, calculated above, is available for primary photon production, and individual primary photons of randomised energies and escape directions are simulated until E_p is exhausted. As each photon has been characterised by energy and escape direction (relative to β-event position), its fate can now be determined as follows.

The absorption spectrum of the solution has been experimentally determined[4] and is used to calculate the absorption coefficient for a photon of that energy. This coefficient is used to generate a randomised distance x travelled through the solution by the photon. The distance y that it must travel in order to reach the curved sides of the vial is also calculated, whereupon:

(i) if $x \geqslant y$ and the photon approaches the vial's outer wall at an angle ψ less than the critical ψ_c in three dimensions, then it escapes towards the detector chamber;

(ii) if $x \geqslant y$ and $\psi \geqslant \psi_c$, then the photon will be totally internally reflected at the curved outer wall and can in theory never escape to the detection chamber.[6a] It will either
(a) escape through the top of the solution or the bottom of the vial, or
(b) be absorbed by the solution (since we assume that the vial walls are perfectly transparent);

(iii) if $x < y$, then the photon will either
(a) be absorbed by the solution, or
(b) escape through the top of the solution or the bottom of the vial.

The photon has therefore three possible fates; first, escape to the detector chamber (which will be discussed shortly); second, escape through the top of the solution or the bottom of the vial whereupon it is absorbed and lost; or third, absorption by the solution, in which case it is either absorbed by a non-scintillating coloured agent and lost (colour quenching) or it is absorbed by the fluor, these two alternatives being simulated in a probabilistic fashion according to the relative absorption coefficient of the fluor at that photon's energy.

Should the fluor absorb a photon, then a secondary photon may be emitted with a probability given by the fluorescence quantum efficiency of the fluor (0.83 for PPO[17]). Should a secondary photon not be emitted, then the fluor has acted as a colour quenching agent in the usual fashion. The point at which the fluor absorbed the photon is calculated, and, as the fluor can be treated as point-source emitter, any secondary photons will be emitted at a known point in the solution.

Secondary photons can be simulated in a manner identical to primary photons since they are emitted from a known point in solution in a random direction three-dimensionally, and have a random energy following the fluorescence spectrum of the fluor (i.e. the energy of a photon emitted by a fluor molecule is independent of the energy of the photon absorbed by the molecule). Therefore, any photon, whether primary or secondary, will meet one of these fates: it is either

(i) lost in the absorbers outside the top of the solution or the bottom of the vial;
(ii) colour quenched in the usual fashion;
(iii) colour quenched by the fluor;
(iv) absorbed by a fluor molecule which subsequently emits a secondary photon; or
(v) it escapes through the curved walls of the vial.

We have already discussed cases (i) to (iv), and will now consider (v). Having left the vial via its curved outer walls, a photon will either (a) pass through the narrow annular air gap between the top or bottom of the detector chamber and the vial, and therefore (vi) be lost from the counting process (see Fig. 2); or (b) it will enter the detector chamber of the instrument, whereupon

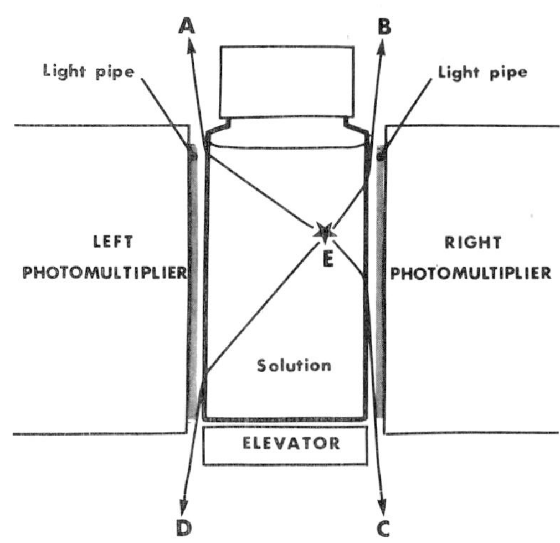

Fig. 2. A source of considerable light loss in the normal LSC instrument. Photons taking paths EA, EB, EC, ED will never enter the light pipe of the instrument illustrated, a Packard Model 3375 Tri-Carb system. They will instead be lost from the detector chamber. About 9% of the light leaving the vial is lost in a Packard 3375, this number being slightly larger for a Searle Isocap 300 instrument.

(vii) it is absorbed in the detector (by the imperfect reflector for example);
(viii) it passes through the photocathode of the photomultiplier without being absorbed or giving rise to a photoelectron;
(ix) it gives rise to a photoelectron at the photocathode;
(x) it is reflected back into the vial (see cases (i) to (v) above);
(xi) it is reflected out of the detector chamber and lost from the counting process (case (vi) above); or
(xii) it is reflected into the opposite side of the detector chamber (see cases (vii) to (xii)).

Consider first case (vi) above. One might expect that a negligibly small proportion of photons would escape through these annular air gaps since they are usually only about 1.0 mm wide. The authors have, however, demonstrated elsewhere[6b, 7] that these losses are disproportionately large because of the cylindrical symmetry of the standard counting vial, being about 9% for a solution 4.0 cm high in a Packard Model 3375 Tri-Carb system or a Searle Isocap 300 instrument.[6b]

Let us now consider the alternative case, that in which the photon enters the detector chamber, this last consisting usually of highly efficient mirrors which are designed to reflect light towards the photocathode of the appropriate photomultiplier (a portion of the light energy entering the photocathode is converted to photoelectrons; these are then amplified in the dynode chain of the photomultiplier and finally registered as an electrical pulse at the anode of the photomultiplier; the anode pulse is analysed by the remainder of the instrument to determine whether or not it arose from a valid 'count', i.e. resulted from an event of appropriate energy in the solution).

Now the design of the detector chamber and associated equipment varies markedly between instruments. For example, of the two instruments available for experimental use in this work, the detector chamber of one, an early model Packard Model 3375 Tri-Carb system, consisted of a rectilinear solid of 'lucite' or 'perspex' mirrored on the top, the bottom and the two sides perpendicular to the photocathodes, and containing a cylindrical hole down which the elevator moved bringing within it the vial. The detector of the other instrument, a Searle Isocap 300, consisted of a complicated curved mirror which was open to the photocathodes at each end and enclosed an air space containing the vial, the vial being loaded by an elevator moving down through circular holes in the top and bottom of the detector.

Because of this wide variation in detector design, and also because of the obvious difficulties in arriving at anything approaching an adequate mathematical description in even a special case, no attempt was made to describe in detail the path taken by photons after leaving the vial; instead, the overall response of the detector was assessed experimentally for light leaving different points on the vial's curved walls[4] (a vial covered completely with black photographic paper except for a small hole at various heights was loaded and counted out of coincidence, then rotated through a small horizontal angle and recounted, etc.; see appendix of Ref. 6a).

These data were then normalised in relation to the best response summed over both photomultipliers, and could thereby be used to determine the overall response of the detector chamber for *all* photons leaving a particular point on the vial's curved walls. Thus the fate of photons leaving the vial could be simulated in a probabilistic fashion according to the overall response of the detector for all photons leaving the vial at that point, with photons simulated as successfully reaching the photocathode giving rise to photoelectrons in a probabilistic fashion according to the photocathode quantum efficiency[4] for photons of that wavelength.

This approach to detector response was therefore adequate for cases (vii, viii, ix and xi) above, but inadequately accommodated case (x) (and therefore also for one possible result of case (xii)), since photons reflected back towards the especially blackened vial were absorbed by the black photographic paper covering almost all its surface. For photons leaving the vial at points midway between the two photomultipliers, this is a serious deficiency in the model since it may be crudely estimated that about 25% of the light leaving the vial is normally reflected back into it when using an early Packard Model 3375 Tri-Carb instrument, this portion being less with the Searle Isocap 300 instrument.

Thus, a photon emitted by the fluor must meet one of the fates described above; it is lost from the counting system in cases (i, ii, iii, vi, vii, viii and xi); it may eventually give rise to a photoelectron and thereby increase the probability of the parent disintegration being 'counted' in cases (iv and xii); it does give rise to a photoelectron in case (ix); and its behaviour is inadequately modelled in case (x) and, to a lesser extent, in case (xii).

The detailed response of the photomultipliers could not be modelled as there is insufficient data concerning pulse spread and losses, and summed photoelectron counts in each photomultiplier are therefore the significant output of the β-event simulated. These summed photoelectron counts were used to simulate coincidence thereby enabling the simulated counting efficiency to be assessed, and digital 'pulse-height spectra' of both the summed and lesser[18,19] kind to be accumulated.

RESULTS AND DISCUSSION

The results presented herein are intended to give the reader some appreciation of the behaviour of the model of the LSC system, and several points in this connection must be mentioned immediately.

First, in validating the model, an extensive laboratory program was conducted[4] to provide the necessary calibration data, and also to provide experimental results with which the forecasts of the model could be compared. It will be shown here, and also in Chapter 3, that, by and large, the model faithfully reproduces the behaviour observed in practice, especially under the more normal quenching conditions.

This does not mean that the authors are suggesting that the model might in some way replace direct laboratory measurements of aspects of the LSC process. This would, in their opinion, currently be quite unrealistic because of the many simplifying assumptions made in defining the model, because some of the data included in the model are necessarily only approximate and also because of the probabilistic nature of the model (see below).

The role of this model is rather to complement and integrate knowledge obtained by direct experimentation. The model is not intended to specify in a completely rigorous numerical fashion the relationships constituting the LSC process. Thus, while the reader will discover many interesting numerical data in the following pages, these data have been presented in order to illustrate the qualitative, not the quantitative, relationships involved. Correspondingly, the main thrusts of our considerations will be the discussion of qualitative, not quantitative, relationships, and these discussions will generally serve to provide a practical background for, and intepretation of, the results produced by the model.

Before turning attention to the model's behaviour for actual β-emitters, let us first consider its behaviour for imaginary sources of monoenergetic β-particles. Table 1 illustrates the behaviour expected for unquenched solutions containing sources of mono-energetic (internal) electrons.

There are several points of immediate interest in Table 1(a). Clearly β-events of very low energy (e.g. < 1.5 keV) will be difficult to measure because they emit so few primary photons (e.g. $<\cong 15$). Some of this emitted energy will be lost by absorption, and of the photons reaching a (bialkali) photocathode, only about 27% will give rise to photoelectrons.

The importance of photocathode response for low β-energies can be appreciated by comparing results for lower pseudo-discriminator settings of a minimum of two and three photoelectrons. Indeed, in the development of LS spectrometers over the past two decades, the improvement of photomultiplier response has been an area of critical attention and considerable progress.[20]

The instrument's detection efficiency increases rapidly with β-energy because the portion of energy available for primary photon production increases rapidly with energy (see above). Quite respectable counting efficiences are obtained for unquenched β-events

Table 1 (a). Simulation of monoenergetic internal β-emitters – comparison of ^{3}H and ^{14}C β-spectra (unquenched sample).

β-energy[a] *e* (keV)	Total no. of primary photons produced	Percent counting efficiencies for a 'pulse' of at least *n* photoelectrons		Portion of β-spectrum of energy ⩾ *e* keV	
		$n = 2$	$n = 3$	^{3}H	^{14}C
1.2	11	12.8	4.5	89.1	98.8
1.3	13	14.3	5.5	88.2	98.7
1.4	14	15.9	6.6	87.2	98.6
1.5	15	18.7	8.4	86.2	98.5
1.6	16	20.9	9.6	85.2	98.4
1.7	17	22.8	11.2	84.2	98.3
1.8	19	25.5	13.2	83.1	98.2
1.9	20	27.9	15.2	82.1	98.1
2.0	21	30.4	17.7	81.1	98.0
2.1	22	31.8	18.9	80.1	97.9
2.2	24	34.4	21.0	79.0	97.8
2.3	25	36.0	23.3	78.0	97.7
2.4	26	38.3	25.8	77.0	97.6
2.5	27	40.7	28.0	76.0	97.5
3.0	34	50.4	38.6	70.8	97.0
3.5	40	59.0	48.6	65.7	96.5
4.0	47	66.8	58.8	60.7	96.0
4.5	54	73.6	66.8	55.8	95.5
5.0	61	78.7	73.3	51.1	95.0
5.5	68	82.8	79.2	46.6	94.4
6.0	75	85.7	82.9	42.2	93.9
6.5	82	89.0	87.0	38.0	93.4
7.0	89	90.6	89.3	34.1	92.8
50	796	100	100	0	44.5
100	1683	100	100	0	8.3
500	9486	100	100	0	0

[a]10,000 events were simulated in each case.

of energy greater than say 7 keV, and events of even higher energy clearly produce so many primary photons that a counting efficiency of 100% is virtually guaranteed. It is also interesting to note that the lower pseudo-discriminator setting becomes less and less important as β-energy increases. This behaviour is of course to be expected.

The last two columns of Table 1(a) illustrate why samples containing ^{14}C behave in practice differently to samples containing ^{3}H. This is because ^{3}H emits β-particles in a much lower energy range than ^{14}C. Thus, according to Table 1(a), only about 34% of ^{3}H β-events are counted with an efficiency greater than 90%, and about 10% of ^{3}H events fall into the very low ($\cong$ 10%) efficiency range. About 90% of ^{14}C β-particles, by contrast, are counted with at least 90% efficiency, and only about 1% fall into the very low ($\cong$ 10%) efficiency range.

These relationships are observed in the laboratory first in that, in an integral window, an unquenched ^{3}H sample generally counts at about 60–65% efficiency, this by comparison with the efficiency of 95%+ normally expected with an unquenched ^{14}C sample.

Second, ^{3}H is much more sensitive to quenching than is ^{14}C. This is to be expected since a relatively large proportion of ^{3}H events occur in the very low energy range, and therefore give rise to very few primary photons. Thus, should even a few photons be lost by colour quenching, then quite marked decreases in counting efficiency will be observed; similarly, in the case of impurity or chemical quenching, even fewer primary photons are produced in the first place.

Table 1(b) serves to illustrate the probabilistic nature of the model's behaviour. Differing counting efficiencies were generated by simulation runs whose only difference lay in the sequence of pseudo-random numbers used to drive the model.

Note that the stochastic variation in results produced by the model might be thought to be analogous with the variation in laboratory results due to experimental error. It would, however, be dangerous to take this analogy too far; for example, while the distribution of experimental estimates of counting efficiencies is known to be Poissonian,

Table 1 (b). Simulation of monoenergetic internal β-emitters—showing the probabilistic behaviour of the model; comparison of results obtained using different sequences of pseudo-random numbers.

β-energy[a] (keV)	Percent counting efficiencies for a pulse of at least n photoelectrons (unquenched samples)			
	n = 2		n = 3	
	Sequence 1	Sequence 2	Sequence 1	Sequence 2
3.0	50.4	50.9	38.6	38.5
3.5	59.0	60.2	48.6	50.1
4.0	66.8	66.5	58.8	58.0
4.5	73.6	74.1	66.8	67.2
5.0	78.7	78.7	73.3	73.5
5.5	82.8	82.9	79.2	79.1
6.0	85.7	85.6	82.9	83.0
6.5	89.0	88.7	87.0	86.6
7.0	90.6	91.2	89.3	89.8
7.5	93.1	92.7	92.0	91.8

[a]10,000 events were simulated in each case.

the distribution of the model's predictions is unknown. The reader is therefore urged once again to use these results in a qualitative rather than quantitative fashion.

In turning now to models of actual β-emitters we will restrict attention to the Searle Isocap 300 instrument, this despite the fact that the Packard Model 3375 Tri-Carb system has been modelled to the same degree of detail. The reason for this restriction, namely the optical qualities of the perspex light pipe installed in our early model machine, is discussed in detail in the following chapter.

Tables 2(a) and (b) show in some detail how ^3H and ^{14}C behave under different conditions of chemical quenching. Tables 3(a) and (b) provide similar details for the colour quenching situation, and chemically- and colour-quenched simulation runs are compared in Table 4. Five thousand events were simulated in all the above cases.

As these results are entirely in accordance with what is to be expected we will not discuss them in detail; the reader is instead recommended to peruse them at leisure. But we will discuss some of the more interesting relationships between comparable simulation runs of chemically- and colour-quenched ^3H and ^{14}C samples. The samples considered are those whose behaviour is detailed in Table 4.

Figures 3 and 4 show how the degree of light trapping due to the curved walls of the vial varies for ^3H and ^{14}C under chemical- and colour-quenching conditions. It must be mentioned immediately that data have only been included for those events in which coincidence is enabled; and, in this case, the total number of total internal reflections (at the curved outer walls of the vial) for *all* photons generated in simulating the event was divided by the total number of photons leaving the vial via its curved walls.

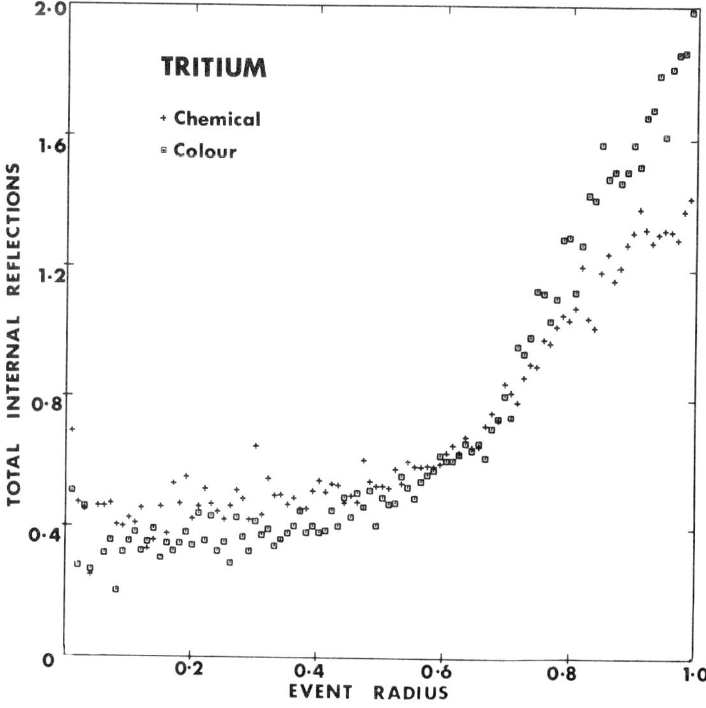

Fig. 3. Light trapping due to the curved walls of the counting vial as a function of the event radius (distance of event from the cylindrical axis) for colour- and chemically quenched ^3H events.

Table 2 (a). Simulation of the unquenched and impurity quenched situations – ^3H.

	Argon-purged $q^a = 0.9$		Air-quenched $q = 0.8$		Impurity quenched to equivalent of 30 μl of CCl_4 in a 20 ml sample $q = 0.5$	
	All events	Coincident events	All events	Coincident events	All events	Coincident events
Average β-energy (keV)	5.6	7.4	5.6	7.5	5.8	8.3
Average number of primary photons	65	86	57	78	36	54
Average number of primary photons trapped due to cylindrical symmetry	23	30	20	27	13	18
Average number of secondary photons	44	57	38	51	24	35
Average number of photons						
colour quenched by PPO	8	11	7	10	5	7
lost in top absorber	9	12	8	11	5	7
lost in bottom absorber	10	13	8	12	5	8
leaving near side of vial	17	24	16	21	10	15
leaving far side of vial	18	24	15	23	10	15
lost in near side of detector	8	10	7	9	4	6
lost in far side of detector	9	13	8	12	5	8

reaching near photocathode[b]	9	13	8	12	6	9
reaching far photocathode[b]	9	12	8	11	5	7
Average number of photoelectrons						
in near photomultiplier	2	3	2	3	1	2
in far photomultiplier	2	3	2	3	1	2
Average summed 'pulse height'	4	6	3	6	2	4
Average lesser 'pulse height'	1	2	1	2	0	1
Proportion of photons reaching detector after m re-emissions						
$m = 0$	0.612		0.618		0.613	
$m = 1$	0.229		0.228		0.231	
$m = 2$	0.094		0.090		0.091	
$m = 3$	0.038		0.037		0.038	
$m = 4$	0.015		0.015		0.015	
$m = 5$	0.007		0.006		0.007	
% Counting efficiency $n^c = 2$	63.9		59.4		45.7	
$n = 3$	58.8		53.3		38.6	

[a] Impurity quenching factor (see text).
[b] Note that photons may cross to the other side of the detector.
[c] Only coincident 'pulses' of at least n photoelectrons are counted.

Table 2 (b). Simulation of the unquenched and impurity situations – ^{14}C.

	Argon-purged $q^a = 1.0$		Air-quenched $q = 0.75$		Impurity quenched to equivalent of 60 μl of CCl_4 in a 20 ml sample $q = 0.4$	
	All events	Coincident events	All events	Coincident events	All events	Coincident events
Average β-energy (keV)	50.2	51.7	49.5	51.4	50.4	53.8
Average number of primary photons	797	821	587	611	319	342
Average number of primary photons trapped due to cylindrical symmetry	282	290	206	214	112	120
Average number of secondary photons	535	551	392	407	213	228
Average number of photons						
colour quenched by PPO	109	112	80	83	43	46
lost in top absorber	114	117	84	87	45	49
lost in bottom absorber	124	128	91	94	49	52
leaving near side of vial	222	223	165	173	90	95
leaving far side of vial	226	229	166	172	90	96
lost in near side of detector	98	101	73	76	39	42
lost in far side of detector	122	126	90	93	49	52

reaching near photocathode[b]	119	113	88	92	48	51
reaching far photocathode[b]	109	112	80	84	44	47
Average number of photoelectrons						
in near photomultiplier	31	32	23	24	12	13
in far photomultiplier	28	29	20	21	11	12
Average summed 'pulse height'	59	61	43	45	23	25
Average lesser 'pulse height'	23	24	17	18	9	10
Proportion of photons reaching detector after m re-emissions						
$m = 0$	0.613		0.616		0.615	
$m = 1$	0.232		0.230		0.230	
$m = 2$	0.092		0.091		0.091	
$m = 3$	0.037		0.037		0.037	
$m = 4$	0.015		0.015		0.015	
$m = 5$	0.007		0.007		0.006	
% Counting efficiency n[c] $= 2$	96.9		95.9		92.8	
$n = 3$	96.4		95.4		91.5	

[a] Impurity quenching factor (see text).
[b] Note that photons may cross to the other side of the detector.
[c] Only coincident 'pulses' of at least n photoelectrons are counted.

Table 3 (a). Simulation of colour quenching – ^{3}H.

	Colour quenching to an equivalent of c μl of coloured solution[a] in a 20 ml sample					
	c = 20.4		c = 73.2		c = 162.4	
	All events	Coincident events	All events	Coincident events	All events	Coincident events
Average β-energy (keV)	5.7	7.6	5.6	7.8	5.6	8.1
Average number of primary photons	56	77	56	80	56	83
Average number of primary photons trapped due to cylindrical symmetry	19	25	16	23	13	18
Average number of secondary photons	36	48	25	35	21	31
Proportion[b] of photons						
colour quenched by PPO	0.131	0.128	0.091	0.089	0.079	0.077
colour quenched by coloured agent	0.047	0.045	0.250	0.240	0.397	0.384
lost in top absorber	0.132	0.130	0.099	0.100	0.076	0.077
lost in bottom absorber	0.141	0.138	0.107	0.106	0.077	0.076
Average number of photons						
leaving near side of vial	15	20	12	18	9	16
leaving far side of vial	15	21	12	17	10	15

reaching near photomultiplier	9	11	7	10	5	9
reaching far photomultiplier	7	10	6	9	5	8
Average number of photoelectrons						
in near photomultiplier	2	3	1	2	1	2
in far photomultiplier	1	2	1	2	1	2
Average summed 'pulse height'	3	6	2	5	2	4
Average lesser 'pulse height'	1	2	0	1	0	1
Proportion of photons reaching detector after m re-emissions						
$m = 0$	0.623		0.691		0.721	
$m = 1$	0.231		0.214		0.200	
$m = 2$	0.087		0.066		0.058	
$m = 3$	0.035		0.020		0.016	
$m = 4$	0.014		0.006		0.004	
$m = 5$	0.006		0.002		0.001	
% Counting efficiency $n^c = 2$	57.7		49.9		40.6	
$n = 3$	51.8		43.3		33.3	

Impurity quenching factor of $q = 0.75$ used throughout.
[a] See following chapter.
[b] Relative to the number of primary photons.
[c] Only coincident 'pulses' of at least n photoelectrons are counted.

Table 3 (b). Simulation of colour quenching – ^{14}C.

	Colour quenching to an equivalent of c μl of coloured solution[a] in a 20 ml sample					
	c = 20.2		c = 74.6		c = 184.3	
	All events	Coincident events	All events	Coincident events	All events	Coincident events
Average β-energy (keV)	49.1	51.2	48.9	51.3	49.9	53.0
Average number of primary photons	607	633	603	635	618	657
Average number of primary photons trapped due to cylindrical symmetry	206	215	186	196	162	172
Average number of secondary photons	409	427	290	305	241	257
Proportion[b] of photons						
colour quenched by PPO	0.139	0.138	0.099	0.098	0.080	0.080
colour quenched by coloured agent	0.012	0.011	0.210	0.208	0.360	0.358
lost in top absorber	0.138	0.138	0.108	0.108	0.084	0.084
lost in bottom absorber	0.149	0.149	0.116	0.116	0.087	0.087
Average number of photons						
leaving near side of vial	171	178	140	148	120	127
leaving far side of vial	171	178	141	142	119	128

reaching near photomultiplier	92	96	75	79	64	68
reaching far photomultiplier	83	86	68	72	58	62
Average number of photoelectrons						
in near photomultiplier	23	24	19	20	16	17
in far photomultiplier	21	22	17	18	15	16
Average summed 'pulse height'	45	47	37	39	31	33
Average lesser 'pulse height'	17	18	13	14	10	11
Proportion of photons reaching detector after m re-emissions						
$m = 0$	0.611		0.680		0.719	
$m = 1$	0.233		0.217		0.202	
$m = 2$	0.092		0.079		0.057	
$m = 3$	0.038		0.023		0.016	
$m = 4$	0.016		0.007		0.004	
$m = 5$	0.007		0.003		0.001	
% Counting efficiency $n^c = 2$	95.6		94.7		93.4	
$n = 3$	94.8		93.8		92.5	

Impurity quenching factor of $q = 0.75$ used throughout.

[a] See following chapter.

[b] Relative to the number of primary photons.

[c] Only coincident 'pulses' of at least n photoelectrons are counted.

Table 4. Comparison of simulations of impurity and colour quenching (results are for coincident events only.)

	^3H		^{14}C	
	$q^a = 0.5$	$q = 0.78^b$	$q = 0.4$	$q = 0.78$
Equivalent volume[c] of CCl_4 (μl)	30	0	60	0
Equivalent volume[c] of coloured agent (μl)	0	115.6	0	184.3
% Counting efficiency $n^d = 2$	45.7	43.3	92.8	93.4
$n = 3$	38.6	36.5	91.5	92.5
Average β-energy (keV)	5.8	5.7	50.4	49.9
Average number of primary photons	36	56	319	618
Proportion of primary photons trapped due to cylindrical symmetry	0.36	0.25	0.35	0.26
Average number of secondary photons	24	22	213	241
Proportion[e] of photons				
colour quenched by PPO	0.136	0.080	0.137	0.080
colour quenched by coloured agent	0	0.370	0	0.360
lost in top absorber	0.144	0.081	0.143	0.084
lost in bottom absorber	0.152	0.085	0.155	0.087
Average number of photons				
leaving near side of vial	10	10	90	120
leaving far side of vial	10	11	90	119
reaching near photomultiplier	6	6	48	64
reaching far photomultiplier	5	6	44	58
Average number of photoelectrons				
in near photomultiplier	1	1	12	16
in far photomultiplier	1	1	11	15
Average summed 'pulse height'	2	2	23	31
Average lesser 'pulse height'	0	0	9	10
Proportion of photons reaching detector after m re-emissions				
$m = 0$	0.613	0.716	0.615	0.719
$m = 1$	0.231	0.204	0.230	0.202
$m = 2$	0.091	0.058	0.091	0.057
$m = 3$	0.038	0.016	0.037	0.016
$m = 4$	0.015	0.004	0.015	0.004
$m = 5$	0.007	0.001	0.006	0.001

[a] Impurity quench factor (see text).
[b] Coloured samples were air-quenched.
[c] In 20 ml of solution.
[d] Only coincident pulses of at least n photoelectrons are counted.
[e] Relative to the number of primary photons.

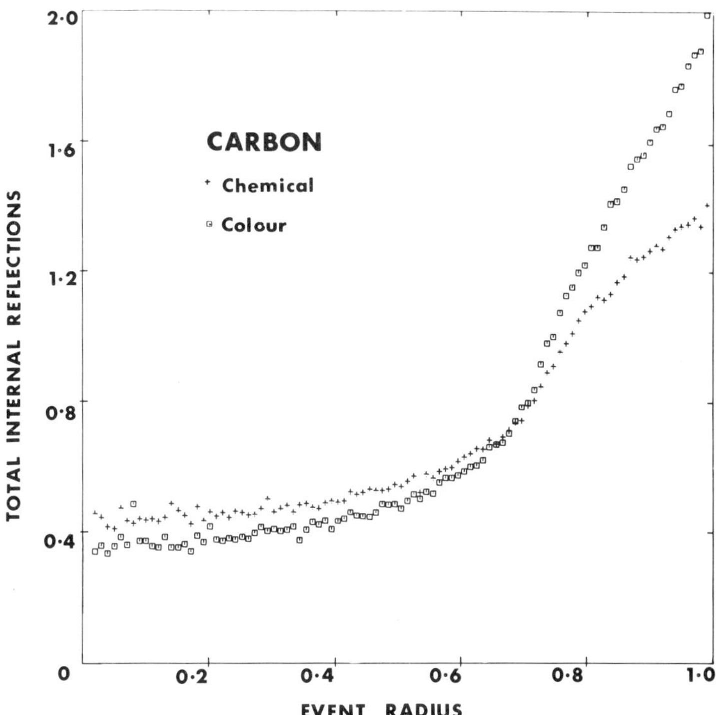

Fig. 4. Light trapping due to the curved walls of the counting vial as a function of the event radius (distance of event from the cylindrical axis) for colour- and chemically quenched ^{14}C events.

The data therefore show the overall light trapping behaviour of a normal cylindrical vial in relation only to those events enabling coincidence (i.e. counts), and may therefore be regarded as uniformly but conservatively estimating the overall light trapping behaviour in relation to all events (i.e. disintegrations). Note, however, that, although assumed in the model, vials are in practice not perfect cylinders,[4] and therefore that our conservative estimate of light trapping behaviour is more likely to represent the real situation than is an estimate in relation to all events simulated.

There are several points of interest in this behaviour. It can first be seen that, overall, the curves obtained for ^{3}H are fuzzier than those for ^{14}C. A total of 5000 β-events was simulated in each of the four cases shown, but ^{3}H showed a counting efficiency of only $\cong$44% by comparison with that of $\cong$93% obtained for ^{14}C. Rather fewer data have therefore been included for ^{3}H, and those data which have been included derive from the many fewer photons produced during the course of a ^{3}H event.

It is, however, more interesting to note that fuzziness increases as radius decreases. This is to be expected because events occur with uniform randomness within the volume of the solution, and therefore about 75% of the events will occur at a radius greater than 0.5 (i.e. proportion of events occurring at radius less than r is proportional to r^2).[6a]

Another feature of Figs. 3 and 4 is that in each case the coloured sample showed more light trapping at higher radii (i.e. $> \cong 0.7$), and less light trapping at lower radii (i.e. $< \cong 0.7$), than its chemically quenched counterpart.

Let us, in considering this behaviour, note first that light trapping at the curved walls of the vial occurs when the angle of incidence, in three dimensions, is greater than the critical angle of the vial–air interface. Now the angle of incidence in three dimensions can be regarded as being composed of horizontal and vertical components, the horizontal one being important in relation to the circular horizontal cross-section of the vial, and the vertical one being important in relation to the rectangular vertical cross-section of the vial and solution.

For photons emitted from points with small radii, the horizontal component of the three-dimensional angle of incidence is largely irrelevant to light trapping behaviour; but, for larger radii, the horizontal component is increasingly influential. A similar argument applies for the vertical component of the angle of incidence; its importance increases as radius increases.

Consider now the overall behaviour at larger radii. In this case, relatively fewer photons will cross to the other side of the vial in a coloured solution, and therefore relatively fewer photons will leave the opposite side of the vial.

Now the horizontal component of the angle of incidence is more important than the vertical one for larger radii, and therefore the total proportion of total internal reflections will be greater in the colour quenching case since:

(i) photons are more likely to escape via the curved walls of the vial if they have only a short distance to travel through the solution; and
(ii) those photons travelling a short distance are more likely to have a relatively large horizontal component in their three-dimensional angle of incidence.

The converse situation applies for small radii since the vertical component of the three-dimensional angle of incidence now becomes more important. The total proportion of total internal reflections will now be less in the colour quenching case because:

(i) photons are more likely to escape via the curved walls of the vial if they have only a short distance to travel through the solution;
(ii) since the horizontal component of the three-dimensional angle of incidence becomes increasingly irrelevant as radius decreases, those photons taking the shorter paths to the curved walls of the vial and then escaping will be those photons travelling near the horizontal plane; and
(iii) similarly those photons taking longer paths will tend to have larger vertical components in their three-dimensional angles of incidence, and will thus be more likely to be absorbed before reaching the curved walls of the vial, there to be totally internally reflected.

The chemical- and colour-quenched curves cross at a radius of $\cong$0.7. The significance of this number can be appreciated by noting that, for radii $< 1/1.474 \cong 0.68$, the horizontal component of the three-dimensional angle of incidence is always less than the critical angle ($\sin^{-1} 1/1.474 \cong 43°$) for the vial–air interface.

Thus, while the horizontal component becomes increasingly important for higher radii, it plays an increasingly irrelevant role for lower radii, with the cross-over in three dimensions, occurring at a radius of $\cong 0.7$.

Clearly, therefore, the light trapping behaviour of the normal LSC system is complicated by its cylindrical symmetry. The seriousness of this light trapping behaviour can be gauged from the improvements in performance observed when cylindrical symmetry is deliberately disrupted by sandblasting the vial's curved outer surface.[21] The authors have proposed elsewhere[5,6b] a spherically symmetrical LSC system of much greater optical efficiency than the cylindrical system, and it should be noted that spherical systems, by comparison with cylindrical ones, enjoy the invaluable advantage of being much more readily understood.

Consider now the variation of counting efficiency with event radius. Figures 5 and 6 illustrate this behaviour for the simulation of chemically and colour-quenched ^3H samples of similar counting efficiencies (see Table 4). The area under the data is therefore approximately the same in each case. Note first that fuzziness decreases as radius increases, a situation which is to be expected because events occur at random points within the *volume* of the solution (see above).

Now note that, despite the general fuzziness of the data (20,000 events were simulated in each case), the colour-quenched case shows higher efficiencies at low radii ($< \cong 0.5$) than the chemically quenched case, the converse situation arising to a more marked degree at higher radii ($> \cong 0.7$). The marginally higher efficiencies at low radii for the colour-quenched case are to be expected since light trapping is marginally lower in this case (see Figs. 4, 5). Similarly, the markedly lower counting efficiencies for the colour-quenched sample at higher radii are to be expected since the colour-quenched case shows markedly more light trapping at higher radii (see Figs. 4, 5).

Similar curves have not been presented for ^{14}C as this isotope shows relatively little departure from a counting efficiency of $\cong 100\%$.

Consider, in this last context, Figs. 7 and 8 which show variation of counting efficiency in relation to β-energy for ^{14}C and ^3H respectively. Results for simulation runs of chemically and colour-quenched samples are overlaid in each case, with ^{14}C

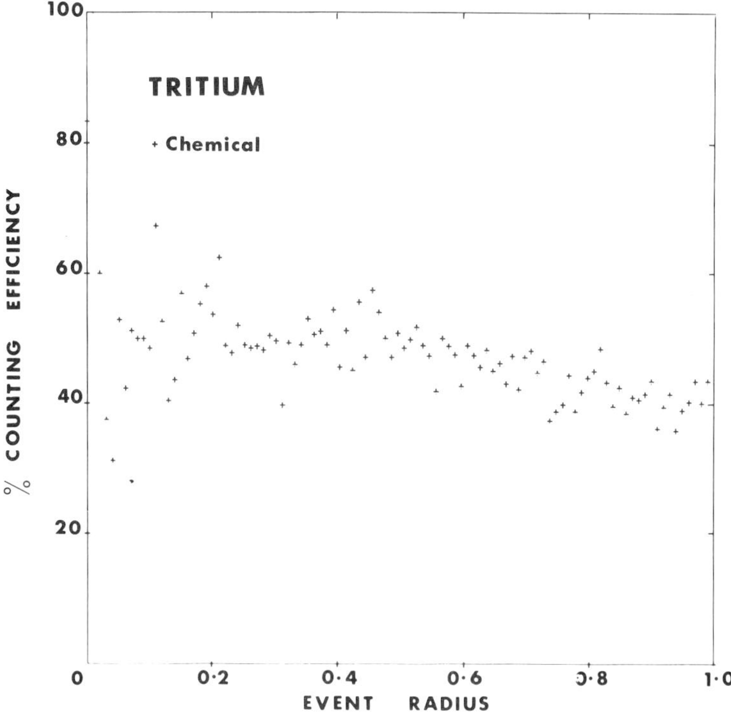

Fig. 5. A plot of average counting efficiency (proportion of events enabling coincidence) for chemically quenched tritium as a function of the event radius (distance of event from the cylindrical axis).

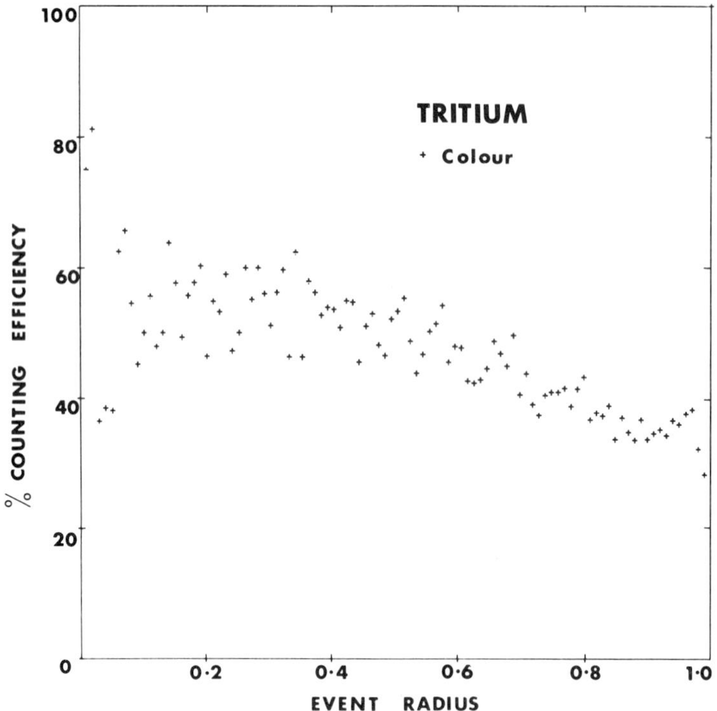

Fig. 6. A plot of average counting efficiency (proportion of events enabling coincidence) for colour-quenched tritium as a function of the event radius (distance of event from the cylindrical axis).

clearly showing a 100% efficiency for almost all events of β-energy greater than about 25 keV (i.e. ≅ 72% of the ^{14}C β-spectrum; see Table 5).

Counting efficiency gradually decreases below 25 keV, with very similar behaviour being shown for both colour- and chemical-quenching. This is to be expected since:

(i) only events of low β-energy produce so few photons that, even under quenching conditions, a counting efficiency of 100% is not virtually guaranteed (see Table 1); and

(ii) because both stochastic simulation runs produced nearly equal overall counting efficiencies the areas under both the chemical- and colour-quenched curves must be approximately the same.

Figure 8 illustrates the counting efficiency/energy relationship for ^3H, and demonstrates more effectively how counting efficiency increases with β-energy in the low range. The relationship is sigmoid and clearly not simple, this being expected since the scintillation efficiency S_E is an increasing but non-linear function of β-energy.

Another feature of interest in Fig. 8 is that fuzziness increases with β-energy. This situation is to be expected since the ^3H β-energy spectrum falls off sharply as β-energy increases, with correspondingly fewer β-events being produced with these energies (see Table 5).

Let us now consider the digital 'pulse height spectra' produced by the model. There are two different approaches to pulse height analysis in LSC, the first and more traditional method being to combine the analogue pulses from each photomultiplier for those events in which coincidence is enabled. This 'summed' pulse then charac-

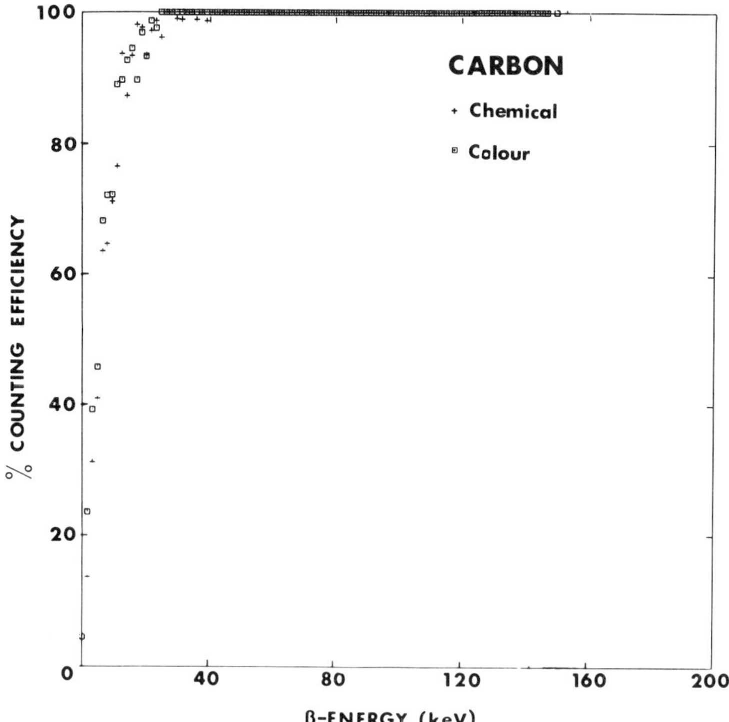

Fig. 7. A plot of counting efficiency (proportion of events enabling coincidence) versus β-energy for colour- and chemically quenched ^{14}C events (integral counting efficiency ≅93%).

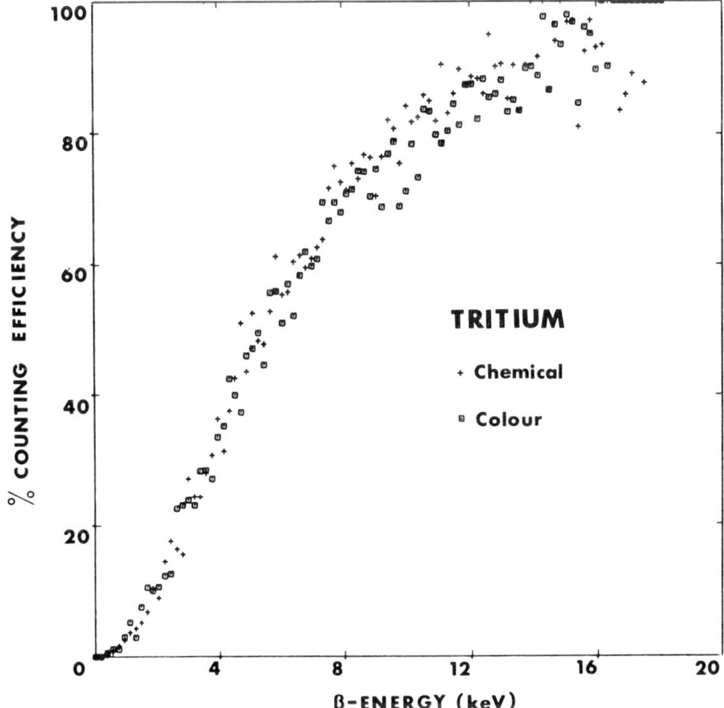

Fig. 8. A plot of counting efficiency (proportion of events enabling coincidence) versus β-energy for colour- and chemically quenched ^{3}H events (integral counting efficiency ≅44%).

Table 5. Fermi distribution[6a] of energies of β-particles emitted by ^3H and ^{14}C.

^3H		^{14}C	
β-energy *e* (keV)	% of β-particles emitted with an energy > *e*	β-energy *e* (keV)	% of β-particles emitted with an energy > *e*
1.0	91.1	10	89.6
2.0	81.1	20	78.1
3.0	70.8	30	66.4
4.0	60.7	40	55.1
5.0	51.1	50	44.5
6.0	42.2	60	34.8
7.0	34.1	70	26.3
8.0	26.8	80	19.0
9.0	20.5	90	13.0
10.0	15.2	100	8.3
11.0	10.8	110	4.8
12.0	7.2	120	2.4
13.0	4.5	130	0.9
14.0	2.6	140	0.2
15.0	1.3	150	0.01
16.0	0.5		
17.0	0.1		
18.0	0.01		

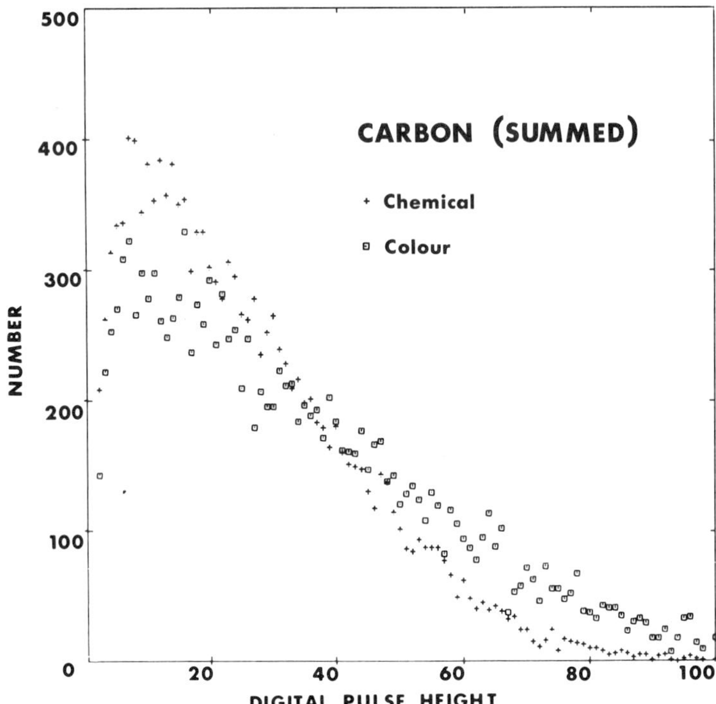

Fig. 9. A digital *summed* pulse height spectrum for colour- and chemically quenched ^{14}C events enabling coincidence. Integral counting efficiency ≅93% in each case. (Note the different vertical scale when compared to Fig. 10.)

terises that (coincident) event, and is further analysed according to discriminator settings, etc. Figure 9 shows digital summed pulse height spectra obtained with the model for colour- and chemically quenched ^{14}C samples counted with similar efficiencies. 15,000 events were simulated in each case.

The second method, known as 'lesser' pulse height analysis, is due to a penetrating insight by Laney.[18,19] The approach in this case is to analyse only the smaller or lesser of the two pulses arising in photomultipliers for coincident events.

Thus, in that only those pulses are analysed which arise from events enabling coincidence, the lesser approach is therefore identical to the traditional one. The difference from the traditional approach arises after coincidence is enabled; whereas the 'summed' method combines the pulses from both photomultipliers, the 'lesser' method, when applied in its pure form, analyses only the smaller of the two pulses from the photomultipliers, the larger pulse being discarded.

Clearly, therefore, the summed and lesser methods will in practice show no difference for counting efficiency in an integral counting window since, in this case, counting efficiency is given by (number of coincident events)/(number of disintegrations or events).

Figure 10 shows the digital lesser pulse height spectra resulting from the same two simulation runs, each of 15,000 events, that produced the data for Fig. 9; the spectra produced in each case are therefore comparable. Figure 9 clearly illustrates the behaviour observed in practice (see, for example, Fig. 9 of Ref. 19) for colour- and chemically quenched samples. The spectrum of the coloured sample has a lower maximum value and a longer 'tail' than its chemically quenched counterpart.

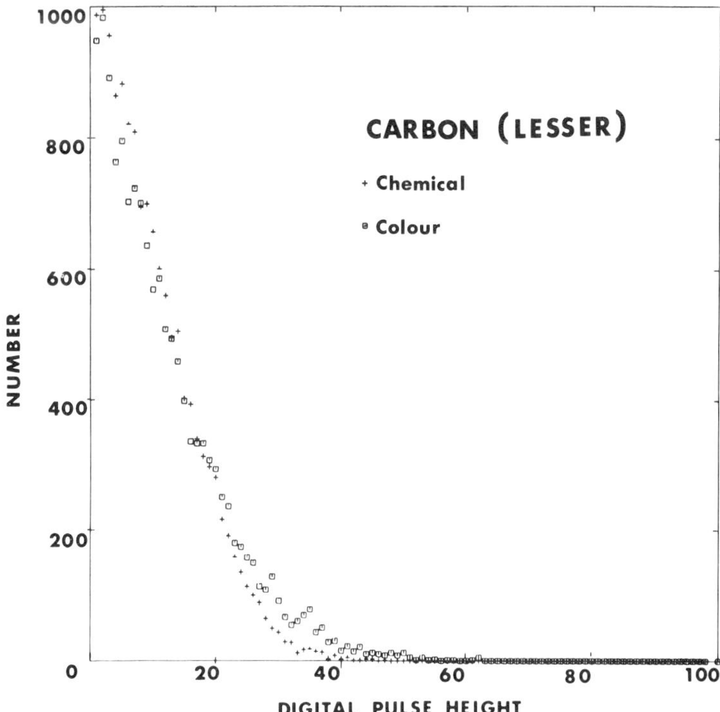

Fig. 10. A digital *lesser* pulse height spectrum for colour- and chemically quenched ^{14}C events. Integral counting efficiency ≅93% in each case. (Note the different vertical scale when compared to Fig. 9.)

This trend is, as expected (see Fig. 9 of Ref. 19), present to a markedly smaller degree in the lesser spectra (see Fig. 10). The behaviour of the model is therefore consistent with the spectra produced in practice for both the summed and lesser systems.

CONCLUSIONS

The above model offers a useful understanding of some of the more essential of the complicated interactions constituting the LSC process as it occurs in practice. While it is clear that definitive accuracy is still lacking in many of the data (e.g. scintillation efficiency and detector response), and other important data are missing (e.g. pulse spread and losses along the dynode string of the photomultipliers in our instruments), the model's behaviour is nonetheless in accordance with that observed in the laboratory. Note that the more practical implications of this study are contained in the following chapter.

ACKNOWLEDGEMENTS

Grateful thanks are given to the Staff of the Computing Centre of the University of Adelaide, without whose assistance this study would have been impossible. We are also indebted to Dr. John Birks, Dr. Donald Horrocks, Mr. Philip Kreveld, Mr. Donald Moore and Mr. Edward Polic for providing invaluable numerical data and advice. Finally we thank Mrs Ermioni Mourtzios and Mrs Helen Simpson for preparing the manuscripts.

REFERENCES

1. J. B. Birks, *The Theory and Practice of Scintillation Counting,* Pergamon, London, 1964.
2. E. Rapkin, *Laboratory Scintillator* **11** (October 1966).
3. L. A. Baillie, *Intern. J. Appl. Radn. Isotopes* **8**, 1–7 (1960).
4. P. E. Stanley and P. J. Malcolm, 'Practical liquid scintillation spectrometry: organising a methodology', this volume, pp. 44–62.
5. P. J. Malcolm and P. E. Stanley, 'Low level counting using liquid scintillation spectrometry: optimizing optical design', paper to be published in the proceedings of an *International Conference on Low Radioactivity Measurements and Applications,* High Tatras, Czechoslovakia, October 1975, Comenius University Press, Czechoslovakia.

6a. P. J. Malcolm and P. E. Stanley, A unified approach to the liquid scintillation counting process: Part I, A stochastic computer model, *Intern. J. Appl. Radn. Isotopes* (in press).

6b. P. J. Malcolm and P. E. Stanley, A unified approach to the liquid scintillation counting process: Part II, Optimizing optical design, *Intern. J. Appl. Radn. Isotopes* (in press).

7. P. J. Malcolm and P. E. Stanley in *Liquid Scintillation Counting: Recent Developments* (eds. P. E. Stanley and B. A. Scoggins), Academic Press, New York, 1974, pp. 77–90.
8. F. E. L. ten Haaf in *Liquid Scintillation Counting,* Vol. 2 (eds. M. A. Crook, P. Johnson and B. Scales), Heyden, London, 1972, pp. 39–48.
9. F. E. L. ten Haaf in *Liquid Scintillation Counting,* Vol. 3 (eds. M. A. Crook and P. Johnson), Heyden, London, 1974, pp. 41–6.
10. F. E. L. ten Haaf and M. L. Verheyke, 'A simple mathematical model of a liquid scintillation counter', this volume, pp. 63–73.
11. M. P. Neary and A. L. Budd in *The Current Status of Liquid Scintillation Counting* (ed. E. D. Bransome), Grune and Stratton, New York, 1970, pp. 273–82.

12. N. Kaczmarczyk in *Organic Scintillators and Liquid Scintillation Counting* (eds. D. L. Horrocks and C-T Peng), Academic Press, New York, 1971, pp. 977–90.
13. D. L. Horrocks, *Applications of Liquid Scintillation Counting,* Academic Press, New York, 1974.
14. D. L. Horrocks in *Liquid Scintillation Counting,* Vol. 3 (eds. M. A. Crook and P. Johnson), Heyden, London, 1974, pp. 1–20.
15. D. L. Horrocks, personal communication.
16. D. Moore, personal communication.
17. J. B. Birks, personal communication.
18. B. H. Laney in *Liquid Scintillation Counting: Recent Developments* (eds. P. E. Stanley and B. A. Scoggins), Academic Press, New York, 1974, pp. 455–64.
19. C. Ediss, A. A. Noujaim and L. J. Wiebe in *Liquid Scintillation Counting: Recent Developments* (eds. P. E. Stanley and B. A. Scoggins), Academic Press, New York, 1974, pp. 91–111.
20. E. Rapkin in *Liquid Scintillation Counting,* Vol. 2 (eds. M. A. Crook, P. Johnson and B. Scales), Heyden, London, 1972, pp. 61–100.
21. B. E. Gordon and R. M. Curtis, *Anal. Chem.* **40,** 1486–93 (1968).

DISCUSSION

For discussion relating to this chapter, see p. 62.

Chapter 3

Practical Liquid Scintillation Spectrometry: Organising a Methodology

Philip E. Stanley

Department of Clinical Pharmacology, The Queen Elizabeth Hospital, Woodville, South Australia 5011, Australia

Philip J. Malcolm

Biometry Section, Waite Agricultural Research Institute, Glen Osmond, South Australia 5064, Australia

INTRODUCTION

Liquid scintillation counting is one of the most commonly used of the ever increasing variety of analytical techniques available to the modern scientist. The complexity of the liquid scintillation counting (LSC) technique is, however, not generally appreciated. It is therefore not surprising that users encounter disproportionate difficulties which are aggravated by them having little or no knowledge of the physical processes involved and matters are complicated further, we believe, by the tendency of manufacturers to make spectrometers which are designed to be easier and easier to use, i.e. for instruments to be equipped with fewer and fewer controls; the user has only to load the samples and return later to remove the printed output. As the results are well presented and neatly formatted he finds it difficult to believe that they may not be valid. Indeed it is only if his results are quite unexpected that his attention is turned to his use of the LSC technique and the problems inherent therein. With fully automatic instruments, the aware user then discovers the difficulties in ascertaining and correcting for spectrometer drift, chemiluminescence, etc. It was with these problems in mind that we decided to design and produce a computer program package to assist users, expert and novice alike, in the selection of counting conditions and the processing of the resultant data. However, in attempting this, we found increasingly that it was necessary to have a unified numerical description of the whole liquid scintillation counting process. Our extensive program package has since been abandoned, at least for the present, and attention has instead been turned to simulating the LSC process on the computer, thereby bringing together data from quite different sources.

We have confined our attention to simulating soft β-emitters such as ^3H and ^{14}C, first since they are the ones most frequently used in the life sciences and second because their physical characteristics encourage a sensitive analysis of the LSC technique.

This chapter deals with the more practical aspects of the project. The preceding chapter dealt in more detail with the simulation itself.

MATERIALS AND METHODS

Instruments

Two scintillation spectrometers were employed in this work. One was an early model Packard Model 3375 Tri-Carb system fitted with a detector comprising two flat quartz-faced bialkali EMI 9635 photomultipliers coupled with optical grease to a perspex light pipe and operated at 20° C ± 0.5 °. The other was a Searle Analytic Isocap 300 instrument fitted with two bialkali EMI 9805/A photomultipliers which have a stippled (frosted) face of borosilicate glass. Its detector assembly was fitted with reflectors and was operated 20°C ± 1°. Both instruments were set to count in an integral window, i.e. all coincident events from 0–2000 keV. The elevators of both instruments were covered with a disc of black photographic paper to act as a 'perfect' light absorber (see the previous chapter).

Absorption spectra were measured using a Varian Techtron Model 635M spectrophotometer and calibrated within 1 nm using a Holmium standard filter (Philips Part No. 700570). The instrument was set for 0.2 mm slit width and operated in the double beam mode. Spectra of samples were read against toluene in 1.00 cm stoppered cells constructed of Suprasil and plotted on a chart recorder. Fluorescence spectra were recorded on an X-Y plotter from an Aminco Bowman Ratio Spectrofluorometer fitted with a photomultiplier having an S-4 response.

Chemicals

Commercial supplies of PPO – 2,5-diphenyloxazole – have been reported to have impurities from various sources.[1] It is also thought that its prolonged exposure to sunlight gives rise to quenching impurities. Consequently, PPO used in this study was twice recrystallised which resulted in an increase of 3°C in the melting point to a value of 71.8°C. A further recrystallisation did not change this value. The absorption spectrum in toluene was similar to that reported[2] as was its fluorescence spectrum at the level of 0.1 μg/ml in toluene.

Toluene (Pronalys Grade, May and Baker, Australia) was fractionally distilled and the portion boiling between 110.2°C and 110.6°C was collected and used in this work.

The scintillation solution was composed of 8.00 g PPO per litre toluene at 20°C. Its refractive index as measured using a simple Abbe refractometer was 1.5018 thus giving a critical angle of 41.75° at a solution–air interface.

Carbon tetrachloride was of spectral grade, obtained from Mallinckrodt, U.S.A., and was used as the chemical quenching agent.

4-Dimethylaminoazobenzene was obtained from Hopkin and Williams, England. A 1.0 mM solution of this compound in the scintillation solution was employed in the colour quenching studies. It was used in preference to methyl orange since both the absorption spectrum and counting efficiency of the treated sample reached a constant value within a few hours; note that with methyl orange, prepared as an alcoholic solution, the samples gave spectra which did not stabilise for at least a day.

Counting vials

A number of brands of scintillation vials were assessed for their uniformity of diameter, height and weight. Of these, Packard Tru-lite vials were chosen since their measurements exhibited the smallest standard deviation. Ten vials, selected at random, had an average 14.236 g (standard deviation (S.D.) = 0.222) of glass, density 2.23 g/ml[3] and refractive index 1.474.[3] The last value gives a critical angle of 42.7° for a glass–air interface. The same vials had an average height to the shoulder of 4.48 cm (S.D. = 0.086), and average diameters (single measurement) at the shoulder, middle and base were 2.716 cm (S.D. = 0.021), 2.718 cm (S.D. = 0.008) and 2.713 cm (S.D. = 0.022) respectively. The average thickness of the curved wall was measured to be 0.1023 cm (S.D. = 0.00435). This thickness was independently estimated by weighing the amount of water necessary to fill the vial to a height of 4.00 cm and assuming the vial had a flat base was found to be 0.096 cm. The average of these two values, 0.099 cm was used in the simulation and the unit of radius was thus taken as 1.260 cm.

All vials were washed with RBS 25 detergent then two rinses of tap water and four rinses of distilled water to ensure maximum cleanliness and to remove any possible quenching agents. Conventional caps were used and these were fitted inside with an aluminium/cork liner covered with a disc of black photographic paper to act as a 'perfect' light absorber (see the previous chapter). The paper was shown to have no quenching agents extractable by the scintillation solution.

Special vial to assess response of detectors

The relative probability that a photon (of given wavelength) reaches the photocathode of a photomultiplier is a complex function of the geometry of the detector and the point and direction at which the photon leaves the curved wall of the vial. To gain some insight into this response, 20 ml (4.00 cm high) of scintillation solution (spiked with about half a millicurie of ^{3}H–toluene) was placed in a vial which was covered entirely with black paper except for a small circular hole (1 mm diameter) 3.95 cm above the inner base of the vial.

It was then placed in the detector of each instrument in turn with the hole directly in front of one photomultiplier. With the coincidence switched off the response of each photomultiplier (above noise) was measured in an integral window. The vial was then unloaded and reloaded so that the small hole pointed at an angle of 15° to the previous one. This was repeated until the vial had been rotated the full 360° (a cardboard protractor was made, taped to the top of the elevator and shielding, and was used to ensure the vial could be accurately placed in all 24 positions from 0–360°). The vial was then fitted successively with a black paper cover having a 1 mm hole at heights of 3.0, 2.0, 0.5 and 0.05 cm above the base (this choice of heights is explained in Ref. 6), and the response plotted for each photomultiplier in each of the 24 angular positions.

Standards

Calibrated tritiated and ^{14}C-labelled hexadecane were purchased from the Radiochemical Centre, Amersham, England, and were certified as having an overall uncertainty of ± 3% and ± 2% respectively. Exactly one litre of the scintillation solution was spiked with a carefully weighed and transferred quantity of the hexadecane standard. The tritiated solution had 155,000 disintegrations min^{-1} per 20.0 ml and the ^{14}C-labelled solution 94,900 disintegrations min^{-1} per 20.0 ml. This approach avoided the errors associated with individually spiking 20.0 ml samples with small weights of the standard. 20.0 ml aliquots were then quickly pipetted into tared scintillation vials, capped, and then reweighed to confirm the volume and the height of the liquid column as 4.00 cm. The (air-quenched) samples were then counted to check their efficiency in an integral window using a preset count of 800,000 or 900,000 counts. Samples which did not have a column height between 3.95 and 4.05 were rejected as were those with counting efficiencies which exceeded the total average disintegrations rate (30 vials) by more than 0.75%.

Quenching studies

Volumes of quenching agent were dispensed into the 20.0 ml aliquots of scintillation solution (after an identical volume of spiked solution had been removed) using a 0.25 ml Gilmont Ultraprecision Micrometer Burette (graduated to the nearest 0.01 μl). In this way the total sample volume was maintained at 20.0 ml and its new disintegration rate was recalculated. Counting of colour-quenched samples was carried out six hours after the addition of the quencher to ensure that the sample counting rate and its absorption spectrum had reached a stable value. Chemically quenched samples were assessed immediately because they had been found to be quite stable.

After the colour-quenched samples had been counted a portion was removed and its absorption spectrum measured as described previously. The spectrum was dissected into about 20 linear portions between 354 nm and 500 nm. These linearised portions were entered into the computer model for the simulation of colour quenching. A spectrum was also measured for an unquenched (argon-flushed) sample for use in the model for assessing secondary photon emission from PPO.

To obtain an 'unquenched' sample an air-equilibrated standard sample was cooled to about -10°C and then dry argon was bubbled gently through it until the counting rate did not increase further after it had been brought to 20°C.

RESULTS AND DISCUSSION

Since counting efficiencies have played an important but not overriding role in validating this model, some considerable effort has therefore been made to produce accurate experimental data. All efficiencies were measured in an integral window, this being simulated in the model by counting all events producing a two-photoelectron coincident event. This was necessary since we cannot satisfactorily simulate discriminator settings due to our lack of information concerning the photomultipliers, amplifiers, pulse perturbations, etc. Counting tritium over an integral window gives, by comparison with the normal case, only slightly higher efficiencies (but with a considerable increase in the background). However, for ^{14}C, since the normal counting channel is from $\cong$ 7.5 keV to 156 keV, this difference is substantial. For example, when counting at about 96% integral the efficiency in the normal window would be 88–90%, and it is interesting to note that the difference corresponds closely to the amount of the β-spectrum for ^{14}C lying below 7.5 keV (7.7%). Compare Table 1 of the previous chapter.

A serious deficiency in our early attempts at modelling the LSC process[4] was the lack of information concerning the directional response within the counting chamber; in describing this response it was necessary to have information on the direction taken by the photon after leaving the vial as well as the point at which it left. This problem has been partially overcome by using the special vial described in the previous section.

Figure 1 (a) (b) show this information in diagrammatic form for the Searle Isocap 300 and the Packard Model 3375 Tri-Carb system respectively. Note that the data are normalised to the maximum response for each instrument. It is apparent that light travelling directly towards one photomultiplier has a small, but non-zero, chance of being detected by the other photomultiplier after multiple reflections within the chamber. The response at about 90° and 270° is less clear-cut but is quite reproducible. Note should be taken of the 'hot-spots' of maximum response on the photocathode. The variation in quantum efficiency across the photocathode is well known,[5] and the current results are consistent with this. Note also the severe fall-off in response for those photons leaving the vial near its base or at the meniscus (note that,

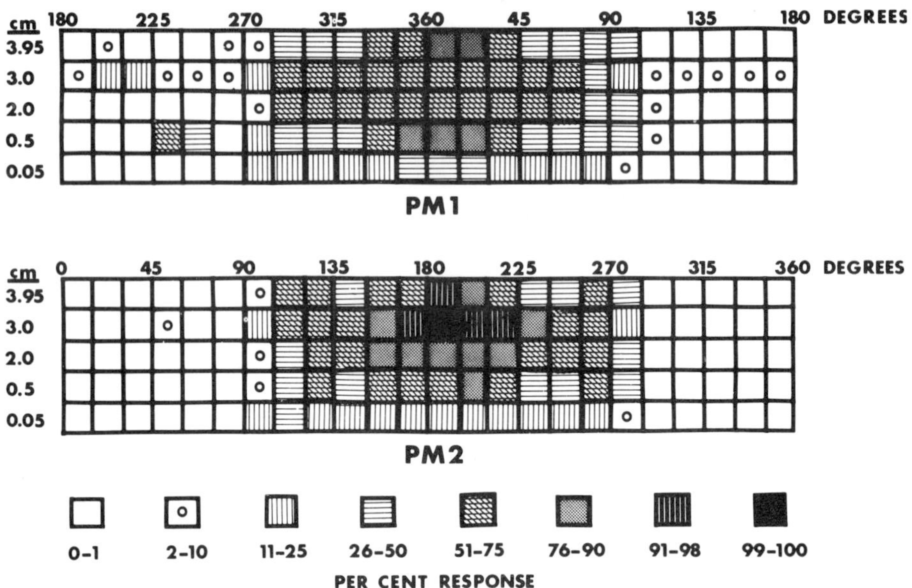

Fig. 1(a). The directional response of the counting chamber of a Searle Isocap 300. See 'Materials and Methods' section for details.

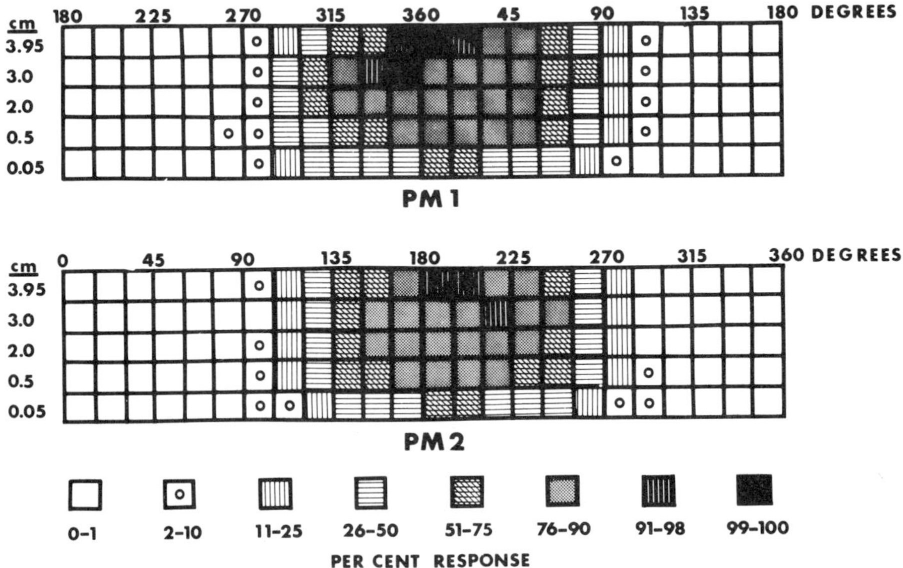

Fig. 1(b). The directional response of the counting chamber of a Packard Model 3375 Tri-Carb system. See 'Materials and Methods' section for details.

Table 1. Comparison percent counting efficiencies obtained by experiment and from the model for conditions of low impurity quenching.

	Impurity quenching factor q in model	Searle Isocap 300				Packard Model 3375 Tri-Carb system			
		Experimental		Modelled		Experimental		Modelled	
		Sample 1	Sample 2	n=2[a]	n=3	Sample 1	Sample 2	n=2[a]	n=3
^{3}H	1	64.2[b]	63.9[b]	67.6	62.4	63.9[b]	63.9[b]	70.6	66.2
	0.9			63.9	58.8			68.9	64.4
	0.8			59.4	53.3			64.8	59.3
	0.75	58.5[c]	59.4[c]	57.4	52.0	58.2[c]	58.3[c]	63.4	58.6
	0.7			55.4	48.7			61.7	56.0
^{14}C	1	96.9[b]	97.1[b]	96.9	96.4	97.8[b]	97.8[b]	97.8	97.3
	0.75	96.3[c]	97.0[c]	95.9	95.4	97.4[c]	97.7[c]	96.4	95.9

[a]A coincident pulse must be of at least n photoelectrons to be counted; n is therefore the 'pseudo-discriminator' setting.

[b]Argon-purged sample.

[c]Air-quenched sample.

as has been shown elsewhere,[6] the solution can be considered effectively of infinite height in respect of the light leaving the curved walls of the vial).

It is imperative to realise that the approach adopted for this measurement has one severe failing: it does not account for those photons which are reflected back into the vial since they are absorbed by the black photographic paper. In the normal case, however, a photon, once back in the vial, may reach the opposite photomultiplier or be reabsorbed by the solution to begin anew the production of photons.

Table 1 contains two sets of data concerning experimental and modelled efficiencies for both ^{3}H and ^{14}C. Several important features emerge. First, compare the differences in the modelled counting efficiency for ^{3}H for 2 and 3 coincident photoelectrons. Such behaviour provides a very sensitive test of the validity of the model, and has frequently been used to gauge its performance during its development. This sensitivity is readily explained by recalling that, in the case of ^{3}H, since the number of photons is small, there is a significantly smaller chance of obtaining a three photoelectron event. Compare this with the situation for ^{14}C where there is very little difference since there are many more photons involved.

Now air quenching causes about a 25% loss in primary photon production,[7] this loss being important for ^{3}H counting efficiencies of both two and three photoelectron events, but, on the other hand, having a negligible effect on ^{14}C.

The reader will no doubt have observed that the tritium data modelled for the Packard instrument are too high when compared with the experimental values, and this is further exemplified in Table 2 (for the meaning of q see the previous chapter). We suspect that the high values recorded on the Packard unit are due, at least to some extent, to the light transmission and scattering characteristics of the perspex light pipe. Figure 2 details some important spectral characteristics of the detector assembly. Note especially the percentage of light lost in a 0.5 inch thickness of normal perspex as measured in the laboratory. It is not known if this is similar to that installed in the early model instrument used; however, it is believed that a special grade of perspex was used in

Table 2. Percent counting efficiencies obtained for different values of the impurity quenching factor.

Impurity quench factor q	^{3}H		^{14}C	
	Searle Isocap 300	Packard Model 3375 Tri-Carb	Searle Isocap 300	Packard Model 3375 Tri-Carb
1	67.6[a]	70.6	96.9	97.8
0.9	63.9	68.9	96.5	–
0.8	59.4	64.8	95.7	96.5
0.75	57.4	63.4	95.9	96.4
0.7	55.4	61.7	–	–
0.6	50.7	56.3	94.5	95.5
0.5	45.7	52.2	93.8	94.6
0.4	37.2	42.7	92.8	–
0.3	29.2	34.3	89.5	90.3
0.2	17.3	21.6	84.5	86.3
0.15	–	–	79.5	81.8
0.1	–	–	70.0	72.5
0.075	–	–	61.3	64.9
0.05	–	–	48.2	52.4

[a]Counting efficiencies for all coincident events, i.e. for an integral counting window.

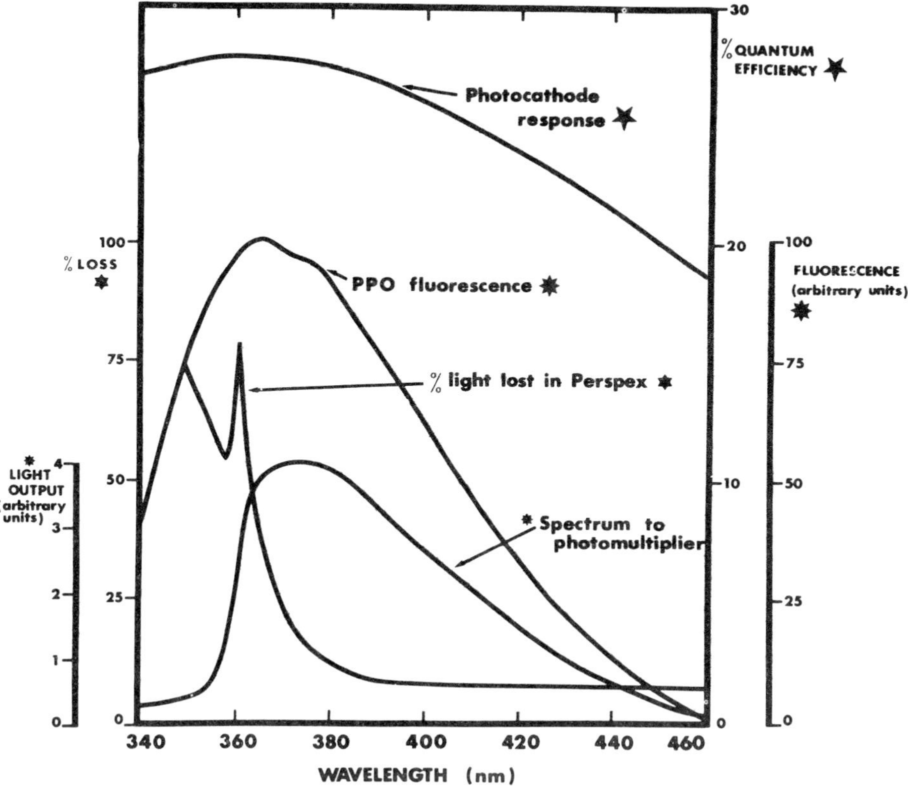

Fig. 2. Some optical characteristics of the scintillation solution as compared to perspex (0.5 in).

spectrometers manufactured after this date – indeed, the later Packard instruments apparently use mirror reflectors. The spectrum to the photomultipliers is the real spectrum leaving the vial and it can be readily seen how the type of perspex measured is poorly matched both to the light emission- and photocathode response-spectra, and would thereby lower the average quantum efficiency observed in practice; this feature, if modelled, would lower the efficiencies predicted by the current model. Because of the complex geometry of the light pipe and the lack of the facilities necessary to characterise it, we decided henceforth to deal only with the Searle Isocap for modelling. The influence on efficiency of the characteristics of individual light pipes in the Packard unit was observed when another light pipe was involved for a short period; this resulted in a decrease of some 3% in the counting efficiency for unquenched ^{3}H.

It is also interesting to note in Table 2 that, after removing 95% of the energy available for the production of photons in ^{14}C, a counting efficiency of about 50% is still recorded whereas, in contrast, an 80% decrease for ^{3}H brings the efficiency into a range that today's users would find only marginally acceptable.

As mentioned above, prediction of the tritium efficiency was a useful guide to the model's validity during its period of development. Table 3 indicates the sensitivity of the model to setting the lower pseudo-discriminator in terms of a coincident event of at least n photons. Counting efficiency decreases drastically as n increases, and q, the impurity quenching factor, decreases. Note also the substantial relative increase in average β-energy necessary not only to produce n coincident photoelectrons, but also

Table 3. Simulation of different 'discriminator' settings for the unquenched and impurity quenched situations (only coincident 'pulses' of at least n photoelectrons are counted).

	$n = 2$		$n = 3$		$n = 5$		$n = 10$	
	% Effic.	keV[a]	% Effic.	keV	% Effic.	keV	% Effic.	keV
^{3}H argon-purged (q[b] $= 0.9$)	63.9	7.4	58.8	7.7	43.3	8.7	14.0	11.1
^{3}H air-quenched ($q = 0.8$)	59.4	7.6	53.3	8.0	37.2	9.1	10.9	11.9
^{3}H impurity-quenched ($q = 0.5$)	45.7	8.3	38.6	8.9	21.2	10.3	2.0	12.8
^{14}C argon-purged ($q = 1$)	96.9	51.8	96.4	52.0	95.0	52.7	90.3	55.0
^{14}C air-quenched ($q = 0.75$)	95.9	51.4	95.4	51.7	93.1	52.8	86.5	56.0
^{14}C impurity-quenched ($q = 0.4$)	92.8	53.8	91.5	54.5	87.2	56.6	75.6	62.4

[a] Average β-energy of events being counted.

[b] Impurity quench factor (see text).

to produce the modelled efficiencies as q decreases. To a much lesser extent the same is true for ^{14}C, since here again the substantially larger number of photons involved play an all important part.

In Table 4 we consider some experimental and modelled data for colour quenching. Notice that for moderate amounts of quenching agent (less than 200 μl) there is reasonable agreement between the experimental and modelled data when $n = 2$ and $q = 0.78$ or 0.75 for air-quenched samples. However, for ^{14}C, there is a considerable divergence in these efficiencies with volumes of colour quenching agent in excess of 200 μl; there are two immediately obvious explanations for this behaviour. First, at these concentrations the 4-dimethylaminoazobenzene may be acting not only as a colour quencher, but also as an impurity or chemical quencher. Second, the model is incorrect, at least in respect of extremely heavy colour quenching (note that the worst case is the sample counted at less than 30% efficiency in a normal ^{14}C window, and is therefore hardly of practical interest).

The first hypothesis, namely that the model can be used to determine the degree of chemical quenching inherent to various coloured agents, is, in the opinion of the authors, while tantalising, as yet by no means substantiated because of the many simplifying assumptions made in defining the model.[8] It is, however, interesting to

Table 4. Comparison of observed and predicted counting efficiency for colour quenching conditions.

	Coloured agent in 20 ml of solution (μl)	Observed counting efficiency (%)	Predicted counting efficiency (%) for a pulse of at least n photo-electrons				Difference between expected and observed efficiencies for
			$n = 2$		$n = 3$		
			$q^a = 0.78$	q=0.75	q=0.78	q=0.75	$n = 2$ $q = 0.78$
^3H	7.6	57.5	60.4	58.7	54.3	53.1	2.9
	20.4	55.4	57.7	58.4	51.8	52.7	2.3
	35.2	53.6	57.0	54.9	50.3	48.1	3.4
	52.4	50.6	54.1	53.0	48.0	46.8	3.5
	73.2	48.9	49.9	48.7	43.3	42.5	1.0
	97.0	44.9	48.2	46.4	41.8	39.7	3.3
	115.6	43.0	43.3	43.8	36.5	37.4	0.3
	139.8	40.0	42.5	40.1	35.9	33.2	2.5
	162.4	38.2	40.6	41.7	33.3	35.1	2.4
^{14}C	20.2	95.2	95.6	95.7	94.8	95.1	0.4
	74.6	94.5	94.7	–	93.8	–	0.2
	124.5	93.0	94.1	94.0	93.4	93.2	1.1
	184.3	91.3	93.4	–	92.5	–	2.1
	235.7	88.3	91.6	–	90.4	–	3.3
	297.4	85.0	90.3	89.9	89.3	88.5	5.3
	353.6	82.5	88.3	–	87.1	–	5.8
	406.0	78.9	88.2	–	87.3	–	9.3
	452.9	76.1	86.3	–	85.0	–	10.2
	503.8	74.9	85.5	83.4	84.1	82.2	10.6

[a] q is the impurity quench calibration factor (see text).

note that a spherically symmetric LSC system, described elsewhere,[9,10] can be mathematically modelled with far greater rigour;[6,9] such a system and its associated model might therefore usefully be applied in distinguishing the chemical quenching component of coloured agents.

Consider Figs. 3 and 4. These data deal exclusively with wavelengths of photons and their absorption within the system. Consequently, no distinction can be made on a photon to photon basis between scintillations derived from either ^{3}H or ^{14}C. The data are produced by the model for colour- and chemically quenched samples counting at similar efficiencies, and include information for all photons simulated. In each case the fluorescence spectrum is built-up by the model from a large number of photons and, whilst their shapes are very similar, they are not identical due to the stochastic nature of the model. From about 354 nm to 360 nm the transmittance through 1 cm of the scintillation solution increases from 1% to 50%. This behaviour is not seen in the figures since the absorption spectra are weighted according to the fluorescence spectrum for PPO.

Note that the absorption spectrum for the chemical quenching (Fig. 3) spectrum applies only to PPO and, above 354 nm, is substantially different to that given for an equivalent colour-quenched case (Fig. 4). As a result, the spectra of photons reaching the photomultipliers are substantially different for chemically and colour-quenched samples. Note also that, due to the various photon losses (at the top and base of the vial, the detector, etc.), the sum of the absorption spectrum and the spectrum to the photomultipliers is always less than the fluorescence spectrum. Note finally that the

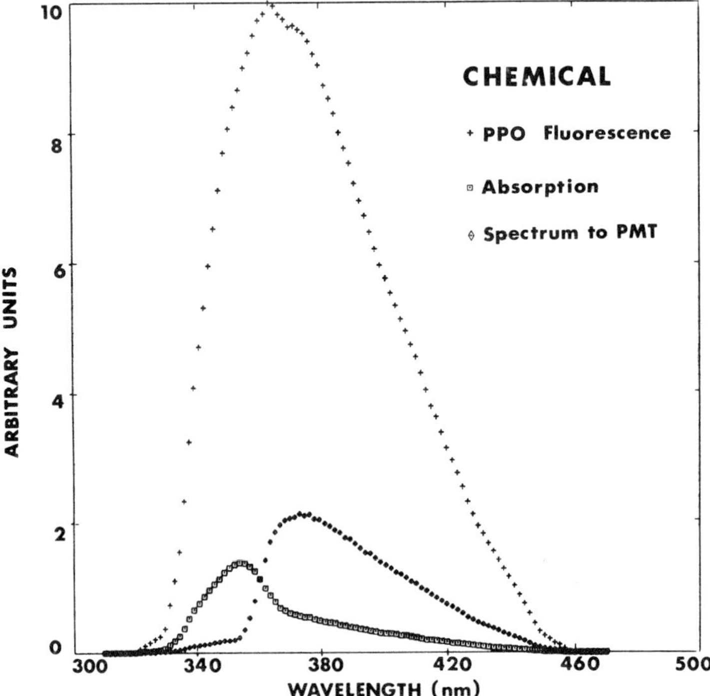

Fig. 3. Spectra generated by the model for the chemically quenched situation. Data derived from 10,000 simulated ^{14}C events. Integral counting efficiency ≅93%.

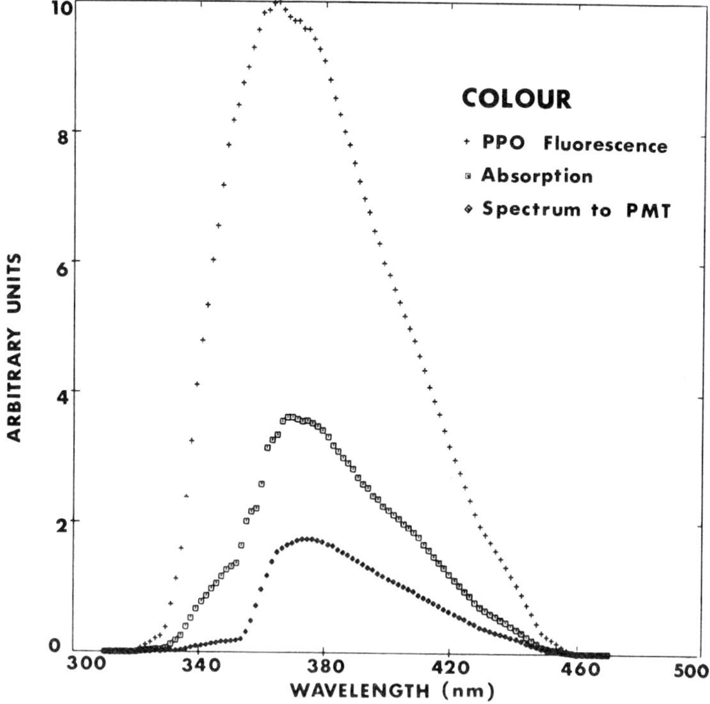

Fig. 4. Spectra generated by the model for the colour-quenched situation. Data derived from 8,000 simulated ^{14}C events. Integral counting efficiency ≅93%.

comparable counting efficiency in the two systems is a function of the number of primary photons and secondary photons simulated (see Table 1 of the previous chapter).

The advantages of lesser pulse height analysis, at least for β-emitters with energies up to 200 keV, have been demonstrated by Ediss *et al.*[11] and Laney[12] who have emphasised its value when assessing efficiencies of either colour- or chemically quenched samples. Various authors have shown that quenching of ^{14}C does in fact give distinctly different channels ratio versus efficiency curves when the summed pulse height is employed. However, if the smaller or lesser of the two pulses which enable coincidence is used then the resultant curves are very similar. Neary and Budd[13] have pointed out that, for ^{3}H, the difference in pulse height is slight and that this is due to the small number of photons produced. However, for more energetic isotopes such as ^{14}C, there is a marked difference since more than 10 times as many photons are involved (see Tables 2–4 of the previous chapter), and this permits a more probabalistic variation for colour than for chemical quenching.

The model reported herein produces results, similar to those made in the laboratory, which are presented in Figs. 5–8. Data produced by the model (and given in Figs. 5 and 6) predict that there should be, for tritium, little difference between the summed pulse height spectrum for chemical- and colour-quenched samples; the same may be said for the lesser pulse height spectrum. Thus, tritium samples, whether colour- or chemically quenched, should give similar pulse height spectra when counted in an integral window for the same counting efficiency. The channels ratio versus efficiency

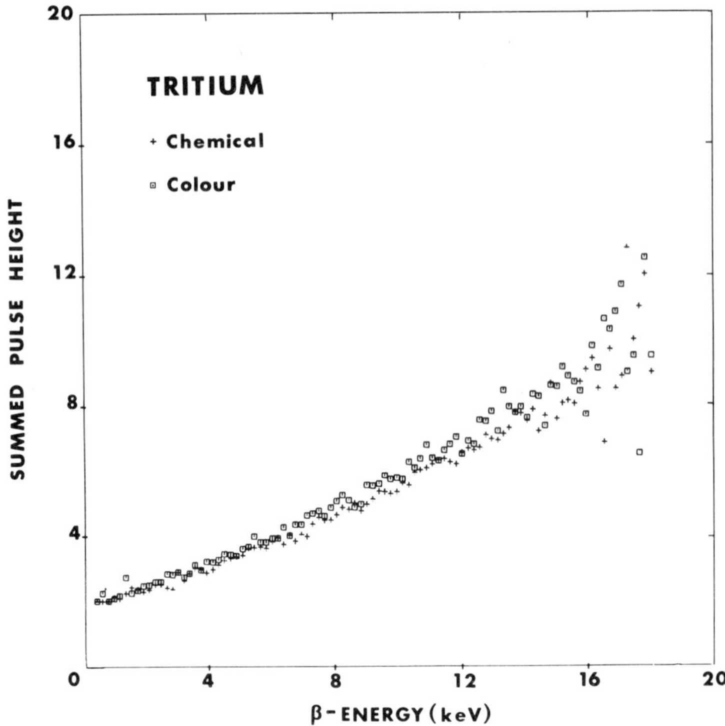

Fig. 5. A plot of average summed pulse height versus β-energy for colour- and chemically quenched tritium events which enable coincidence (note the different scale for the vertical axis when compared with Fig. 6). Integral counting efficiency ≅44%.

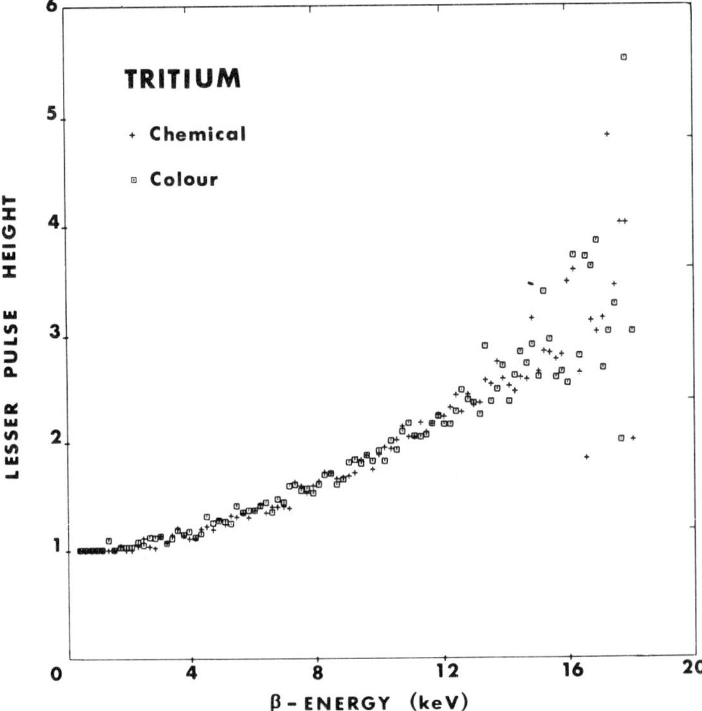

Fig. 6. A plot of average lesser pulse height versus β-energy for colour- and chemically quenched tritium events which enable coincidence (note the different scale for the vertical axis when compared with Fig. 5). Integral counting efficiency ≅44%.

curve should therefore be similar. The fuzziness in the distribution for β-energies above 15 keV is due in part to relatively few (1.3%) events occurring at energies in excess of this.

The situation of ^{14}C is quite different, as can be seen in Figs. 7 and 8, and, as has been mentioned several times before, this is mainly due to the larger number of photons involved. The position of the scintillation of β-event within the vial, particularly in relation to the curved wall, now becomes of importance and this is discussed in detail elsewhere.[6,9] Figure 7 clearly shows the difference between the summed pulse heights for colour- and chemically quenched ^{14}C samples counting at 93% (integral), and this confirms that the channels ratio versus efficiency curve would be substantially different. However, in considering the lesser pulse heights for the same samples (Fig. 8), the data for chemical and colour quench lie closely together. This is consistent with the channels ratio versus efficiency curves being similar. Note also that increased fuzziness at high energies, especially for the colour-quenched sample, is due in part to relatively few (8.3%) β-energies occurring above 100 keV.

It is likely for β-emitters which are much more energetic than ^{14}C, ^{32}P for example, that the summed pulse height spectra for colour- and chemically quenched samples will diverge even more, but the situation cannot be treated here since such an emitter cannot be considered to produce a scintillation at a point source.

The number of greater and lesser pulses for photons of different wavelengths are plotted in Figs. 9–12 and included here is the data for both ^{3}H and ^{14}C samples which have been either chemically or colour quenched. (Note that the optical behaviour is independent of the degree of chemical or impurity quenching.) The peak of each

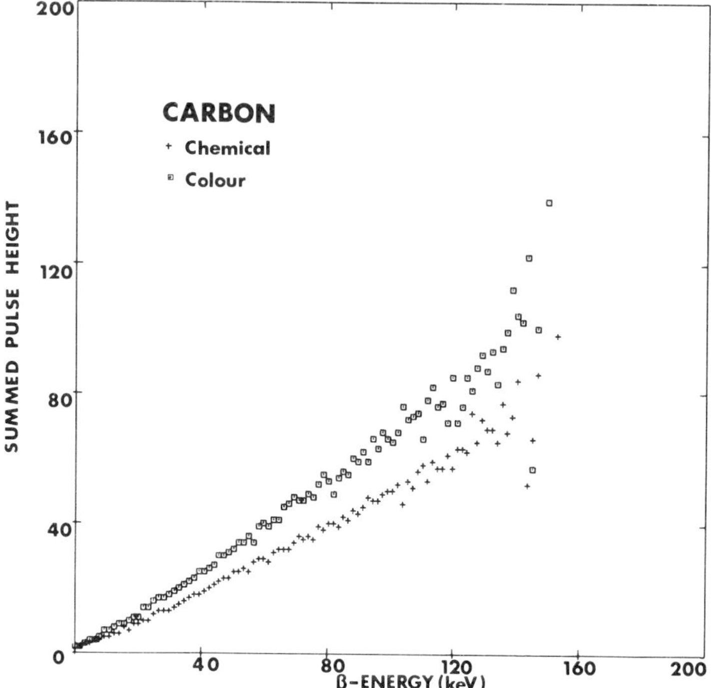

Fig. 7. A plot of average summed pulse height versus β-energy for colour- and chemically quenched ^{14}C events which enable coincidence. (Note the different scale for the vertical axis when compared with Fig. 8). Integral counting efficiency ≅93%.

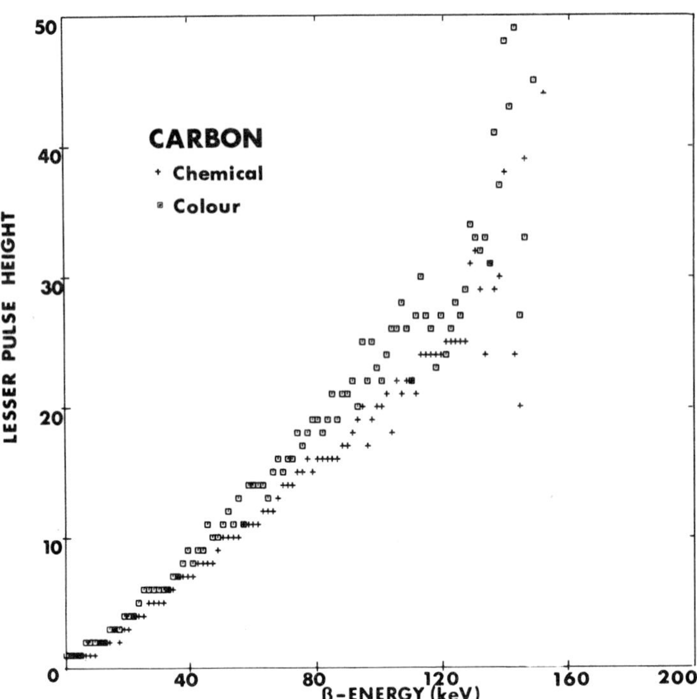

Fig. 8. A plot of average lesser pulse height versus β-energy for colour- and chemically quenched ^{14}C events (note the different scale for the vertical axis when compared with Fig. 7). Integral counting efficiency ≅93%.

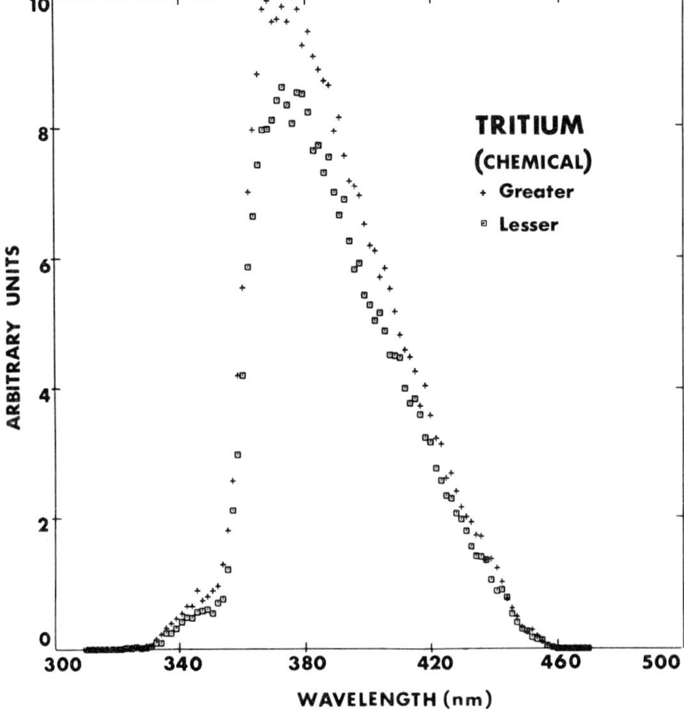

Fig. 9. The relationship between the photon spectra received by the greater and lesser photomultipliers for chemically quenched tritium events (integral counting efficiency ≅44%).

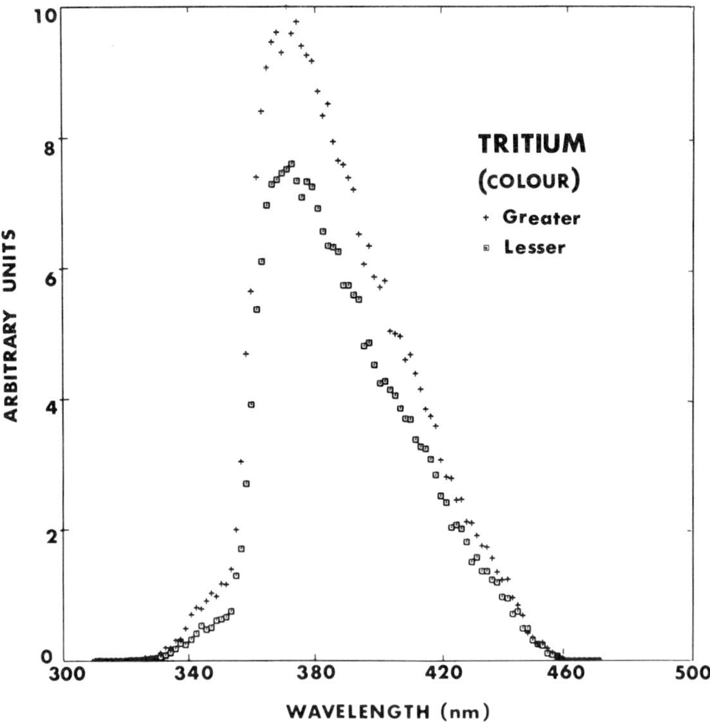

Fig. 10. The relationship between the photon spectra received by the greater and lesser photomultipliers for colour-quenched tritium events (integral counting efficiency ≅44%).

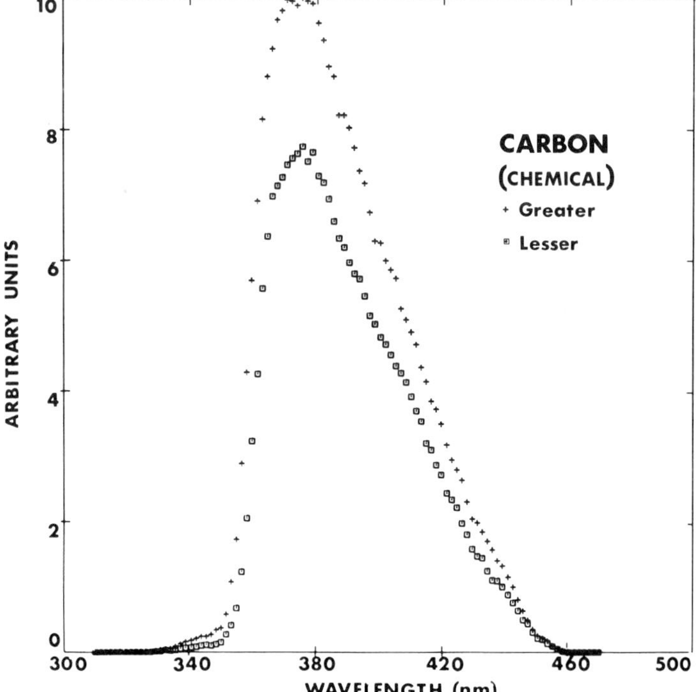

Fig. 11. The relationship between the photon spectra received at the greater and lesser photomultipliers for chemically quenched ^{14}C events (integral counting efficiency ≅93%).

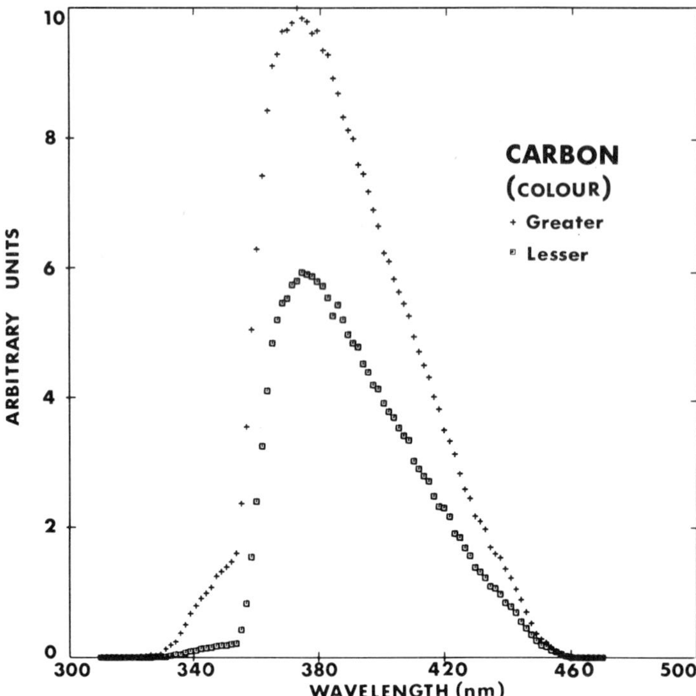

Fig. 12. The relationship between the photon spectra received at the greater and lesser photomultipliers for colour-quenched ^{14}C events (integral counting efficiency ≅93%).

curve is reached at about 380 nm. Again these results are for those events enabling coincidence and for the small number of photons leaving the vial in the case of ^{3}H there is, as expected, only a small difference between the greater and lesser plots. For ^{14}C, however, this difference is more marked since there are more photons involved. There is also, for the colour-quenched situation, a noticeable step in the 'greater' distribution below about 355 nm. Relative transmittance is less for wavelengths below this value for chemical since *relatively* more photons absorbed by PPO ($\lambda < 355$ nm) travel a shorter distance to the photomultipliers. This step, again due to the small number of photons, is just discernable in the case of the colour-quenched ^{3}H sample.

CONCLUSION

In providing numerical data for the mathematical model, it has been necessary to examine carefully, and modify where necessary, the methodology applied in the laboratory even for a system as simple as the PPO–toluene one which has been simulated. The importance of purity of the scintillation chemicals was reaffirmed and time-dependent changes in the optical spectrum for colour-quenched systems have been noted. The effect of small changes in vial geometry has been previously reported[14] and this led us to use vials of the best uniform dimensions to provide optical uniformity from vial to vial. We also report the directional response within the detector chambers of two spectrometers. Perhaps, though, of most importance is the apparent validity of the model in providing numerical data which is consistent with that found in the laboratory and as has been described in the previous chapter, how it has provided an insight into the complex dynamic interactions (such as total internal reflections) which occur within the system.

In addition, the model itself should be of considerable practical value in allowing workers to make better use of the liquid scintillation spectrometer as well as to assist both beginner and expert alike in their understanding of the LSC process. Furthermore, its use could provide an economical alternative to constructing and testing new designs. Finally, the model should not be overlooked as a tool for fundamental research into the liquid scintillation system; indeed, it has already led the authors to propose a spherical counting vial and detector[9,10] in which a tritium counting efficiency in excess of 80% might well be expected.

ACKNOWLEDGEMENTS

Acknowledgements given in the previous chapter apply also to the present one. In addition, one of us (Philip E. Stanley) wishes to thank the Department of Agricultural Biochemistry, Waite Agricultural Research Institute, The University of Adelaide, for making available the Packard Model 3375 Tri-Carb system.

REFERENCES

1. K. Painter and M. Gezing, in *Liquid Scintillation Counting: Recent Developments* (eds. P. E. Stanley and B. A. Scoggins), Academic Press, New York, 1974, pp. 183-7.
2. J. B. Birks, *Solutes and Solvents for Liquid Scintillation Counting,* Koch-Light Laboratories, Colnbrook, 1969.
3. D. Moore, personal communications.
4. P. J. Malcolm and P. E. Stanley, in *Liquid Scintillation Counting: Recent Developments* (eds. P. E. Stanley and B. A. Scoggins), Academic Press, New York, 1974, pp. 77-90.
5. D. E. Persyk and T. T. Lewis, in *Liquid Scintillation Counting,* Vol. 3 (eds. M. A. Crook and P. Johnson), Heyden, London, 1974, pp. 21-7.
6. P. J. Malcolm and P. E. Stanley, A unified approach to the liquid scintillation counting process: Part I, A stochastic computer model, *Intern. J. Appl. Radn. Isotopes* (in press).
7. J. B. Birks, personal communication.
8. P. J. Malcolm and P. E. Stanley, 'Systematically understanding the liquid scintillation counting process: a stochastic computer model', this volume, pp.15–43.
9. P. J. Malcolm and P. E. Stanley, A unified approach to the liquid scintillation counting process: Part II, Optimizing optical design, *Intern. J. Appl. Radn. Isotopes* (in press).
10. P. J. Malcolm and P. E. Stanley, 'Low level counting using liquid scintillation spectrometry: optimizing optical design, paper to be published in the proceedings of an *International Conference on Low Radioactivity Measurements and Applications,* High Tatras, Czechoslovakia, October 1975, Comenius University Press, Czechoslovakia (in press).
11. C. Ediss, A. A. Noujaim and L. I. Wiebe, in *Liquid Scintillation Counting: Recent Developments* (eds. P. E. Stanley and B. A. Scoggins), Academic Press, New York, 1974, pp. 91-101.
12. B. Laney, in *Liquid Scintillation Counting: Recent Developments* (eds. P. E. Stanley and B. A. Scoggins), Academic Press, New York, 1974, pp. 455-64.
13. M. P. Neary and A. L. Budd, in *The Current Status of Liquid Scintillation Counting* (ed. Edwin D. Bransome, Jr.), Grune and Stratton, New York, 1970, pp. 273-82.
14. P. E. Stanley, in *Liquid Scintillation Counting,* Vol. 2.(eds. M. A. Crook, P. Johnson and B. Scales), Heyden, London, 1972, pp. 285-92.

DISCUSSION

J. B. Birks: The fluorescence parameters (spectrum lifetime and quantum yield) of a scintillator solute molecule are independent of the wavelength of excitation. Hence the fluorescence spectrum of a molecule excited by self-absorption is identical with its normal emission spectrum. The fact that the energy of the emitted photon may exceed that of the absorbed photon does not conflict with the conservation of energy. The extra energy is acquired by the excited molecule as thermal energy from the solvent molecules. This process of thermal equilibration occurs in about 10^{-12} s, which is much less than the solute excitation lifetime of about 10^{-9} s.

F. E. L. ten Haaf: Concerning the difference between 'summed' and 'lesser' pulse height analysis, have you any quantitative figures on efficiencies of samples produced via the 'lesser' versus the normal method? The shape of the spectrum will influence channels ratio and other methods used to determine efficiencies by comparison to a standard series.

P. E. Stanley: This was not considered.

J. B. Birks: Concerning the validity of computer modelling of the quenched samples, is there any theoretical justification for lesser pulse height analysis? Its apparent purpose is to make impurity quenching and colour quenching look similar. Impurity quenching obeys a linear Stern–Volmer law, while colour quenching depends exponentially on the scintillator pathlength. The greater of the two coincident pulses experiences the lesser quenching and it would appear to be a more promising candidate than the pulse. The arbitrary discarding of over half the precious experimental data, as in lesser pulse height analysis, requires more justification than it has yet received. Perhaps Dr. Laney would care to comment?

E. F. Polic: How were the chemical colour standards chosen to be 'identical' for the spectral comparison?

P. J. Malcolm: The instrument was set to count integrally with the lower level discriminator set at the coincidence threshold. Counting efficiencies for both the ^{14}C chemical-quenched and colour-quenched samples were set at 93% by careful addition of quenching agents.

Chapter 4

A Simple Mathematical Model of a Liquid Scintillation Counter

F. E. L. ten Haaf and M. L. Verheijke

N. V. Phillips' Gloeilampenfabrieken, Eindhoven, The Netherlands

INTRODUCTION

Although it is possible to understand each of the basic processes taking place in a liquid scintillation counter, the number of parameters involved, the complex way in which they interact with each other and the difficulty in assessing some of them quantitatively make it virtually impossible to give an accurate mathematical description of the complete liquid scintillation counting process. However, even imperfect mathematical models may eventually become helpful in understanding certain aspects of this process and possibly lead to improvements in measuring methods or instrument design. In recent years, various models have been proposed.[1-3]

The mathematical model presented in this study is a 3-dimensional extension of the model discussed earlier by one of the authors.[4] It is used to generate probability distributions representing pulse height distributions from differently quenched liquid scintillation samples. These are compared with experimentally obtained pulse height distributions. Calculated channels-ratio calibration curves, based on the pulse height distributions produced by the model, are compared with experimentally obtained channels-ratio curves.

THE MATHEMATICAL MODEL

In order to reduce the complexity of the model a number of simplifying assumptions are made. Secondary effects, such as re-emission of light following an absorption, and effects caused by the vial wall and cap are neglected. The bottom of the vial and the detector chamber are considered to be perfect mirrors. No allowance is made for the spectral distribution of light emission and absorption. The relation between pulse height and β-energy is assumed to be linear. Statistical fluctuations in any parameter concerned are not taken into account.

We now consider a scintillation event caused by a β-particle with an energy E at an arbitrarily chosen point P (ρ, α, y) in the sample, assuming that its light is emitted evenly in all directions (see Fig. 1). In the absence of quenching, we suppose that one half of the light goes to photomultiplier 1, whereas the other half is collected by

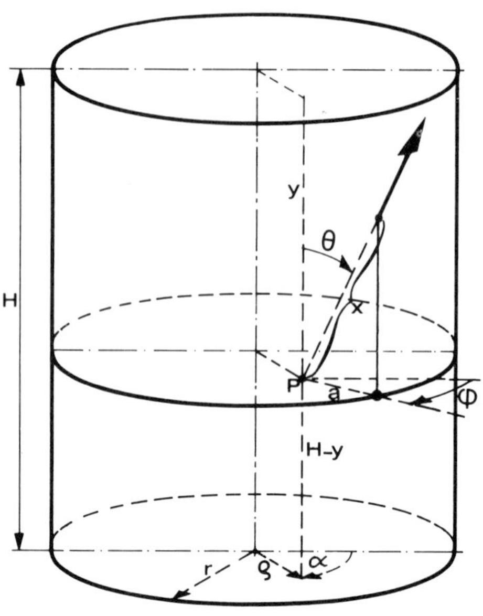

Fig. 1.

photomultiplier 2. The output pulse heights of both photomultipliers will then be equal and proportional to E:

$$h_1 = h_2 = \tfrac{1}{2} cE \tag{1}$$

In the presence of quenching, h_1 and h_2 will generally be different:

$$h_1 = f_1 cE \tag{2a}$$

$$h_2 = f_2 cE \tag{2b}$$

For f_1 and f_2 we find:

$$f_1(\rho, \alpha, y) = \frac{q}{4\pi} \int_{\phi = \pi/2}^{3\pi/2} \int_{\theta = 0}^{\pi} \exp(-\mu x) \sin\theta \, d\theta \, d\phi \tag{3a}$$

$$f_2(\rho, \alpha, y) = \frac{q}{4\pi} \int_{\phi = -\pi/2}^{\pi/2} \int_{\theta = 0}^{\pi} \exp(-\mu x) \sin\theta \, d\theta \, d\phi \tag{3b}$$

Here q is the chemical quench factor ($0 < q \leqslant 1$), μ is the optical attenuation coefficient of the sample material ($\mu \geqslant 0$) and x is the length of the light path through the sample. The value of x is a function of the coordinates ρ, α, y, θ and ϕ, and of the height H and radius r of the sample:

$$\left.\begin{aligned} &x = y \sec\theta \quad \text{for } 0 \leqslant \theta \leqslant \arctan\frac{a}{y} \\ &x = a \operatorname{cosec}\theta \quad \text{for } \arctan\frac{a}{y} \leqslant \theta \leqslant \pi + \arctan\frac{a}{y-H} \\ &x = (y-H)\sec\theta \quad \text{for } \pi + \arctan\frac{a}{y-H} \leqslant \theta \leqslant \pi \\ &\text{where} \\ &a\sqrt{r^2 - \rho^2\sin^2(\alpha-\phi)} - \rho\cos(\alpha-\phi) \end{aligned}\right\} \tag{4}$$

If linear pulse summation is applied, the height of the pulse presented by the detector to the analyser as a result of a scintillation event at P(ρ, a, y) is

$$h_s(\rho, \alpha, y) = \frac{qcE}{4\pi} \int_{\phi=0}^{2\pi} \int_{\theta=0}^{\pi} \exp(-\mu x) \sin\theta \, d\theta \, d\phi \tag{5}$$

This pulse, however, will only be counted if both h_1 and h_2 are above the one-photoelectron level.

In order to find the pulse height distribution for a given β-emitter divided homogeneously in a sample with given quenching parameters, the integrations (3) and (5) have to be performed for a sufficient number of points in the sample and a representative number of values of the β-energy distribution function. We chose 4400 points, divided evenly throughout the sample and properly weighed for the volume they represented. However, the actual number of points required for calculating each pulse height distribution could be reduced to 550 on symmetry considerations. The time required by the computer to calculate one pulse height distribution was about 4 min.

For the β-energy distribution we used the expresssion

$$p(E) = \mathrm{K}F(E + 511)(E_{max} - E)^2 (E^2 + 1022E)^{\frac{1}{2}} \tag{6}$$

where $p(E)$ is the probability that a β-particle has an energy E (in keV). The function F, which is a correction factor for the influence of the electrical field of the nucleus, was disregarded and made equal to 1. The constant K was used to normalise the function in such a way that

$$\int_0^{E_{max}} p(E) dE = 1 \tag{7}$$

It is well known from practical measurements that instrumental quench correction methods, such as the channels-ratio method and the external standard method, are sensitive to the shape of the pulse height spectrum. It therefore seemed interesting to calculate channels-ratio curves from pulse height distributions generated by the model and to compare these with experimental data. In order to represent quenchers with mixed chemical and colour quenching properties in a form suitable for use in the model, we assumed

$$\left.\begin{aligned} q &= \exp(-\gamma Q) \\ \mu &= \gamma M \end{aligned}\right\} \qquad (8)$$

where γ fulfils the function of a concentration.

Values for the integral efficiency and channels-ratio were obtained from five series of calculated pulse height distributions, each for a different pair of values of Q and M, according to the following table:

Curve	Q	M
A	1	0
B	0.75	0.25
C	0.5	0.5
D	0.25	0.75
E	0	1

The channels-ratio curves obtained by means of these values are shown in Fig. 2. The calculated distributions were based on a value of 158 keV for E_{max}. The curves can, consequently, be associated with ^{14}C.

A few calculated pulse height distributions are shown in Fig. 3. They are the distributions that produced the 'calibration points' indicated by the dotted circles in Fig. 2.

EXPERIMENTAL RESULTS

Experimental data were obtained by means of a commercial liquid scintillation counter (Philips PW 4540). For comparison with the calculated results we prepared several series of progressively quenched ^{14}C samples. All samples contained 50,000 disintegrations min^{-1} ^{14}C in the form of ^{14}C-hexadecane. They were quenched with CCl_4, Azobenzene and Sudan red, respectively, and of each we prepared three series with different scintillator concentrations. Figures 4 to 6 show the ratio of the integral counting efficiency of the quenched samples to that of the unquenched sample of each series.

The carbon tetrachloride shows its typical chemical quenching properties by producing markedly different results at different scintillator concentrations. Sudan red is practically insensitive to changes in the scintillator concentration, demonstrating its predominant colour quenching properties. Azobenzene seems to assume an inter-

efficiency
CALCULATED
CALIBRATION CURVES
CARBON 14
q=1
μ=0
A
B
C
D
E
μ=6
channels ratio

Fig. 2.

differential
intensity
(arbitrary linear scale)
CALCULATED
PULSE HEIGHT DISTRIBUTIONS
OF CARBON 14
97.3% (unqenched)
60.1%
59.5%
28.0%
26.3%
predominant
chemical
quenching
predominant
colour
quenching
counting level
ratio level
2log pulse height

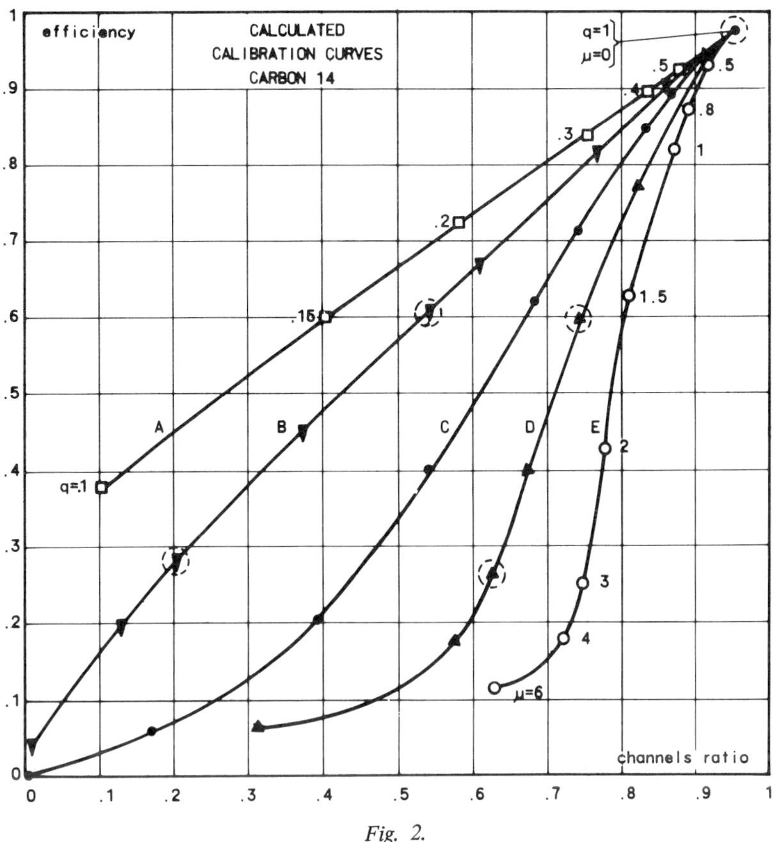

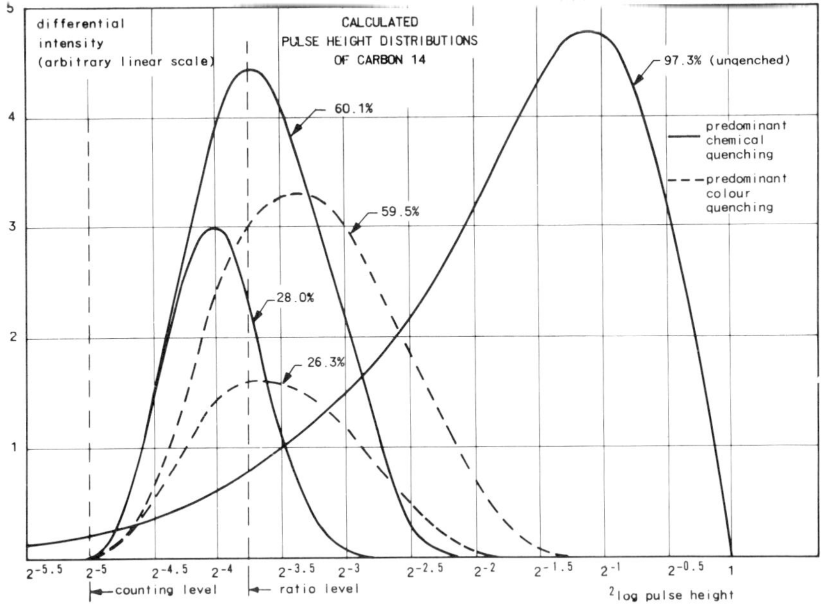

Fig. 3.

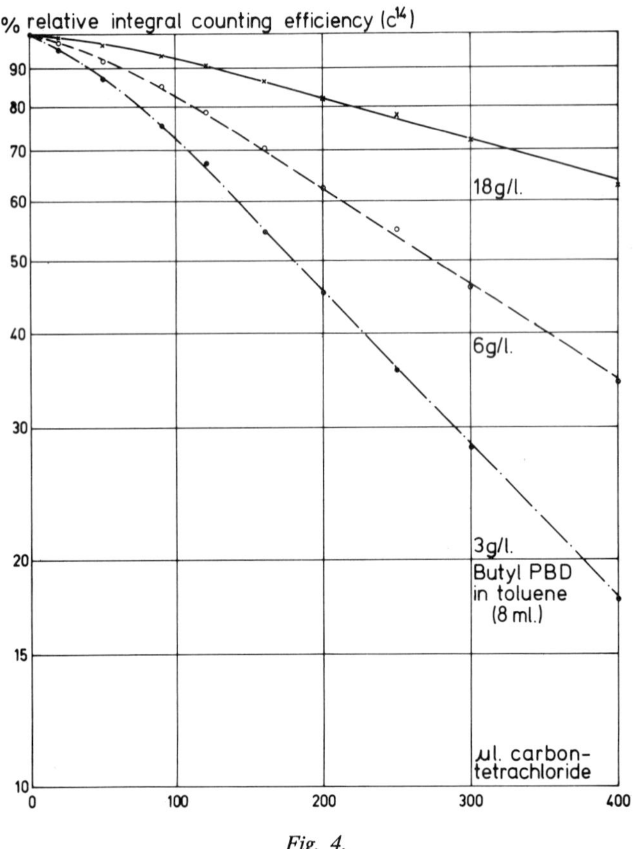

Fig. 4.

mediate position. These considerations are in agreement with Fig. 7, which shows experimental quench curves obtained with the three quenching agents previously mentioned. The scintillator was 6 g l^{-1} butyl-PBD in toluene. The curves represent integral counting efficiency versus sample channels-ratio.

Of a few samples (indicated by circles in Fig. 7) pulse height distributions were determined by means of an experimental logarithmic pulse height analyser. They are shown in Fig. 8. As each channel of this analyser represents a pulse height increment $H + \Delta H = \sqrt{2}$, the horizontal scale of Fig. 8 is equivalent to that of Fig. 3.

DISCUSSION AND CONCLUSION

The experimental, as well as the calculated, pulse height distributions (shown in Figs. 3 and 8, respectively) show the typical differences in shape attributable to different types of quenching. It should be noted that the presentation on a logarithmic pulse height scale has two advantages. One is that it is easier to represent a wide range of pulse amplitudes in a single figure. The second advantage is that a multiplication in the pulse height direction amounts to a simple translation. This makes it easier to compare changes in spectral shape due to different quenching conditions.

When comparing the calculated with the experimental channels-ratio curves (Figs. 2 and 7), one is tempted to conclude that, apparently, carbon tetrachloride and

% relative integral counting efficiency (c14)
Butyl PBD
in toluene
(8 ml.)
18 g/l.
6 g/l.
3 g/l.
5 g/l.
Sudan III
in toluene
μl. Sudan III
solution
% relative integral counting efficiency (c14)
Butyl PBD
in toluene
(8 ml.)
18g/l.
6g/l.
3g/l.
40g/l.
azobenzene
in toluene
μl. azobenzene solution

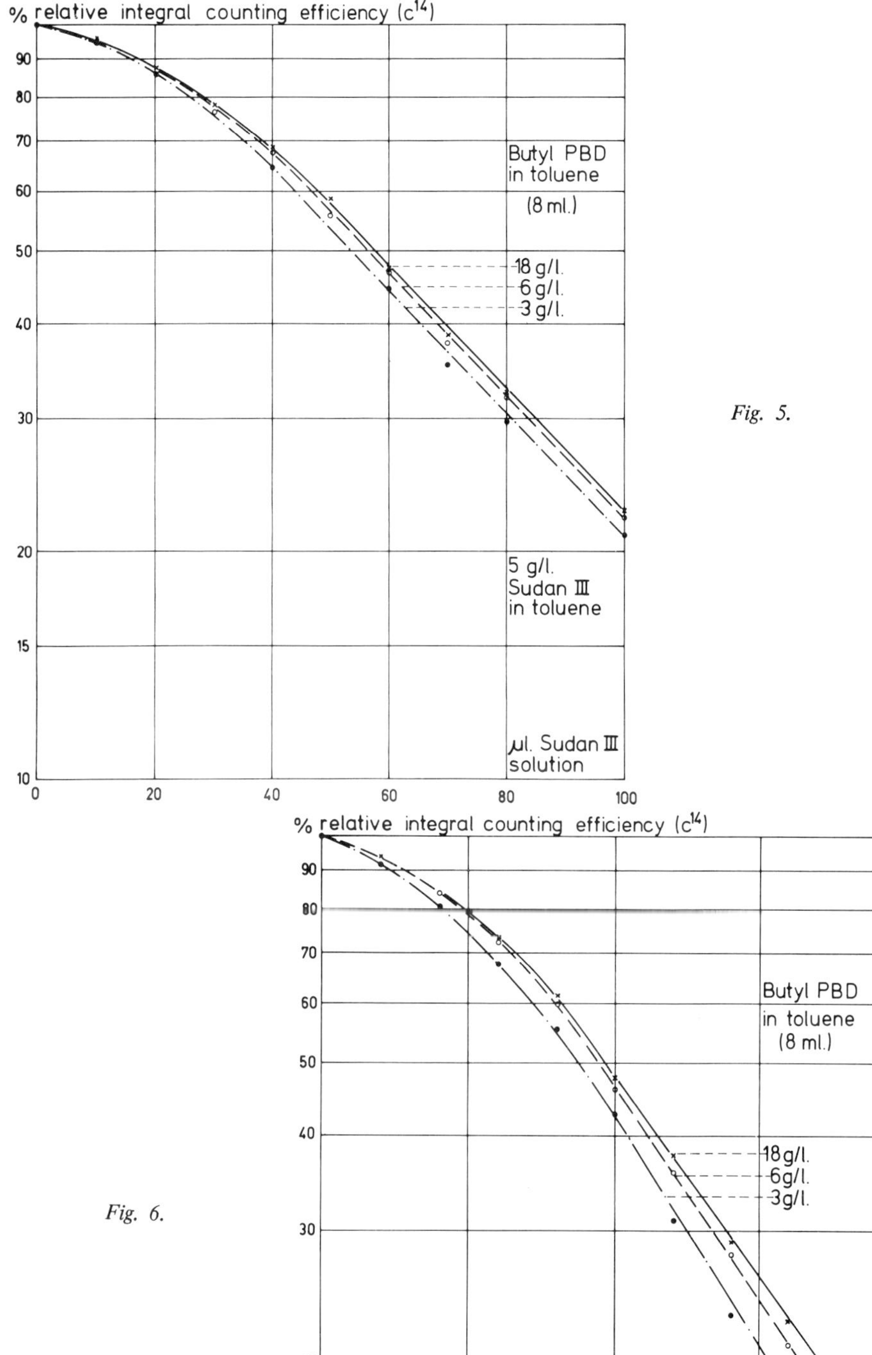

Fig. 5.

Fig. 6.

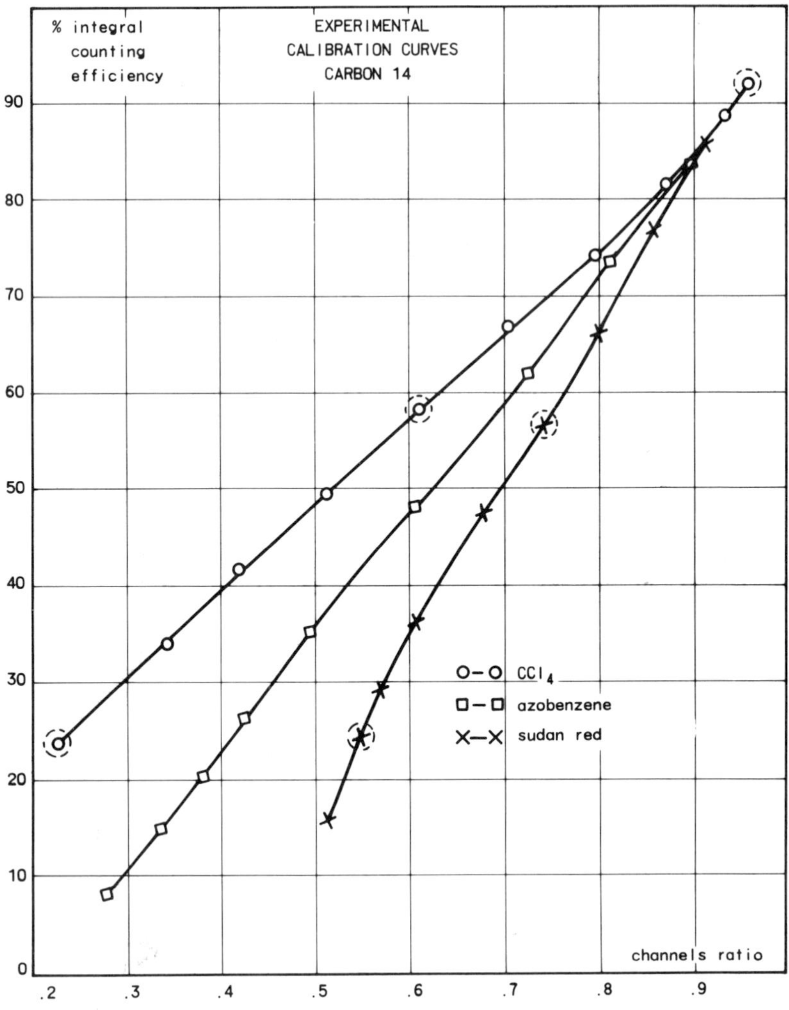

Fig. 7.

Sudan red do not represent truly extreme conditions. But the discrepancies are more than likely due to the simplifying assumptions that were made when the mathematical model was set up. However, the model correctly 'predicts' the trend found in the experimental data, where colour quenchers produce a more rapid decrease of efficiency with channels-ratio than chemical quenchers.

It should be mentioned that we purposely considered integral ^{14}C-efficiencies in order to obtain substantially different results for different types of quenching that could easily be measured and demonstrated. However, such measuring conditions would have a detrimental effect on measurement accuracy in practical liquid scintillation counting. Fortunately, this effect can be strongly reduced by using a counting window. The effects mentioned are most pronounced at low pulse heights and can be appreciably reduced by means of the lower level of the counting window.

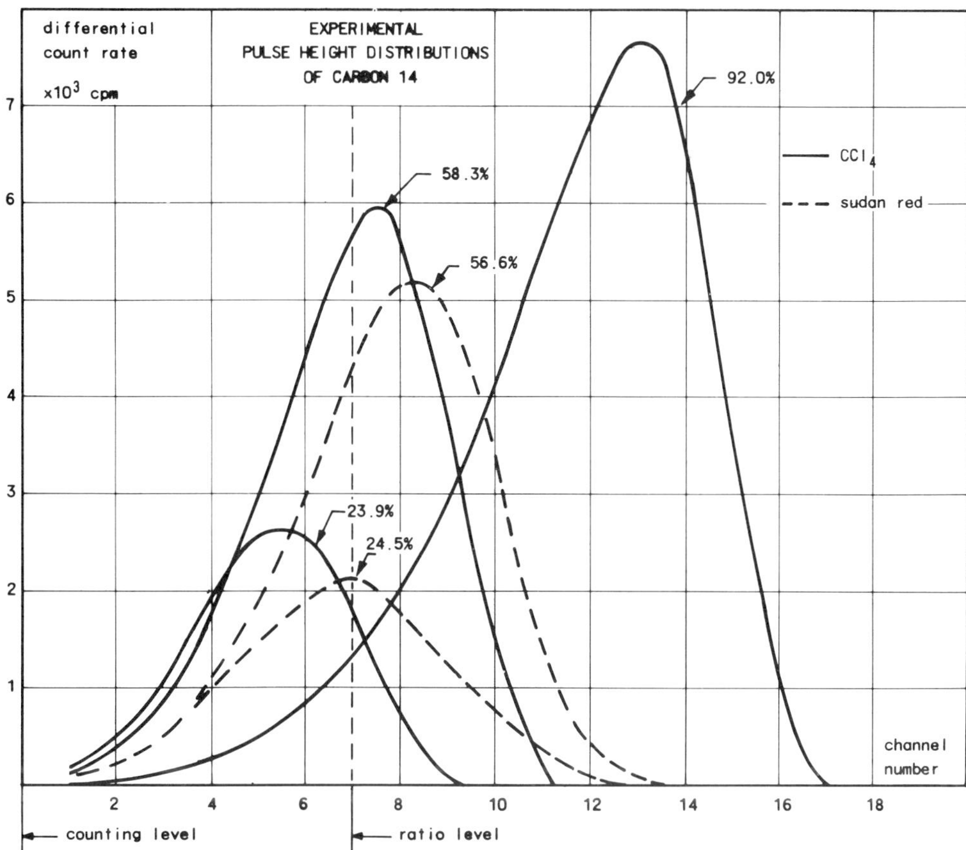

Fig. 8.

If external standardisation with a high-energy γ-source is used, the effect can even be practically eliminated as shown in Fig. 9. A comparison of the integral ^{14}C-efficiencies with those obtained in a standard ^{14}C-window shows that, although the former may seem to offer better statistical accuracy, they are extremely sensitive to colour quenching, whereas the efficiency curves obtained for the window are practically insensitive to colour over a fairly wide range. Normally this range will be even more extended, as practical samples will hardly ever show so much variation in quenching properties as the two quenching agents used to obtain the curves shown.

It is an unfortunate aspect of liquid scintillation counting that optimalisation of measuring conditions has to be based, to a large extent, on trial and error methods. We feel that it would be worthwhile to refine the mathematical model sufficiently to be able to obtain more or less quantitative results from it. It could then become a useful tool for improving liquid scintillation methods and instrumentation. At that stage it would almost certainly become rewarding to study external standardisation by means of the model.

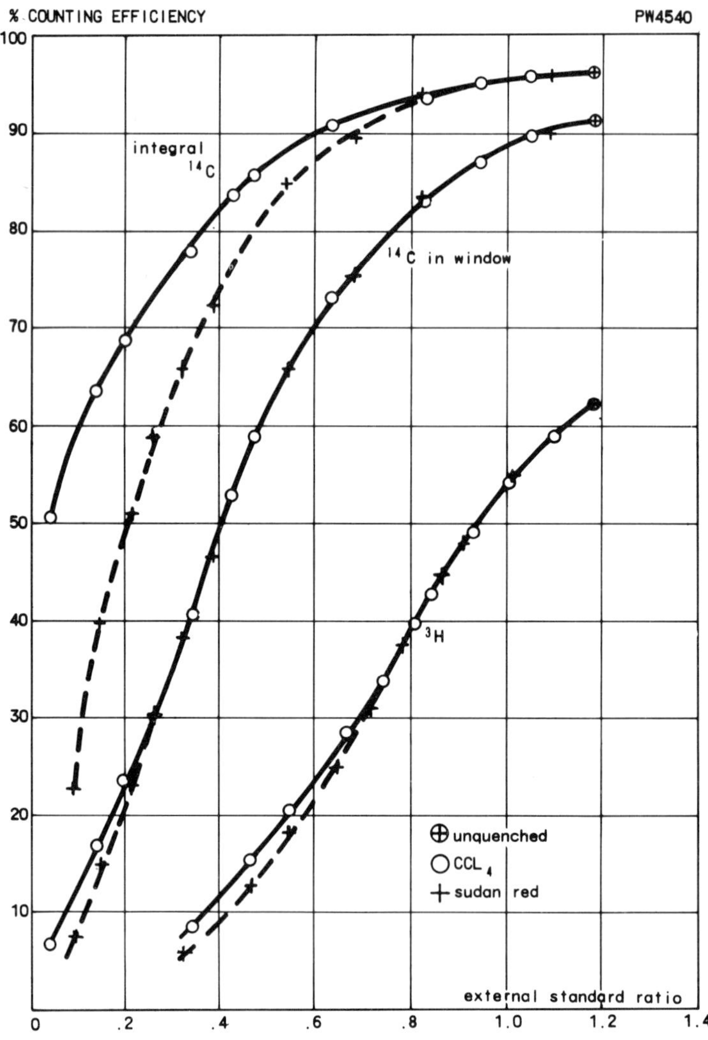

Fig. 9.

REFERENCES

1. M. P. Neary and A. L. Budd, in *The Current Status of Liquid Scintillation Counting,* Grune and Stratton, New York, 1970, pp. 273-82.
2. N. Kaczmarczyk, in *Organic Scintillators and Liquid Scintillation Counting,* Academic Press, New York, 1971, pp. 977-90.
3. P. E. Stanley and P. J. Malcolm, 'Practical liquid scintillation spectrometry: organising a methodology', this volume, pp. 44–62.
4. F. E. L. ten Haaf, in *Liquid Scintillation Counting,* Vol. 2 (eds. M. A. Crook, P. Johnson and B. Scales), Heyden, London, 1972, pp. 39-48.

DISCUSSION

L. A. Currie: By applying a log-transform to the abscissa (pulse height axis) you were able to compare relative pulse heights by simple translation (x-axis). Have you also applied such a transform to the ordinate (y-axis, pulse height frequency) in order to similarly compare the relative shapes of the pulse height frequency with distributions for unquenched, chemically quenched and colour-quenched samples?

F. E. L. ten Haaf: We did not do this, but it sounds a good idea. We might try it in the future.

Chapter 5

Two-Parameter Pulse Height Analysis in Liquid Scintillation

B. H. Laney

Nuclear Chicago, Des Plaines, Illinois, U.S.A.

INTRODUCTION

Virtually all liquid scintillation spectrometers since 1950 have utilised dual-phototube coincidence detectors to reduce background count rate.[1] Various methods of combining the output signals from each detector have been used for pulse height discrimination against background.[2–7]

This chapter will show how analysis of the relative pulse height from each detector is utilised to provide additional information about the sample specimen and photodetectors. Two-parameter pulse height analysis[8] is used to examine the relative response from each photodetector. Correlations between the response from each detector are used to resolve some of the ambiguities which arise in single-parameter observations. The added dimension provides a new means of discriminating between background events and those caused by scintillations in the sample, by rejecting regions of the 2-parameter plane. Various transformations of the 2-parameter spectrum into a single-parameter spectrum to improve background discrimination, spectral resolution and/or efficiency will be described.

OPERATIONAL OVERVIEW OF A COINCIDENCE SPECTROMETER

A generalised coincidence spectrometer is depicted in the block diagram of Fig. 1. Pulses from each of the photodetectors are simultaneously applied to a high-speed (typically less than 50 ns) coincidence detector, and also to a processing element H which transforms the 2-parameter X and Y pulse height information from each phototube into a single-parameter pulse height h. A pulse height analyser (PHA) sorts and counts the pulses. The coincidence detector enables pulse height analysis only when a signal is received from both detectors at the same time ($\sim$50 ns). Gating by the coincidence detector can alternatively be accomplished prior to the H transform, i.e. analogue gating.

The H processing element can be analogue or digital. For example, H may be a voltage, current or pulse width proportional to the magnitude of h, or it may be a serial or parallel digital word proportional to the magnitude of h. More than one transform can be used. Depending upon H, the pulse height analyser can utilise

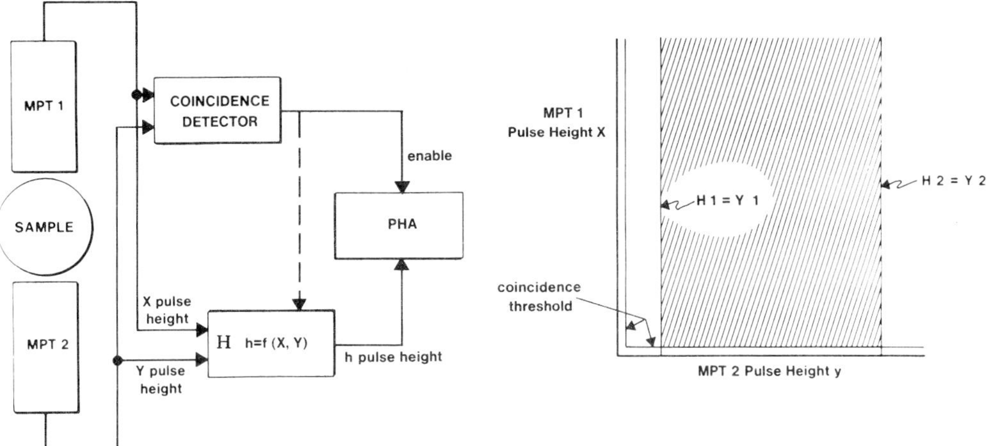

Fig. 1. Generalised coincidence spectrometer. *Fig. 2.* Unsummed pulse height analysis, $h = y$.

analogue or digital discriminators in either a single or multichannel analyser configuration.

SINGLE-PARAMETER TRANSFORMS

Examples of three commonly used methods of pulse height analysis are depicted in Figs. 2, 3 and 4.

Figure 2 shows a 2-parameter pulse height field of 'unsummed' pulse height analysis. Phototube 1 is only used as a coincidence '*gate*'. Its pulse height is irrelevant, providing it exceeds the coincidence threshold. Events which produce a y pulse height greater than H1 but less than H2 are accepted in the pulse height window. This method of pulse height analysis was demonstrated by Hiebert and Watts,[2] and was typical of early commercial LS spectrometers.[6]

'Summed' pulse height analysis is shown in Fig. 3. In order for an event to be within the acceptance window, the sum of the x and y pulse heights must exceed level H1 and be less than level H2. Both pulse heights are linearly added prior to pulse

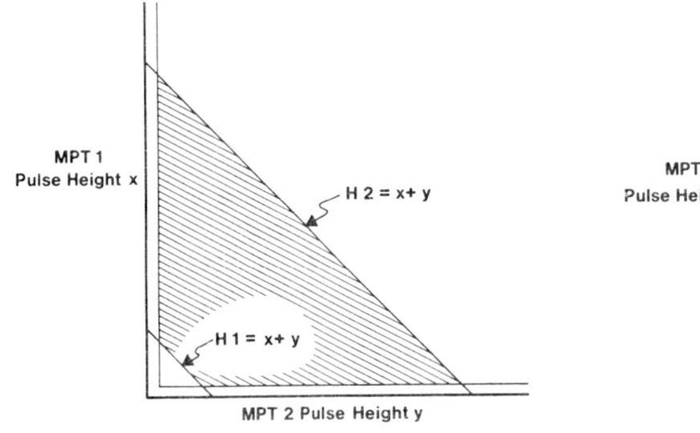

Fig. 3. Summed pulse height analysis, $h = x + y$.

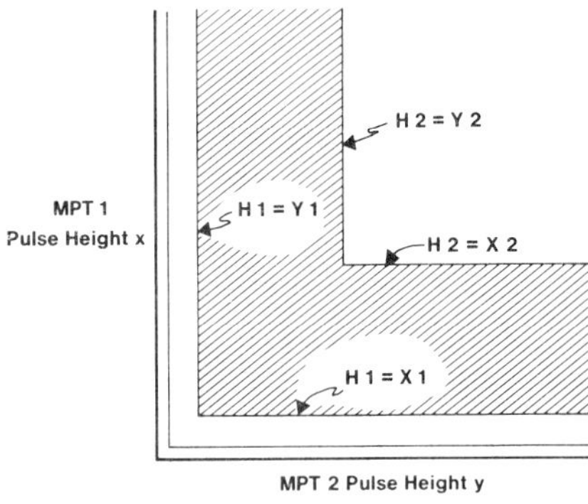

Fig. 4. Lesser pulse height analysis, $h = x < y$, $h = y$ if $y < x$.

height discrimination. In addition, each must exceed the coincidence threshold to be accepted. This method of pulse height analysis was demonstrated by Swank,[3] and is still used in many commercial systems.

In the 'lesser' pulse height analysis shown in Fig. 4, both x and y pulse heights must each exceed level H1 and be less than level H2 to be accepted. Thus, the lesser of the two pulse heights determines acceptance and rejection. The coincidence detector only determines the minimum level for detection. The effective coincidence threshold follows the lower level discriminator setting whenever it exceeds the threshold of the coincidence detector. This method of pulse height analysis was demonstrated by Utting,[5] and is also used in many commercial systems.

When 2-parameter information is transformed into the single parameter h, it is effectively re-mapped from a plane to a single line.[9] Thus every point on the line H1 in Figs. 2, 3 and 4 is represented by the single magnitude H1, as shown in Fig. 5. Since there are an infinite number of functions $h = f(x,y)$, a spectrum in the 2-parameter field can have many distributions in the single-parameter mapping. This may be used to explain why the pulse height spectrum of an isotope, particularly those with low energy, differs from its β-spectrum.

MATERIALS AND METHODS

The sources and instrument settings used to analyse spectra are given in Table 1. Sources contain 15 ml PPO–POPOP–toluene unless specified otherwise. Figures 6, 7 and 8 were obtained using a Searle Analytic Mark III system, while all others were with a Mark II, two Model 8928 interfaces, a Model 25609 multichannel analyser and two 100 MHz ADC's. EMI type 9750QB phototubes were used to obtain all data shown. Except for background measurements, spectral distributions are the same with RCA type 4501-V1 phototubes. Two-parameter data was read out on perforated paper tape and processed off-line. A 20-level digital grey scale conversion described by Macleod[10] was used as the log of normalised count rate, with the peak count rate being set to the maximum density. This method provides a means for obtaining both graphical and digital data simultaneously. The approximate normalised count rate at any point can be determined from the characters printed. Slight distortions in the visual image are caused by the use of a character font different from the one specified by Macleod.

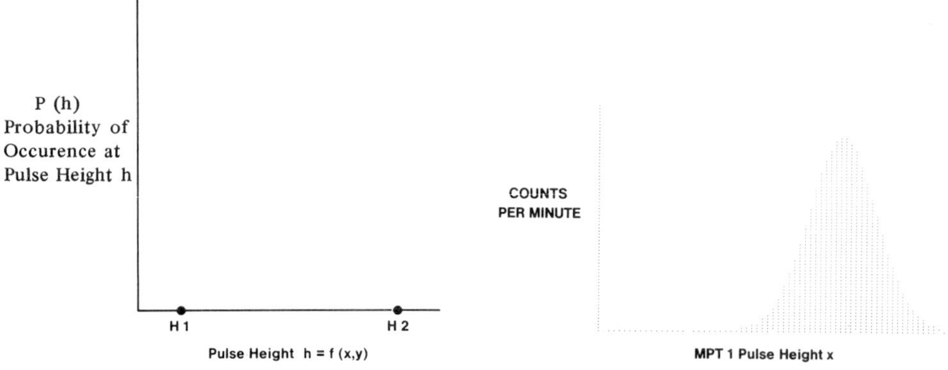

Fig. 5. Single-parameter pulse height analysis.

Fig. 6. Americium-241 unsummed pulse height spectrum.

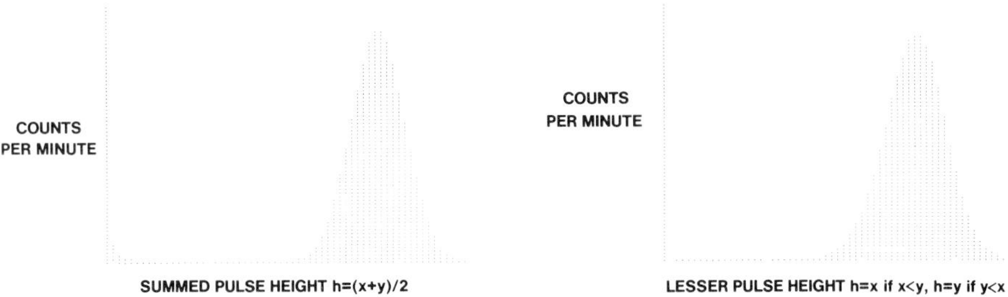

Fig. 7. Americium-241 summed pulse height spectrum.

Fig. 8. Americium-241 lesser pulse height spectrum.

Table 1. Sources and instrument settings.

Figure	Source	Scintillator relative pulse height	Instrument attenuation factor
10	^{241}Am	0.7	32
6	^{241}Am	0.7	16
7	^{241}Am	0.7	16
8	^{241}Am	0.7	16
11	^{36}Cl	1.0	64
12	^{3}H	1.0	1
13	^{14}C	1.0	8
14	Blank	1.0	8
15	None		8
17	None	-	8
19	^{14}C	0.1	2
20	^{14}C	0.1	2
21	^{14}C	0.2	8
22	^{14}C	0.2	8
18	Chemi.	-	1

ANALYSIS OF ISOTOPE SPECTRA

The spectrum of ^{241}Am is shown 3-dimensionally in Fig. 9. The α-peak is clearly visible in the centre. This method of visualisation is acceptable for well-defined α-peaks, but is poor for less pronounced β-emitters. Americium-241 is shown as a 2-parameter density gradient in Fig. 10 using the method of Macleod.[10] Correlation of the magnitudes from each detector is indicated by conformance to a 45° line. Deviation from the 45° line is caused by non-uniform sample geometry and cathode sensitivity, as well as by the statistics of photon collection and electron multiplication.

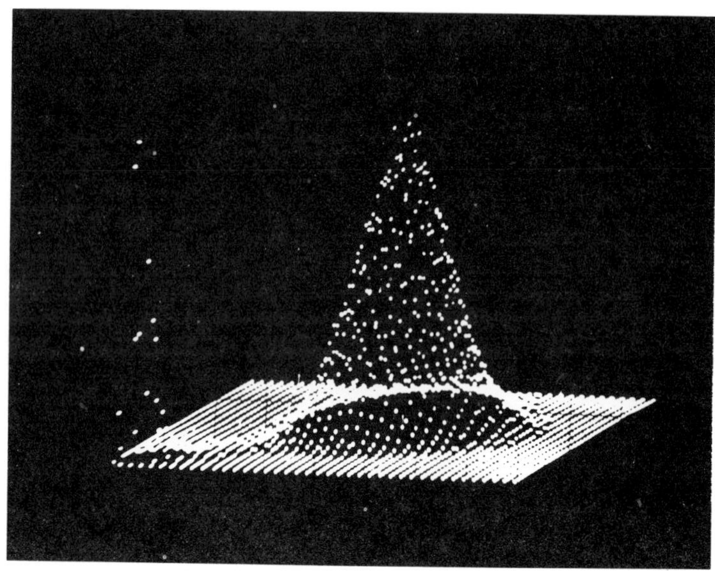

Fig. 9. Three-dimensional spectrum of ^{241}Am.

Applying unsummed, summed and lesser transforms to ^{241}Am results in the single-parameter spectra shown in Figures 6, 7 and 8. The analyser was set to stop at the same peak channel counts for all three spectra. The unsummed spectrum is broader and has poorer resolution than either the summed or lesser spectra. The summed spectrum has slightly better resolution than the lesser spectrum.

Chlorine-36, ^{3}H and ^{14}C spectra are shown in Figs. 11, 12 and 13 respectively. Because ^{36}Cl produces more photons per event, it has the best resolution as measured by its close proximity to a 45° line. In contrast, ^{3}H has poor correlation between the pulses from each phototube due to poor collection statistics associated with the small numbers of photons generated by the weak β-emitter. Thus, the greater the pulse amplitude, the greater the potential energy resolution.

When ^{14}C is quenched to 60% efficiency, the 2-parameter spectral shape nearly coincides with that of unquenched ^{3}H. In the absence of colour quenching, spectral resolution increases as the number of photons detected increases. Thus, the spectral resolution of a quenched sample containing a high-energy emitter can be poorer than for an unquenched but lower energy isotope.

ANALYSIS OF BACKGROUND SPECTRA

Blank background and empty chamber spectra are compared with a ^{14}C spectrum at the same amplification in Figs. 14, 15, and 13 respectively.

The blank background spectrum can be analysed in three parts. The high intensity region near the origin is primarily due to radioactive contamination in the glass vial. The region of intensity along the 45° line is primarily caused by Compton interactions in the liquid scintillator from external environmental and cosmic radiation. The region of intensity along the axes results from light emissions from the phototubes.

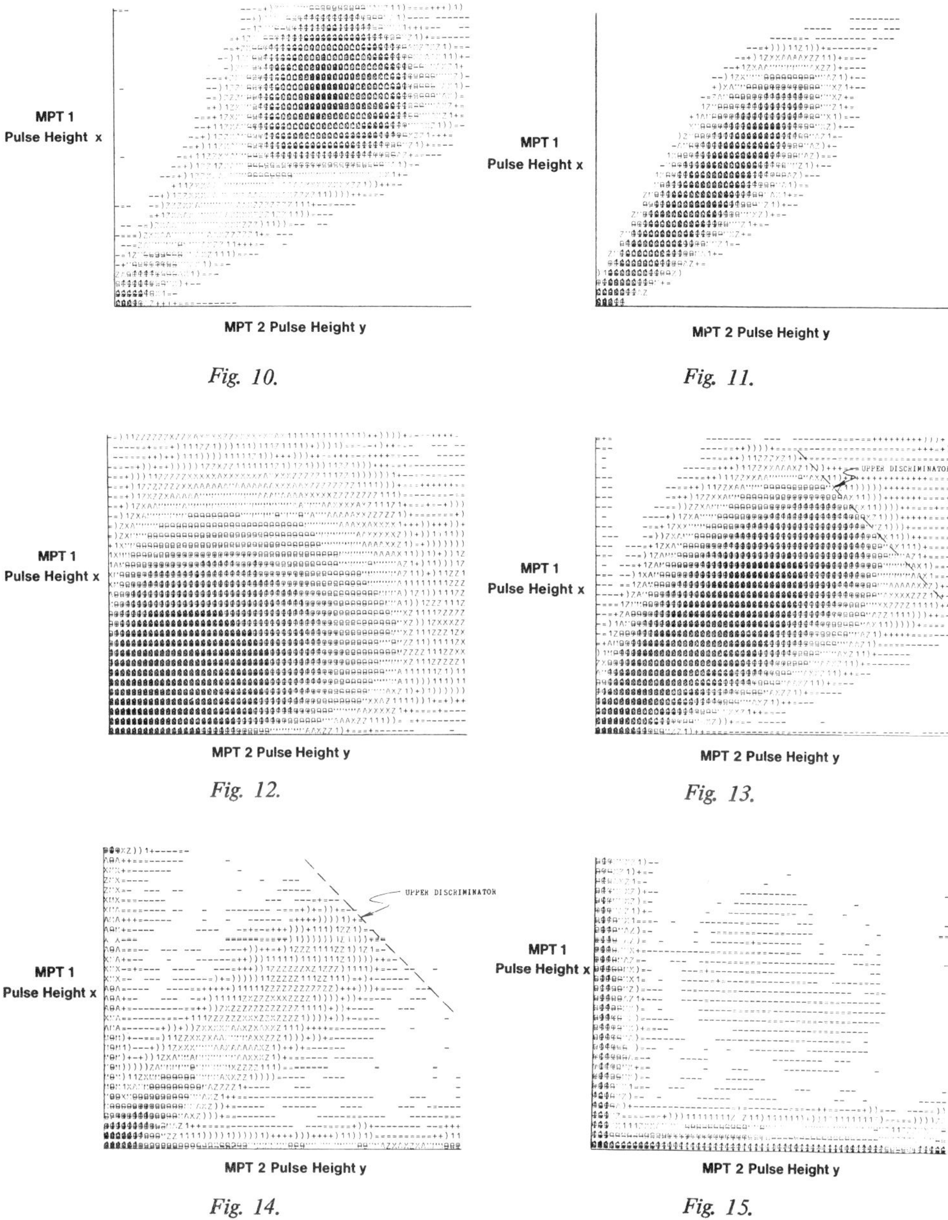

Fig. 10. *Fig. 11.* *Fig. 12.* *Fig. 13.* *Fig. 14.* *Fig. 15.*

Fig. 10. Two-parameter spectrum of ^{241}Am.
Fig. 11. Two-parameter spectrum of ^{36}Cl.
Fig. 12. Two-parameter spectrum of ^{3}H.
Fig. 13. Two-parameter spectrum of ^{14}C.
Fig. 14. Background 2-parameter spectrum.
Fig. 15. Empty chamber 2-parameter spectrum.

Removing the background sample from the chamber virtually eliminates the first two sources of background (Fig. 15). The circle of weak intensity in the centre of Fig. 15 is believed to be caused by scintillations in the phototube face plates. Except for scintillations in the glass caused by external radiation, the empty chamber background is almost entirely from within the phototubes.

Comparison of the ^{14}C and background spectral distributions clearly suggests areas along the axis which can be rejected with negligible loss in ^{14}C detection efficiency. Placement of the lower level discriminator has much more control over rejecting background events than does the placement of the upper level discriminator.

Of the three methods of pulse height analysis described, the lesser method is superior for separating low pulse height background from ^{14}C events. An unsummed lower level discriminator will only reject events along one axis. A summed lower level discriminator only rejects events near the origin. A lesser lower level can be set to reject most of the background along both axes, including the region near the origin.

Since the coincidence threshold is a lesser discriminator, raising the coincidence threshold is very effective for removing background from a ^{14}C counting window. Decreasing the high voltage to both phototubes effectively raises all discrimination levels relative to the pulse height spectrum. This explains why decreasing the high voltage reduces the background more than raising the lower level discriminator of a summed instrument. With lesser pulse height analysis, raising the lower level discriminator produces the same reduction in background as decreasing the high voltage provided the phototubes are not overstressed in the former instance.

The situation is quite different for assaying ^{3}H. Because a significant number of ^{3}H events occur at the lowest detectable level near the origin, increasing the lower level discriminator significantly reduces the detection efficiency and overall sensitivity. Consequently, background events are best eliminated from a ^{3}H counting window by proper placement of the upper level discriminator.

Of the three methods of pulse height analysis, the summed method is superior for separating background from ^{3}H events. An unsummed upper level discriminator will only reject events along one axis. A lesser upper level discriminator does not reject the background along either axis. A summed upper level discriminator rejects most of the events along both axes. An even better upper level discriminator is defined by the 'greater' of the two pulse heights, as demonstrated by Hiebert and Hayes.[4]

The best method of discriminating against the background contributions along the axes is with a crosstalk discriminator.[7] A separate system is used to reject these events which is independent of the method of pulse height analysis. Consequently, efficiency and sensitivity are similar with either of the three methods of pulse height analysis.

A new technique is used to analyse the sources of light emitted from the phototubes. Since the photocathodes are designed to absorb most of the photons impinging on the surface, it is clear that only a fraction of them will pass through. Both photocathodes, therefore, form an optical attenuator to the light generated in one phototube as seen by the other. Consequently, the pulse height distribution between the two phototubes is highly unbalanced when light is generated within one of the phototubes. The 2-parameter spectrum of the empty chamber count rate shows this high concentration of events along the axes.

Recognising this uneven distribution provides a mechanism for determining which phototube produced the scintillation. In Fig. 16, the x-y plane is divided into two regions on each side of a 45° line. Events occurring in Region 1 have a larger pulse amplitude from phototube 1 than from phototube 2, and are thus assumed to originate

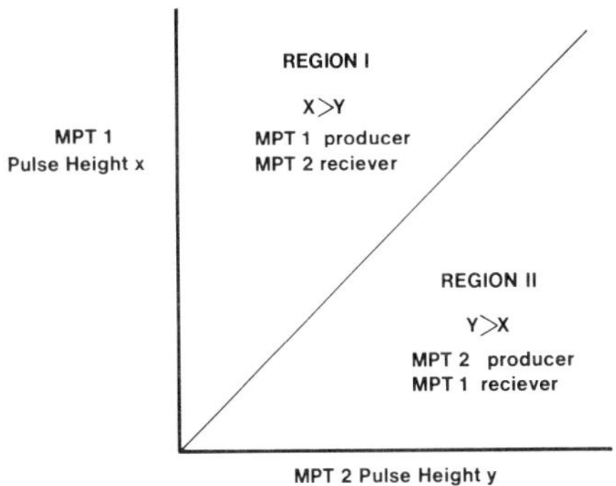

Fig. 16. Transformations for analysis of MPT light emissions.

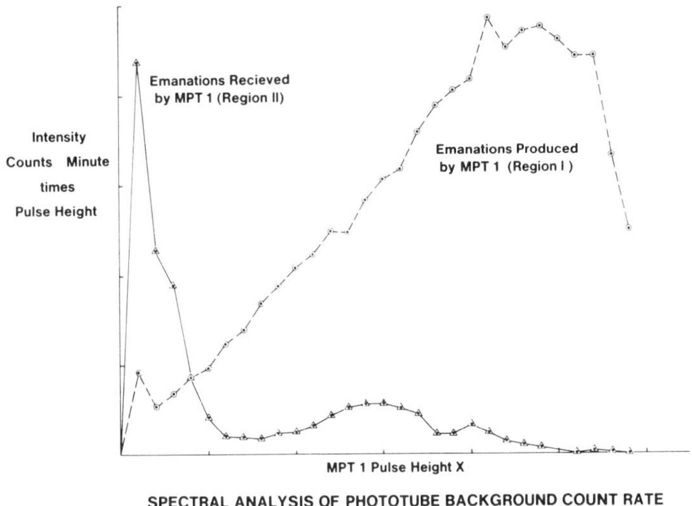

Fig. 17. Spectral analysis of phototube background count rate.

from phototube 1. Similarly, events recorded in Region II are assumed to originate in phototube 2. By adding all the events in Region I at each pulse height x, the spectrum of the emissions produced by phototube 1 can be constructed. Adding all the events in Region II at each pulse height x constructs the spectrum of emissions received by phototube 1 from phototube 2. The spectrum produced and received by phototube 1 is shown in Fig. 17. Similar integrations in the y direction will construct the spectrum of emissions which phototube 2 generated, and which phototube 1 detected from phototube 2.

Chemiluminescence and phosphorescence are other sources of background count rate. The accidental coincidence spectrum characteristic of excessive monophotonic events is shown in Fig. 18. The single photon pulse height can be determined from this spectrum. Events along the axes are probably artifacts caused by pulse pile-up in the 100 MHz ADC's used to digitise the phototube signals.

ANALYSIS OF QUENCHING

Chemically quenched and colour-quenched samples containing ^{14}C are compared in Figs. 19 and 20. Both samples have the same total counting efficiency. Events from the colour-quenched sample have a larger deviation from the 45° line than for events from the chemically quenched sample. More events have a large pulse height difference because the sample itself is an optical attenuator to light. Consequently, scintillations produced near one phototube will produce a strong signal from that phototube and a weak signal from the far phototube.[11,12]

The uneven distribution of light to each of the phototubes produces spectral distortions which can cause errors in the estimation of efficiency.[13] The lesser method of pulse height analysis has been shown to produce significantly smaller errors in efficiency than summing, as determined by the ratio method of quench correction.[14] This is because the coincidence detection efficiency is most influenced by the phototube which receives the least amount of light.[15] With summing, the signal from the phototube which receives the most amount of light contributes proportionately more to the spectra as colour quenching increases. Thus, efficiency determined from a summed pulse height spectrum is overestimated in the presence of colour quenchers.

Two samples containing urine in different surfactant cocktails are compared (Figs. 21 and 22). Using 2-parameter pulse height analysis, we can see that the Triton X-100 sample exhibits strong colour quenching even though both samples visually appear identical. Colour quenching is even visible in the 2-parameter spectrum of PCS. This spectrum would be more rounded if only chemical quenching were present.

CONCLUSIONS

Two-parameter pulse height analysis is a useful technique for analysing the signals from a liquid scintillation detector. Data presented in an array of α-numeric characters provide both digital quantitation and a graphical density gradient.

Background events generated by Compton interactions in the sample solution, radioactive contamination in the vial walls, scintillations from within the phototubes and luminescence are more easily distinguishable in the 2-parameter plane. Phototube crosstalk can be identified and electronically discriminated from scintillations within the sample.

Alternative methods of combining the signals from both phototubes prior to single-parameter pulse height analysis are readily interpreted from a 2-parameter spectrum. Spectral distribution as a function of isotope quencher and detector response can be analysed. Chemically quenched samples can be distinguished from colour-quenched samples. Although a summed spectrum has slightly better pulse height resolution than the lesser spectrum, the latter is more accurate for efficiency determinations, particularly when colour quenchers are present.

ACKNOWLEDGEMENT

The author wishes to thank Mr. Harry Engberg for performing the measurements, and Mr. J. Marshall Dudley for programming and processing the 2-parameter spectra.

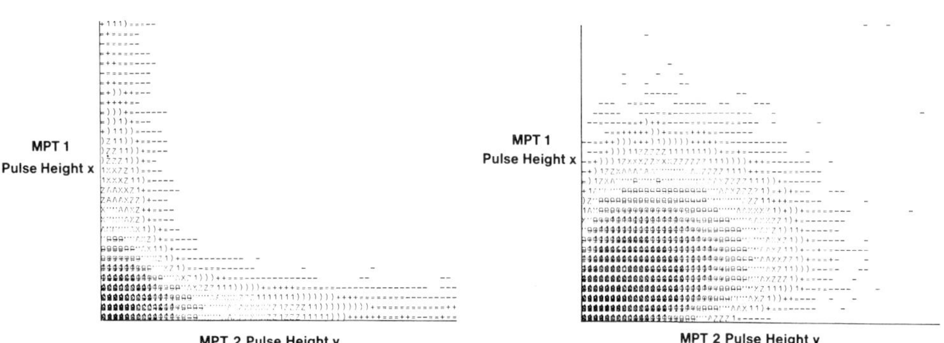

Fig. 18. *Fig. 19.*

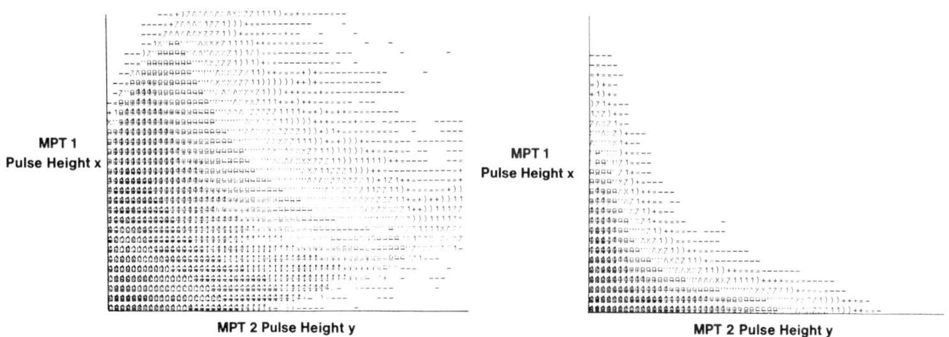

Fig. 20. *Fig. 21.*

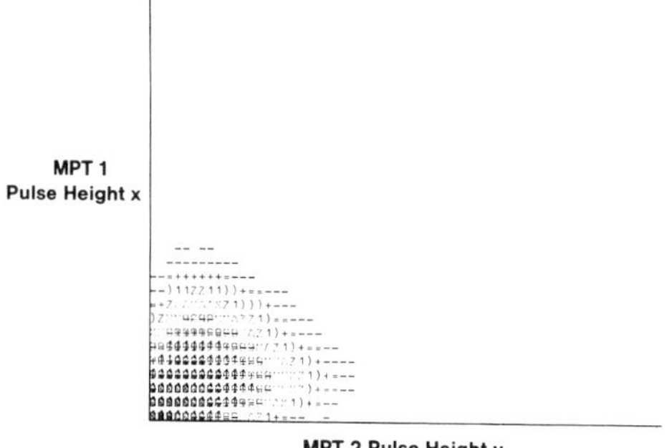

Fig. 22.

Fig. 18. Chemiluminescence 2-parameter spectrum.
Fig. 19. Carbon-14 chemically quenched to 70% efficiency.
Fig. 20. Carbon-14 colour quenched to 70% efficiency.
Fig. 21. Carbon-14 urine in Triton X-100/toluene 1/2.
Fig. 22. Carbon-14 urine in PCS.

REFERENCES

1. G. T. Reynolds, F. B. Harrison and G. Salvini, *Phys. Rev.* **78**, 448 (1950).
2. R. D. Hiebert and R. J. Watts, *Nucleonics* **11**(12), 38 (1953).
3. R. K. Swank, in *Liquid Scintillation Counting* (eds. C. G. Bell and F. N. Hayes), Pergamon Press, Oxford, 1958, p. 23.
4. R. D. Hiebert and F. N. Hayes, in *Liquid Scintillation Counting* (eds. C. G. Bell and F. N. Hayes), Pergamon Press, Oxford, 1958, p. 41.
5. G. R. Utting, in *Liquid Scintillation Counting* (eds. C. G. Bell and F. N. Hayes), Pergamon Press, Oxford, 1958, p. 67.
6. L. E. Packard, in *Liquid Scintillation Counting* (eds. C. G. Bell and F. N. Hayes), Pergamon Press, Oxford, 1958, p. 50.
7. B. H. Laney, in *Organic Scintillators and Liquid Scintillation Counting* (eds. D. L. Horrocks and C. T. Peng), Academic Press, New York, 1971, p. 991.
8. M. Birk, T. H. Braid and R. W. Detenbeck, *Rev. Sci. Instrum.* **29**, 203 (1958).
9. A. E. Taylor, *Advanced Calculus,* Ginn & Co., New York, 1955, p. 254.
10. I. D. G. Macleod, *IEEE Transactions on Computers* **C-19**(2), 160 (1970).
11. M. P. Neary and A. L. Budd, in *The Current Status of Liquid Scintillating Counting* (ed. E. D. Bransome, Jr.), Grune and Stratton, New York, 1970, p. 273.
12. F. E. L. ten Haaf, in *Liquid Scintillation Counting*, Vol. 2 (eds. M. A. Crook, P. Johnson and B. Scales), Heyden, London, 1972. p. 39.
13. A. Noujaim, C. Ediss and L. Wiebe, in *Organic Scintillators and Liquid Scintillation Counting* (eds. D. L. Horrocks and C. T. Peng), Academic Press, New York, 1971, p. 705.
14. C. Ediss, A. A. Noujaim and L. I. Wiebe, in *Liquid Scintillation Counting: Recent Developments* (eds. P. E. Stanley and B. A. Scoggins), Academic Press, New York, 1974, p. 91.
15. B. H. Laney, in *Liquid Scintillation Counting: Recent Developments* (eds. P. E. Stanley and B. A. Scoggins), Academic Press, New York, 1974, p. 455.

DISCUSSION

F. E. L. ten Haaf: Have you done a spectrum analysis on acetone as a colour quencher?

B. H. Laney: No.

J. B. Birks: In my paper I discussed two types of impurity quenching; that due to electron capture and that due to exciplex formation by excited solvent or solute molecules. There is a further type of impurity quenching, known as static quenching (the other type is dynamic quenching) which is due to the formation of donor–acceptor (charge-transfer) complexes between the quencher and solvent or solute molecules in the unexcited ground-state. Such complexes are readily formed by impurity molecules which have a high electron affinity. The complexes commonly have a charge-transfer absorption band at longer wavelengths than either of the molecular components, which extends over the solute fluorescence spectrum. They thus function both as an impurity quencher and a colour quencher. Further details will be found in my book *Photophysics of Aromatic Molecules* (Wiley–Interscience, London and New York, 1970).

Chapter 6

Design of a Micellar-Based Technique for Recycling Liquid Scintillation Glass Vials

N. G. L. Harding and J. Dixon

University of Glasgow, Department of Pathological Biochemistry, Royal Infirmary, Glasgow, Scotland

INTRODUCTION

Liquid scintillation vials have become, with progressive financial stringencies, an item to be erased from the budget of a laboratory provided an acceptable technique could be developed for recycling used glass vials.

The current cost of glass vials is around ten vials per dollar; so an average research programme where vial consumption is of the order of several thousand per year, discards about one thousand dollars-worth of vials per year. In order to excise this budgetary loss, a reliable recycling technique was required for cleaning a used vial and placing it in circulation again. A prerequisite was that an investigator should not need to check the background count of the used vial in the required isotope channel. Once a certain count-independent method for achieving this had been established it should prove possible to introduce a standard recycling procedure into the laboratory with ready and unanimous acceptance among users of a wide variety of isotopes.

At the outset we anticipated that a major difficulty would be to convince users that the following criteria were acceptable without actually counting each recycled vial:

(i) the background count is not increased by the washing procedure;
(ii) chemiluminescence characteristics of the vial remain unchanged;
(iii) the efficiency of the isotope assay procedure is constant;
(iv) external standardisation and internal quench estimates remain unaffected;
(v) improperly washed vials are readily detected and rejected without the need for counting.

PRELIMINARY SURVEY

When we began this programme some years ago there was little detailed information available about the structural relationships between the glass in a scintillation vial and the 'average' residue left after a used vial has been emptied of the majority of its radioactive contents. To investigate this we used a standard glass vial 'contaminated' with a residue of naphthalene-based scintillator containing a variety of inert (non-radioactive) 'contaminant' molecules. At the same time, by using this standard system we were able to investigate available commercial washing reagents in a preliminary study. Results can be summarised as follows:

(i) Commercial washing powders and liquids were variable in their ability to remove standard contaminants from the glass wall. Some were unable to remove precipitated protein; others could not remove the hydrophobic components of the scintillator; others did not remove all traces of radioactive labels introduced into the 'contamination'.

(ii) Some of these washing preparations were at a grave disadvantage in that they left residues upon the vial wall. These were later found to interfere with internal quench estimation procedures. Some preparations yielded such ingrained deposits that even subsequent machine washing and acid-soaking failed to regenerate the original optical surface.

(iii) With some materials there was not a consistent spectrum of cleaning ability: those found not to be satisfactory for hand washing of the vials were also unsatisfactory for machine washing; other materials able to remove certain types of contaminant by hand washing also proved unsatisfactory in the washing machine.

This preliminary study of commercially available preparations confirmed the need for a fundamental investigation of the problem rather than an attempt to modify an existing preparation to give a simple one-step vial-recycling technique.

This basic approach had several advantages, in particular enabling complete control over the composition of washing reagents and the structure–contaminant relationship of ingredients of a scintillator and samples. Accordingly, the programme started at a fundamental level and aimed to produce a simple one-step washing procedure for use in the laboratory.

As a first step we investigated simple procedures for removing standard oleaginous contaminants (Table 1), and confirmed that neither simple anionic nor cationic systems were individually capable of performing the required tasks.

Table 1. Effectiveness of different treatments at removing a standard greasy soil from scintillator vials.

Isopropanol, 2 min	Oily residue and fingerprint remains	Slight wetting
Isopropanol, 2 min + ultrasound	Cleaning complete	Slight wetting
Alkylsulphonate ion	Partial cleaning	No wetting
Quaternary ammonium ion	Partial cleaning	No wetting

Vials were treated with a standard scintillator residue and dried. Soiled vials were treated as shown, at room temperature, and examined for residues, fingerprints and wetting by distilled water.

An impressive finding was that although ultrasonication in isopropanol proved effective for these contaminants, it was ineffective for rapid removal of all scintillant complexes. In addition, these preliminary experiments confirmed that there was an empirical relationship between 'wettability' of the washed surface and the extent of cleaning in the test system.

An important question arising from this first project was whether there was a relationship between the structure of anionic and cationic surfactants and their ability to remove contaminants from the glass wall. To avoid an essentially boundless survey of compounds the following principles directed our attention to several classes of reagent for further study.

STRUCTURAL BASIS OF CONTAMINATION

Two factors set the rate and determine the extent of contamination, namely the glass wall and the impinging molecules.

The anatomy of the glass wall is a subject of some doubt. The precise composition of the wall will vary with the type of glass used, and in turn this will determine the solution properties of the glass. For example, silica, borosilicate, aluminosilicate and high-lead glasses have excellent water resistance, in contrast to the poor water resistance of high-alkali, phosphate and borate glasses. For the purposes of the present paper, it is useful to assume that the wall consists of an outer, hydrated layer and an inner, structural layer based upon the tetrahedral silicon atom. Normally the hydrated layer of the wall will have hydroxyl or alkoxyl moieties in contact with the internal fluid layer. These are illustrated in Fig. 1.

The nature of the impinging molecules is complex as diverse types and amounts find their way as samples and scintillators into a scintillation vial. Several factors thereby complicate the chemistry of contamination: labelled molecules introduced into glass vials will react at varying rates with the vial wall; sample preparation and counting may be performed at a variety of temperatures, both above and below ambient resulting in varying rates of interaction with the vial wall; finally, as the composition of the vial wall may vary according to the batch of the glass, contaminant properties may not always be consistent.

I
Si O O $\xrightarrow{H_2O}$ OH O—Fe(OH)(OH) O O Fe—OH → >Fe.OOC.$C_nH_{2n}CH_3$

II
Si O—Na O Si $\xrightarrow{H_2O}$ >Si—OH O >Si $\xrightarrow{Na^+OH^-}$ >Si(OH)(OH) O$^-$···Na^+ >Si

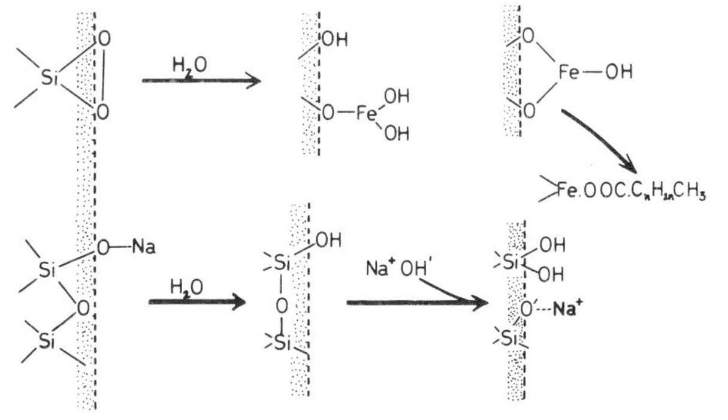

Fig. 1. Glass vials: hydration and derivatisation of wall silicon atoms. The shaded area refers to the glass wall.

Despite this spectrum of difficulties, contaminants are able to bond to the vial wall by rather a restricted number of mechanisms. As a result it will be predicted that any failure to remove a contaminant by a properly designed reagent could be ascribed to a rate-dependent failure to attack a large lump of contaminant adhering to the glass wall rather than to an aberrant physicochemical interaction. Clearly, any washing reagent must therefore be able to reverse the common modes of contaminant bonding. For convenience we have classified the three modes of contaminant bonding, though other and more complex classifications come to mind. As an operational tool the following classifications have proved fundamental to the design procedure.

Covalent bonding

This is likely to be an unusual mechanism in the formation of contamination complexes with the glass wall. Nevertheless, it has some important aspects, of which one is illustrated in Fig. 1. A reaction between iron or manganese and the wall hydrate layer leads to covalent bonding of an iron atom in alternative modes. Coupling to both oxygen atoms of the silicon tetrahedron forms a ferric hydrosilicate in which a reactive hydroxyl group is attached to the iron atom. This hydroxyl can in turn be displaced by other residues, a carboxylate derivative is shown in Fig. 1, to yield large metallo-organic complexes. The stains which can be left on glass surfaces by iron-rich solutions are familiar, and those who have tried to remove them will know that they are removed with appreciable difficulty. This type of contamination is important in two ways: first, the iron atom, if radioactive, presents a radioassay problem for the cleaning process; second, if non-radioactive, the iron is troublesome in acting as an intermediate for formation of metallo-organic contaminants.

Ionic processes

For convenience, both ionic bonds and ion-exchange processes can be considered together. An underlying mechanism is illustrated in Fig. 1. In the presence of hydroxyl ion, the wall hydrate layer is susceptible to attack at the oxygen atom which joins two silicon atoms. This structure is cleaved, leaving an anionic oxygen atom able to partake in ionic processes. At the same time, hydroxyl density in the wall hydrate layer appears to increase as the remaining silicon atom sequesters an hydroxyl residue. Thus, as the wall gains negative charge, counter-ions would be expected to be absorbed from aqueous solutions. Anions such as borate, carbonate and phosphate are known for their ability to adhere to glass, though fortunately dissociation is relatively simple under non-etching conditions.

Etching is an important part of the contamination procedure, and as already described seems to be a common feature of many washing solutions. In such cases etching is often accompanied by a deposition of virtually insoluble white material. The dependency of vial erosion upon hydroxyl ion is shown in Fig. 2, which also shows the erosion of a commercial laboratory washing reagent and compares these with a current vial recycling reagent specifically designed. The etched surface of the vial is greatly increased in surface area and contains pits for entrapment of contaminants by the various binding mechanisms. Reproducible low-background counts during vial recycling has been a guiding principle requiring a washing procedure free from the 'etching pitfall'.

Hydroxyl ion is very important in the physical chemistry of the glass surface owing to the ability to break bonds in the hydrate layer and create etched zones.

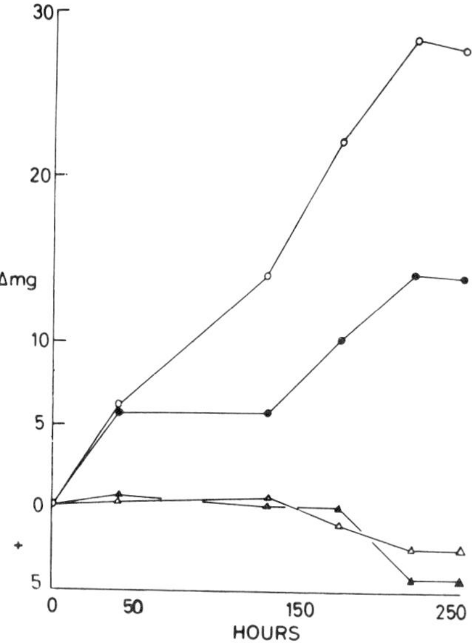

Fig. 2. Erosion of scintillator vials by hydroxyl ion at room temperature. Note that loss of weight (mg/vial) on the ordinate is expressed above zero and weight gain below. O = pH 10.5; ● = pH 8.5; △ = aqueous solution of a micellar reagent at pH 4.0 and at tenfold concentration pH 4.0 (▲).

In the presence of high concentrations of hydroxyl ion, a progressive erosion of the hydrate layer would be expected as the process shown in Fig. 1 extends into and beyond the normal hydrate layer of a glass wall. This is conveniently termed etching. During these studies we also observed that those vials etched with certain preparations during washing became unsuitable with some quenching reagents. This problem is described in further detail below.

Yellowing of quenched standards

A surprising finding during the preliminary screen of available products was that of interference between the washing procedure and subsequent performance of the vial when used with quenched standards. This is illustrated in Table 2, which shows that the phenomenon was restricted to a particular product (W3, W3A) in the presence of carbon tetrachloride. A visual yellowing of the scintillator eventually occurred in these vials and was associated with a progressive fall in light output of the scintillator. The phenomenon was not dependent upon the presence of tritium and did not occur with acetophenone-quenched scintillator.

Electrostatic bonding mechanisms

For convenience a variety of different mechanisms is classified within this general heading, as in practice similar approaches would be used to dissociate material bound to the glass wall by this type of mechanism. An example is the vial which has been allowed to dry when containing a naphthalene-rich scintillator. A familiar incrustation of naphthalene and scintillator is obtained, and is not easy to remove from a glass surface by aqueous solvents. However, addition of toluene rapidly results in dissolution of the complex. There are a series of complicated interactions which describe such a deceptively simple washing procedure and it is beyond the scope of the present paper

Table 2. Yellowing of quenched standards by washed vials.

Wash reagent	NE 233	$+CCl_4$	$+C_8H_8O$	
W1				^3H omitted
W2				
W3		y		
W3A		y		
W1				Plus 10^5 dpm ^3H
W2				
W3		y		
W3A		y		

Standard glass vials were washed with a variety of commercially available washing reagents (W2, W3, W3A) and with a micellar wash reagent (W1) based upon the polyethersilane core mentioned in the text and legends to Figs. 9 and 10. The dried vials were dosed with a standard scintillator (NE 233) and left unquenched or dosed with the quenching reagents shown. Identical groups of vials were left without added radioactivity or contained the amount of tritiated toluene shown. Washing reagents W3 and W3A produced yellowing of the scintillator fluid upon standing in daylight for a few days (y).

to consider these in detail. It is sufficient here to mention that non-polar solvent-solute interactions are a fundamental design requirement for a washing procedure of the type specified. The problem is to obtain such an effect in an aqueous environment.

In terms of the bulk of interacting material within the vial this general class of mechanism probably predominates, and may of course be superimposed upon a covalent mechanism. For example, the contaminant iron atom shown in Fig. 1 may eventually terminate in a long hydrophobic alkyl chain which would then prove a powerful attractant by means of hydrophobic interactions with other molecules. This illustrates the problem of developing a suitable strategy for dissecting such mixed-bond complexes. A reagent capable of dissociating the hydrophobic complex would not necessarily cleave the covalent bond, though once the covalent bond had been cleaved, the complex could be removed intact provided the solubility characteristics were appropriate.

DESIGN PRINCIPLES

Both the preliminary survey of existing materials and a subsequent theoretical consideration of the problem of mechanisms of contamination yield the following points summarised for the design procedure:

(i) many commercial cleaners give rise to time-dependent etching and leave white deposits in the glass wall. These later interfere with quench techniques (Table 2).

(ii) hydroxyl ion is an important determinant of certain mechanisms which bond molecules to the glass wall (Fig. 1).

(iii) etching of the glass wall depends upon the concentration of hydroxyl ion and other anions and is to be avoided (Fig. 2).

(iv) contaminants are bonded to glass by a variety of mechanisms involving covalent, ionic and electrostatic forces.

(v) both polar and non-polar solvent–solute interactions are required for effective removal of contaminants.

(vi) a relationship between wetting of a glass surface and cleaning has been established for long-chain aliphatic compounds.

(vii) this cleaning was only partial when both anionic and cationic derivatives were tested individually as surfactants (Table 1).

A MICELLAR REAGENT

The preliminary work defined some reagent specifications: acid cleaning alone was inadequate unless prolonged; base washes produced etching; simple non-ionic, anionic and cationic reagents were ineffective to a varying degree unless reinforced by thermal or sonic energy.

A complex reagent was therefore required, compatible with both aqueous and oleaginous systems, capable of dissociating contaminants affixed by all classes of bonding mechanism and free from corrosive effects upon the glass wall.

Micellar theory offered the required approach and enabled us to consider a further question: could the vial wall be protected against subsequent contamination by incorporating into the micelle a reagent itself capable of reacting covalently with the wall? Such reagents would be organosilicon derivatives or metallo-organic compounds.

The concept of protection of the vial wall

In principle, the idea is simple. A covalent derivative of the vial wall offers the opportunity for steering properties of the glass surface towards those desirable on both theoretical and practical grounds. Conversion of heterogeneous reacting sites in the glass to sites with homogeneous properties is possible in several ways. The problem is to decide which properties to emphasise. A wall with hydrophobic properties is obtained by either of two principal classes of reaction, illustrated in Fig. 3.

Scheme I (Fig. 3) indicates that in the simplest conversion an homogeneous set of hydroxyl groups is generated in the hydrated layer. The attendant problems of so doing have already been inferred. Scheme II shows the use of an organochromium

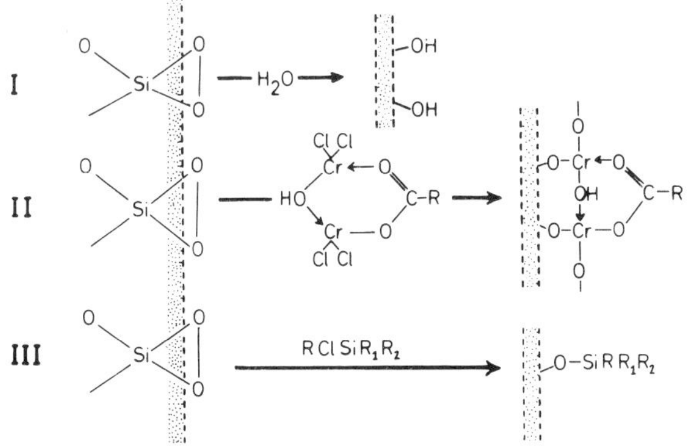

Fig. 3. Principal classes of derivatives of wall silicon atoms. These derivatives are referred to in the text.

derivative as an intermediate coupling agent. Methacrylic acid co-ordinated to chromium yields films with high flexural strength.[1] For our purpose, methacrylic acid was disadvantageous in view of its further reactivity, although published evidence confirms that the chromium-oxygen terminals become integral with the glass structure. Rather than undertake a programme based upon chromium co-ordinate chemistry, the organosilicon approach commended itself. Scheme III illustrates the manner in which a chlorosilane derivative reacts with the glass wall. In this way alkyl silicon polymers can be formed on the glass with covalent linkage. Such chemistry has a long history dating from Johannson's work in 1938 on the interaction of organochlorosilanes with glass for improving adhesion of lacquers.[2]

Silicon alkyl derivatives have the advantage of optical transparency at the required wavelengths and so seem ideal. Yet by emphasising the adhesive properties of such films, Johannson's work apparently contradicts the present purpose. Resolution of the paradox lies in the properties of the organic substituent. For example, aminopropylsilane and methylsilane derivatives exhibited different properties as shown in Fig. 4 and Table 3. In these experiments vials treated with different procedures (W1-S2) were lavaged with $^{14}CO_3^{2-}$, rinsed and counted after addition of micellar scintillator. Compounds W1-W3 are commercial laboratory washing detergents, which give considerable and apparently unpredictable scatter in their ability to elute carbonate. When the glass is treated with a methylsilane derivative (S1), excellent elution of carbonate occurs, with good clustering of individual results. As expected, introduction of an anion-exchange function, using a ω-aminopropylsilane, resulted in retention of some carbonate after rinsing (S2).

Having conferred hydrophobic properties upon the wall, it is important to confirm that the reagent is compatible with a range of scintillators (Table 3). In general, appropriate silicon derivatives of the vial wall do not affect the scintillation process. Indeed, a scintillator purchased from one source (Scintillator 2) started with a very high background count in the tritium channel which was diminished after derivatisation of the glass, though the precise reasons for this effect are unknown.

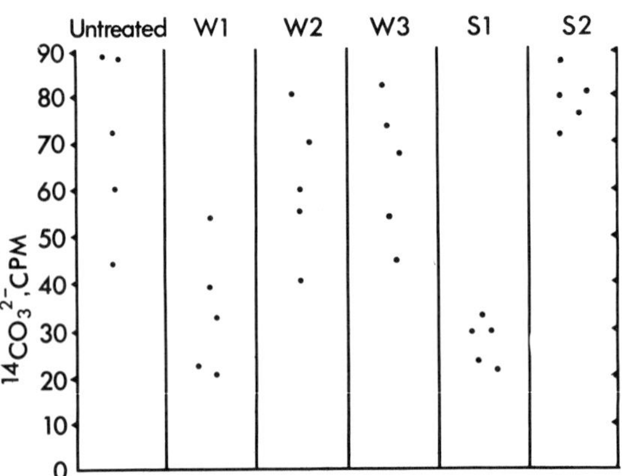

Fig. 4. Retention of carbonate by washed and derivatised standard glass vials. Vials were washed with a micellar reagent (W1), with commercial reagents (W2 and W3) or derivatised with a methylsilane (S1) or γ-aminopropylsilane (S2). Treated vials were rinsed under standard conditions with $^{14}CO_3{}^{2-}$, distilled water, and counted following addition of a micellar scintillator.

Table 3. Scintillator-wall interaction after hydrophobic wash.

Scintillator	PRE-WASH				POST-WASH			
	cpm in channel			Efficiency	cpm in channel			Efficiency
	^{3}H	^{14}C	$^{3}H + ^{14}C$		^{3}H	^{14}C	$^{3}H + ^{14}C$	
1	19	12	29	41.7	20	14	34	40.8
2	115	13	128	44.9	63	13	86	40.6
3	22	14	36	43.3	24	10	36	43.7
4	19	12	31	43.3	19	8	28	44.1
5	27	14	41	41.6	33	9	45	41.5
S1	23	13	32	42.1	20	14	36	41.9
S2	22	12	29	40.9	24	10	33	42.0

Scintillators 1–5 refer to commercially available solutions purchased from standard suppliers, except 1, which was a gift. Vials were filled with scintillator and counted (PRE), contaminated with ^{3}H, washed with a micellar recycling reagent and counted in the channels shown, using ^{3}H-toluene as an internal standard. S1 and S2 are referred to in the legend to Fig. 4.

The silane reagent persisted through successive use cycles of the vial so the next problem to solve was the introduction of a chlorosilane derivative at the appropriate moment of the wash cycle. Micellar theory indicates that this should be possible provided critical points in the carriage and unwrapping of the package are identified so that sensitivity of the procedure to solution conditions can be minimised by an appropriate choice of chlorosilane derivative and carrier micellar molecules.

MICELLES AS REAGENT CARRIERS

Critical stages in carriage of the reagent to the vial wall become apparent from further theoretical considerations. In structural terms one can envisage linear micelles (Fig. 5) protecting their cargo of silane and delivering it by ordered or random interaction with the glass surface. In contrast, spherical micelles do not have the opportunity for multiple linear contact. Figure 6 illustrates the manner in which a spherical micelle could be constructed to deliver its cargo, either from a reservoir in the core of the micelle or from a coaxial domain in the ternary model. A further point, not illustrated in these models, is the need to interact with the contaminants in the wall prior to initiating deposition of the hydrophobic coat. To some extent this directs choice of the micelle type towards complicated structures so that the correct reaction sequence can be maintained.

Unlike the enzymologist we at least have the opportunity wilfully to design a reaction mechanism so that the problems can be clarified by considering the kinetics of a hydrophobic deposition sequence. Figure 7 shows a micellogenic molecule (A) coaxially protecting a core reagent (B) in solution (S) within a boundary (G). In this model, there are three abortive dead-end complexes, AG, AB and BS, the desired complex being BG. Assuming that G represents a single site capable of reacting with A, B and S, then BG could be formed in a linear displacement sequence of the type shown in Fig. 7. The forward velocity, V5, becomes critical in relation to the rate of reaction of B with S. If the latter greatly exceeds the former, then inadequate amounts of the desired complex BG will be formed.

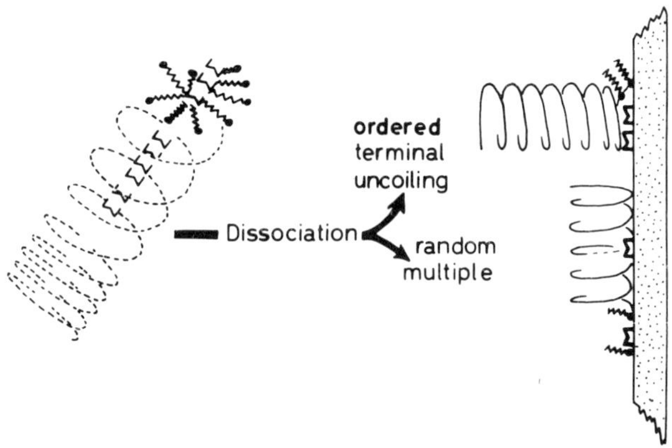

Fig. 5. Models of linear micelles acting as reagent carriers. Σ = a derivatising reagent; as with Fig. 1, the shaded area refers to the vial wall.

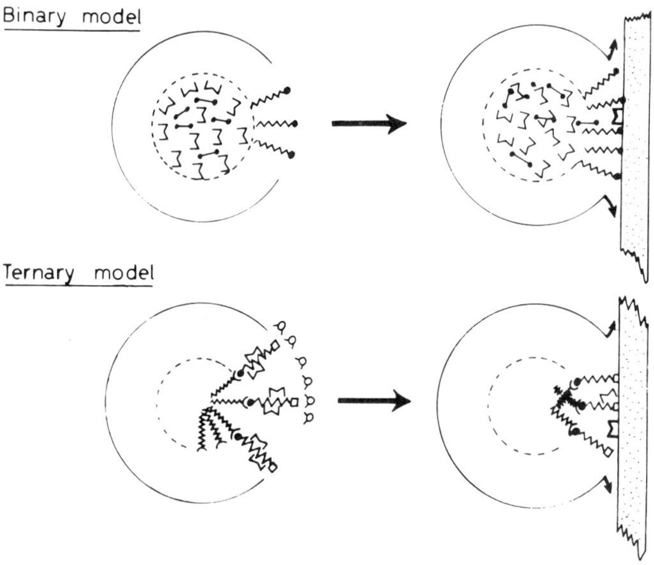

Fig. 6. Models of spherical micelles acting as reagent carriers. For simplicity the micellar and reagent molecules are depicted in a similar way to those in Fig. 5.

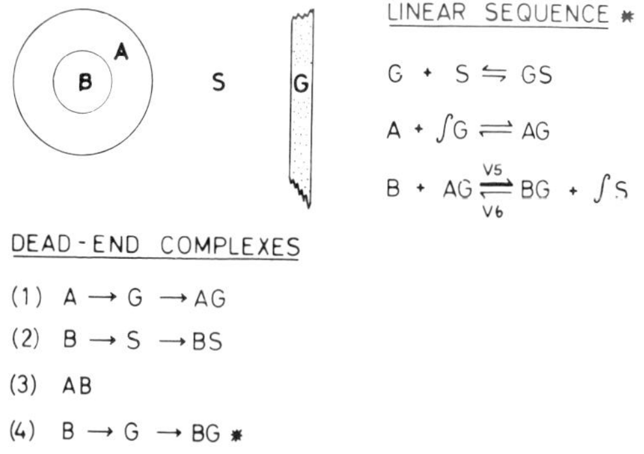

Fig. 7. Kinetic models of reaction between a micellogenic molecule (A), core reagent (B) and glass substrate (G) in a solution (S). Unwanted and wanted (*) dead-end complexes are shown together with a linear (*) sequence leading to the wanted complex.

Another restraint imposed by the model is the reaction sequence, whereby if A and S interact with G the resultant complexes will dissociate in the presence of unreacted B. This sequence has the advantage that A and S could perhaps be used to fulfil the other function required of the reagent, namely that of facilitating removal of unwanted contaminants.

A FUNCTIONAL COMPLEX

Formulation of a micellar complex depended upon resolution of two central issues: satisfactory detergency of contaminants and ability to deliver a reactive molecule to exposed residues on the glass wall. The former was solved by studies outlined previously whilst the latter proved to be especially complicated by virtue of the factors mentioned in the previous section. By finding that polyglycol ethers formed complexes with silane reagents, a skeleton became available upon which a complex micellar reagent could be constructed with the desired properties.

Such a complex has the desired properties as seen from the following studies. Figure 8 shows for two scintillators that the background count (A1, B1) of new vials clusters at a satisfactory low value after a single wash. The efficiency of these vials is at the expected value for tritium (A2, B2). After rewashing, the background (A3, B3) and efficiency (A4, B4) are essentially unchanged. During these experiments a group of vials was isolated with a moderately high background count (C1) but essentially normal efficiency (C2). After wash-recycling, the background fell to the expected value (C3) and the efficiency remained at the previous value. The reason for these unused vials clustering originally at the relatively high background value is not known.

To examine the comparative effectiveness of the complex reagent, groups of 16 vials were selected for a range of background counts from 10–70 counts min^{-1} in the tritium window (Fig. 9). These vials were recounted after washing with a recycling reagent (W1) and with standard commercial laboratory glassware washing preparations (W2 and W3). Although all reagents tended to reduce the background count, the

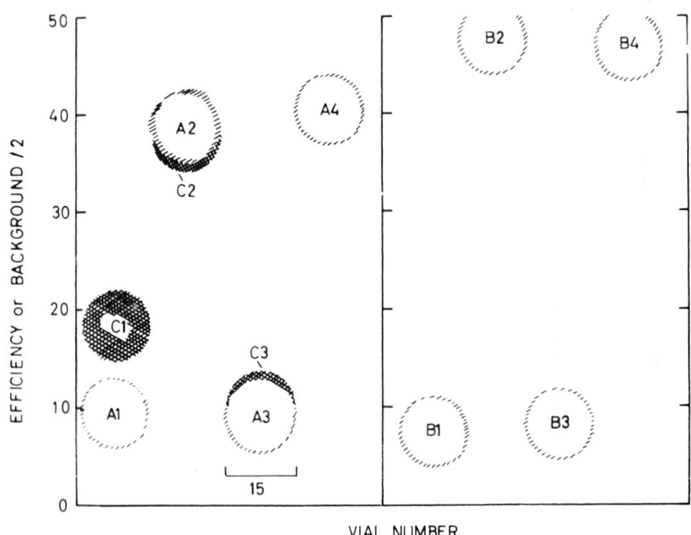

Fig. 8. Clustering of background counts and efficiency of washed vials. A batch of standard glass was selected for background counts in the region of 10 counts min^{-1} (A and B) or 18 counts min^{-1} (C) in the ^{3}H-window, treated as described in the text with a silane complex during washing, and finally assayed with a micellar scintillator (A and C) or hydrophobic, xylene-based scintillator (B).

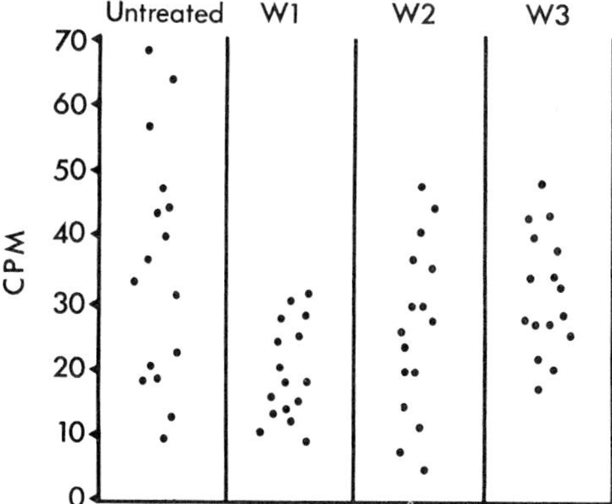

Fig. 9. Reduction of intrinsic ^{3}H-background of used vials by washing. Each point represents one used vial. Vials were assayed with a micellar scintillator after extensive water rinsing (untreated), washing with a micellar washing complex (W1) and standard laboratory washing solutions (W2 and W3).

recycling yielded the lowest counts and the most compact clustering. These properties are conserved over at least a tenfold range of reagent concentration during washing (Fig. 10), where it is also seen that the kinetics of this process are rapid at all concentrations of the reagent. It would be expected that erosion profiles of the vials show minimal erosion by a recycling reagent of this type. This is confirmed in Fig. 2, where groups of vials were soaked for up to 250 h. Vials soaked in sodium hydroxide show intense erosion, those soaked in a standard laboratory glassware washing detergent also show a surprising amount of erosion. In contrast, those soaked in a recycling reagent (NE555) show little change, perhaps a slight gain in

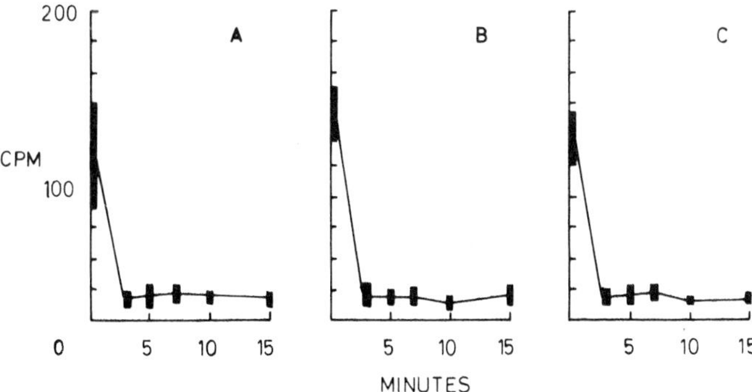

Fig. 10. Compact clustering of ^{3}H-background counts over a tenfold range of concentration of a micellar wash reagent. Used vials were washed in batches, for the times shown, in 1% (w/v;A), 5% (w/v;B) and 10% (w/v;C) aqueous solutions of a polyethersilane micellar complex washing reagent. Undried vials were assayed with a micellar scintillator in the ^{3}H-channel.

mass, by the end of the experiment. Encouraged by these findings we have been able to expand the properties of these micellar vial recycling reagents by formulation, along the lines described, of polynomial micelles containing several classes of molecule each conferring a specific function upon the micelle for use in separate phases of a single washing cycle. This not only permits specific elution of an isotope from a vial, but also enables desirable mechanical properties to be conferred upon the glass, such as scratch resistance.

ACKNOWLEDGEMENTS

We express gratitude to colleagues who have willingly and generously supplied many of the molecules studied in this programme. Nuclear Enterprises Ltd. are gratefully thanked for support of the programme.

REFERENCES

1. J. V. P. Torrey, *Modern Plastics* **30**, 154 (1952).
2. E. G. Rochow, *An Introduction to the Chemistry of Silicones,* 2nd Edition, Chapman and Hall, London, 1951.

DISCUSSION

C. McEvoy: My research shows that efficient rinsing in cold tap water, using a device of my own invention, is a sufficient method of decontaminating most vials. Are you suggesting that this scintillator should be used because of its easy decontamination in all cases or only in specific applications in which radionuclides are known to be strongly adsorbed onto the vial wall?

N. G. L. Harding: In our carbonate elution assay we have shown that elution of carbonate is variable, particularly in etched vials. With 'fresh' vials, elution is also variable, although with properly washed vials I find that a distilled water rinse will elute carbonate. The type of scintillator is also important, for we find that a vial used with a micellar scintillator is easier to wash than when used with non-micellar types, particularly if proteinaceous samples are used. Finally, chemical modification of the vial wall enables us to have a set of vials in the laboratory with known properties of a uniform minimal capacity for isotope adsorption.

B. W. Fox: We use a commercial sonicator successfully in routine washing of vials – with minimal incidence of contamination.

N. G. L. Harding: I agree that this is an extremely effective procedure for washing a vial, particularly if one sonicates in the presence of isopropanol as we show in the paper. However, in a multi-isotope, multi-procedure laboratory, the problem can be complicated, particularly if users presoak vials before sonic washing. Large sonicators are expensive and so we decided to develop a washing procedure, essentially independent of a supply of sonic and thermal energy. At the same time the opportunity presented itself for coating a vial wall to obtain an homogeneous set of vials with the desired properties.

B. Bakay: We use oven-cleaned vials satisfactorily in our laboratory.

N. G. L. Harding: I agree that oven cleaning is effective for removing organic isotopes though I am not sure of the long-term effects upon recycling, and the procedure is expensive to run so one might do equally well to purchase new vials.

B. W. Fox: This procedure only applies to organically found isotopes, e.g. ^{3}H and carbon.

B. Bakay: Yes.

N. G. L. Harding: Metal residues will combine with heated glass.

SECTION II

RECENT ADVANCES IN SAMPLE PREPARATION

Chapter 7

Recent Advances in Sample Preparation for Liquid Scintillation Counting

B. W. Fox

Paterson Laboratories, Christie Hospital, Manchester, England

There are three basic questions to be considered when sample preparation for liquid scintillation counting is to be undertaken:

(i) Has the elaborate nature of the technique, apparently necessary, destroyed accuracy?

(ii) Has the simplicity of the technique to be used destroyed sensitivity?

(iii) Is the intended counting system homogeneous or heterogeneous in character?

HOMOGENEITY AND HETEROGENEITY IN SCINTILLATION COUNTING

It is important to recognise whether the scintillation system and the β-emitter are totally within a single phase (homogeneous) or whether they are distributed unevenly between two phases, either solid–liquid or liquid–liquid (heterogeneous). For correction of quenching it would appear that in the majority of cases *sample channels ratio* may be used. The use of internal standard (other than the β-emitter itself as standard) or external standards should, however, be used with considerable care in heterogeneous systems.

Ideally, one should strive to obtain an homogeneous counting system in which the β-emitter is in direct thermodynamic contact with and within the same phase as the scintillant system itself. This is, however, ideal, and many biochemical experiments, due either to the nature of the final sample to be measured or sheer numbers, cannot be undertaken without the use of heterogeneous procedures. Mueller[1] has described an *operational definition* of a true solution from the point of view of scintillation counting as a mixture of components in which any one of those components is counted with the same efficiency, if tagged with a given radionuclide. This definition would allow certain regions of colloidal compositions to be treated as though they were effectively homogeneous. However, where absolute levels of radioactivity are required, or where mixed isotope counting is to be undertaken, homogeneous systems should be made the prime aim.

An apparently elaborate procedure may often be used effectively to concentrate β-emitters, prior to assay. Bulk impurities are first cleared away to reduce impurity quenching. An example of such an elaborate preparation is the formation of osazone derivatives of sugars from biological fluids, which due to their high insolubility will effectively concentrate radionuclide-tagged sugars. However, the yellow and orange colours of the osazone create colour quenching. A colourless derivative was thus described by Steel *et al.*,[2] the glucotriazole into which the osazone may be converted quantitatively. However, although colourless, this derivative is very insoluble, but is readily solubilised in a boric acid mixture described by Jones and Henschke.[3] This mixture consists of naphthalene 60 g, PPO 4 g, POPOP 0.2 g, methyl alcohol 100 ml, ethylene glycol 20 ml, *p*-dioxane 880 ml and boric acid 25 g. Up to 100 mg of the triazole may be dissolved in 5 ml of this mixture. However, at this stage of elaboration one would begin to question again the exact requirements of the experimenter. Would thin layer or column chromatography be more appropriate, quicker and perhaps more quantitative?

The final choice of sample preparation must be a judgement arrived at as a result of a large number of individual decisions. One can reduce these decisions to a few important questions:

(i) is the isotope insoluble in hydrocarbon solvents?

(ii) is the sample to be measured of low specific activity (i.e. say less than 1000 counts min^{-1}/0.2 ml or 40 mg) or of higher specific activity (i.e. greater than this value)?

(iii) are there *few* (less than 20) or *many* samples to be measured?

(iv) does one need absolute activity *or* relative radioactivity assays between samples?

(v) is there likely to be several millilitres available of each sample or only a very small volume (1.0 ml or less)?

The answers to these kind of questions can critically determine the sample preparation techniques appropriate to the sample being measured.

HOMOGENEOUS SYSTEMS

If the isotope is soluble in toluene, there is no great problem, and it is worth remembering that many lipids, long chain fatty acid salts of metals, noble gases, hydrocarbons, both liquid and gas, certain alcohols, especially the longer chain series, and ethers are soluble in toluene and are best assayed in a simple solution of PPO in toluene. A recent example of the employment of this fact has been described by Vikari[4] for the estimation of labelled cholesterol in plasma, where steroid was partitioned directly into a 15 ml mixture of a toluene-based scintillation mixture containing 37.5% ethylene glycol monomethyl ether and 1 ml of water. The upper scintillant phase contains all the cholesterol and the lower most of the impurities. No bleaching procedures are necessary and chemiluminescence is negligible. Being an homogeneous system, external standardisation can also be freely applied.

Another example of the use of partitioning into the toluene phase of the labelled material to be measured has been described by Sankaran and Pogell,[5] in the determination of methyl transferase enzyme activity, in which the labelled product is extracted directly into the toluene-based scintillant within the vial, which is also used for the reaction.

A β-emitter which partitions into the aqueous phase, however, such as a polar organic substance, biological macromolecule or as a polar salt, provides a solution in which the sample preparation will be largely determined by the specific activity of the solution as a whole. Since many scintillant mixtures have been devised to accept water and aqueous solutions in low proportion (i.e. up to 10% by volume) and still retain homogeneous counting characteristics, then clearly for high specific activity solutions these are the methods of choice.

A useful summary of the assessment of the optimal conditions for a few commonly employed blending solvents was given by White.[6] He concluded that 2-ethoxyethanol was superior to many other alcohols tried for their ability to blend without introducing excessive quenching. A useful mixture suggested by White consisted of a mixture of 100 g naphthalene, 7 g PPO, 0.6 g POPOP in a mixture of 1000 ml toluene and 300 ml 2-ethoxyethanol. This mixture will accommodate up to 1% of aqueous solutions. For higher proportions of watery solutions (i.e. up to 4–5%), the naphthalene must be omitted, but for higher proportions still, a dioxane-based scintillant, such as that of Bray's,[7] using methyl alcohol and ethylene glycol as combined anti-freeze and blender must be used.

In many modern counters, the use of ambient temperature conditions would appear to do away with the necessity for the use of anti-freeze in Bray's mixture. We investigated whether the ethylene glycol played a significant role in the counting efficiency, and more particularly the blending properties, of this scintillant, and different volumes of salt solution of different concentrations were examined in Bray's mixture with and without this component. The results are shown in Fig. 1.

A surprising feature was the fact that a low proportion of salt solution to scintillant caused precipitation (of salt?) to occur, even at low concentrations, whereas at higher proportions of salt solution, more could be incorporated, without precipitation occurring. By about 34% aqueous solution, precipitation of naphthalene occurred, even by water alone. The ethylene glycol did allow a higher concentration of salt to be incorporated between 0 and 10%, but made very little difference above that level.

For relatively small volumes of aqueous solution of higher specific activity, this type of scintillant would therefore appear to be the method of choice, since the strict homogeneity of the final scintillant mixture will allow for confidence in the use of external and internal standardisation techniques, as well as the sample channels ratio technique.

It is worth noting at this point that the accuracy of quench correction procedures properly applied have approximate coefficients of variation ranging from internal standard ratio 1.3%, sample channels ratio 2.5% to external standard ratio 3.5%. Automatic dispensing of internal standards at the 10 μl level allows for variations of between 1.5 and 2.0% between samples.

An alternative sample preparation technique for solid material such as tissue, or biological fluids like plasma and blood containing ^{3}H, ^{14}C or ^{35}S, is of course combustion. Although eminently suitable for the assay of these isotopes, the method is only really suitable where the number of counts obtained from the sample is sufficient to be statistically reliable after combustion. The automation and convenience of the procedure has much to commend it since the final counting mixture is in the best possible form for absolute determinations. Yet another simple train combustion device has recently been described by Baba *et al.*[8] A bomb combustion method for up to 10 g of biological and environmental samples in cases where the specific activity is low has also been described by Moghissi *et al.*[9]

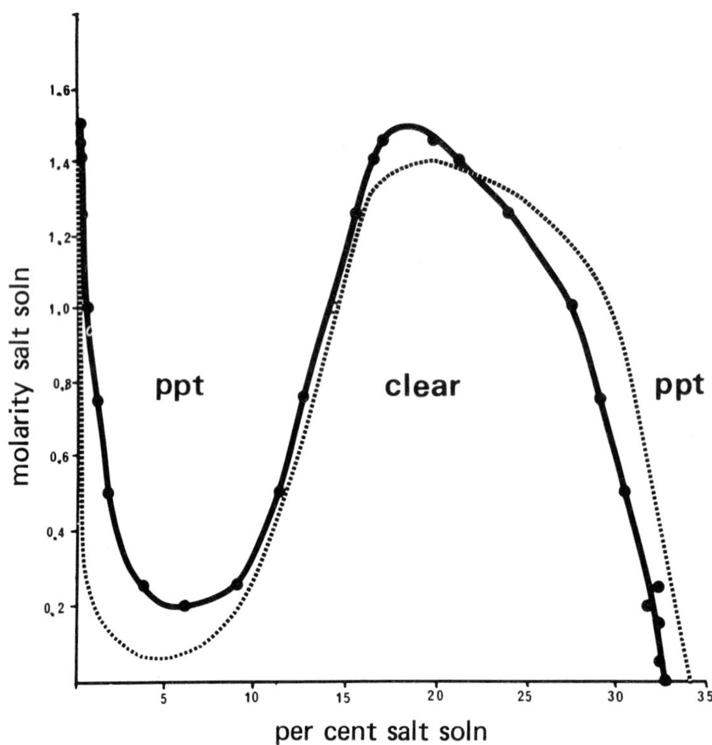

Fig. 1. The effect of adding increasing volumes of sodium chloride solution of different concentrations to Bray's scintillant. The solid line indicates the boundary above which precipitation occurs. The dotted line is a similar plot with Bray's scintillant in which the ethylene glycol has been left out.

SOLUBILISATION METHODS

Solubilisation methods involving only quaternary ammonium salts have continued to provide a suitable procedure for the assay of tissue. The chemiluminescence associated with many of the quaternary ammonium solubilisation procedures has been traced to the lipid component of the tissue being dissolved. Laine-Boszormenyi and Fallot[10] suggested that a lipid extraction prior to assay would reduce the level of chemiluminescence considerably, provided, of course, the label itself is not in this fraction. The incompatibility between certain primary solutes such as PBD and its derivatives observed by Dunn[11] has been confirmed by Painter and Gezing,[12] but not by Wetter and Dyck.[13] Clearly there appear to be other parameters in the local use of these agents which may modify the results.

HETEROGENEOUS SYSTEMS

In many situations in biochemical research, large numbers of samples, often of low specific activity, require to be assayed. The experiment often only becomes possible if counting systems and sample preparation procedures are used which are designed to provide reproducible comparative data between samples. The use of solid supports, such as paper, thin layer materials, glass fibre or membranes, has been used extensively and it is often desirable to solubilise material from the filter for more accurate assays in homogeneous counting conditions. Jonsonbaugh *et al.*[14] examined this problem with membrane filters and showed that the cellulose acetate membranes (EHWG type) did not produce colour quenching, whereas the mixed ester (HAWG type) did with solubilisers.

Assay of substances on solid supports is a common technique, but quantitation is unreliable unless the β-emitter is completely eluted from the disc into the scintillant mixture before measurement. Care should be taken that complete and not just partial solubilisation has taken place, since the latter will introduce considerable variability in the results. An example of this is in the measurement of monomeric and polymeric carbohydrates on paper chromatograms, described by Sandford and Watson,[15] who also gave optimal conditions for the elution and assay in blended xylene-based scintillation mixtures.

A further example of the use of differential elution in assaying enzyme activity has been described by McKenzie and Gholson[16] for methionine adenosyl transferase on phospho cellulose paper. The enzyme reaction is stopped with 10% perchloric acid and is neutralised by the addition of a fixed volume of 3M potassium carbonate, 0.5M triethanolamine mixture and after washing is assayed in an eluting scintillant mixture.

Counting of energetic β-emitters, such as ^{32}P, on discs should be avoided,[17] since losses occur through the vial base due to the shorter path length of the scintillant available compared with the range of the β-particle.

Suspension counting of BaCO–^{14}C has usually been employed for $^{14}CO_2$ radioassay. Larsen[18] has pointed out that up to 6 mg of $BaCO_3$ can be dissolved in 1 ml of 0.03M EDTA tetra-sodium salt, which can be assayed in a colloid counting system.

The heterogeneity and loss of 4π counting by adsorption of materials on to the vial walls have been studied by Wigfield[19] and Wigfield and Srinavasan.[20] The measurement of adsorption shift, which varies from 0.0 with non-adsorbed components to 0.4 for fully adsorbed compounds, is determined from an equation containing two constants, one based on external standard ratios and the other on the sample channels ratio. Using this concept, a satisfactory calibration curve can be obtained which takes into account both homogeneous and heterogeneous aspects of the scintillation counting process. This is basically similar to the problems encountered in many heterogeneous counting conditions and there appears to be no reason why the principle involved here should not be more widely applicable in these systems.

An interesting observation associated with wall adsorption effects was made recently by Crouthamel and Van Dyke,[21] who showed that the adsorption of ^{14}C-labelled polyethylene glycol on the walls of a glass vial could be completely prevented by the use of detergent-based scintillant mixtures.

POLYETHYLENE VIALS

There have been a number of reports on the problems arising from the use of polyethylene vials in liquid scintillation equipment having different types of γ-emitting isotopes as external standards. The observation of discrepancies was reported by Rauschenbach and Simon,[22] who attributed them to the use of low-energy (^{137}Cs) sources in some instruments and high-energy (^{226}Ra) sources in others. This effect coupled with absorption of the fluor into the walls produced abnormal quenching values. This observation has essentially been confirmed by Gogan and Gogan.[23] Horrocks[24] has recommended that when sample counts are high enough the *sample channels ratio* should be used, but *external standard ratios* of either type may be used provided a careful choice of the channels used for the external standardisation is ensured.

COLLOID SCINTILLATION COUNTING

The use of colloidal systems in scintillation counting has predictably increased enormously over the last two or three years. The two-phase nature of the system has not only increased the possible uses to which it has been put, but has increased the possibilities of reporting 'improved' or 'preferred' recipes. It is not surprising, therefore, that lack of easy verification of the claims has resulted in many publications that have not added significantly to either our knowledge of the system or to its efficient use.

I would like, therefore, to summarise some of the aspects of this powerful counting system, and to endeavour to understand some of the ways in which the problems of counting instability and the partial heterogeneity exhibited may be modified and possibly controlled.

The prime value of the colloid scintillation counting system is to incorporate a larger volume of salt or macromolecule-loaded aqueous solution into an heterogeneous association with an efficient scintillator system. Its main advantage does not lie in its absolute efficiency, but in its ability to get maximum counting rates out of a low activity sample that could otherwise be immeasurable in an homogeneous system demanding lower volumes of sample in the mixture. This is an important criterion in determining whether or not to use the system. For solutions of high specific activity, such that 0.2–0.4 ml of solution would give statistically adequate counts, there is no advantage to be gained in using a colloid system at all. Indeed, low sample volumes in colloid systems (i.e. 5–10%) could result in considerable instability in counting.

In an attempt to establish some criteria for a proper comparison of these systems, there are certain conditions that should be met before an advantage can be claimed for one colloidal system over another:

(i) does the modification allow increased water acceptance without excessive loss of efficiency?

(ii) does the modification increase the stability of the counts obtained between repeated assays over a 48-h period?

(iii) does the modification increase the variety of solutions capable of assay?

Between any detergent, hydrocarbon solvent and aqueous solution there are clearly an infinite number of combinations. Unless it is possible to describe the properties of a combination uniquely in these terms, then clearly we shall continue to be fed with an increasing number of new, so-called improved combinations.

The word solubilisation was coined by McBain *et al.*[25] in the Forties to describe the way in which certain dyes may be introduced or suspended in water by long chain amphiphilic salts or detergents. A similar phenomenon concerning liquids in liquids, such as aniline by short chain amphiphilic salts, was described by Neuberg[26] earlier and referred to as 'hydrotropy'. Although the latter would appear to be a much more descriptive term for the events which occur in colloidal scintillation systems, the term solubilisation has gained wide usage.

In the solubilisation of water-insoluble liquids (in our case toluene or xylene) in water, or vice versa, by means of detergent, a characteristic and theoretically important liquid crystalline phase can be achieved at certain specific concentrations of the components. The liquid crystalline state can vary from a thick grease to a firm gel or more correctly a *smectic* phase. In this region of composition the molecular components are arranged in layers like packs of extremely thin, highly polished playing cards. Through polaroid filters, the light is characteristically orientated in these regions. Under the microscope, using polaroid optics, one can see the general organisation of the liquid components.

The establishment of this carefully aligned and organised molecular architecture will thus require time for assembly and this time will presumably vary with the component proportions, some combinations being capable of more rapid assembly than others. This is a technically difficult parameter to observe and to quantitate optically. Its implications are, however, important to liquid scintillation counting in this system and although we do not understand precisely how the energy of a β-particle is conducted from the disintegrating isotope to the hydrocarbon layer, this clearly happens and probably via the detergent molecule itself.

Triton X-100 itself may act as a primary solvent in its own right (Turner[27]) and it seems likely that for the colloid scintillation process to operate successfully the detergent molecule itself may require some electron-transferring properties. Figure 2 illustrates a possible arrangement of the molecules of a toluene:Triton X-100:water system at its optimal point for counting. The relative numbers of molecular species are correct but the orientation as shown is purely speculative.

The molecular architecture in the vicinity of the β-particle will clearly influence the counting efficiency at that point. In particular, the more intricate and complex the architecture, the longer it may take to achieve its final organised state and this will depend on mixing efficiency and thermal history of the sample. A convenient way of visualising this settling down phenomenon, therefore, would be to observe the changing count rate of selected combinations with time. Such a parameter is the coefficient of variation of the counts taken five or six times during a 48-h period.

We have used this parameter to examine the stability of counts at various points in the phase diagram of any combination. The method of construction has been described previously by the author.[28]

We have used the xylene:Triton X-100:water system containing PPO as our study system. Xylene appears to result in an equally efficient system to that of its toluene equivalent, but does show a greater degree of instability (Fig. 3).

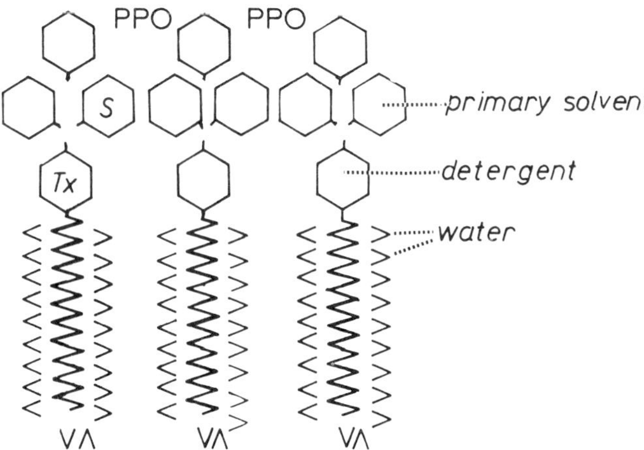

Fig. 2. Represents the number-relative of molecules of the scintillant components at the optimum counting region in the toluene:Triton X-100:water system. The primary solute is assumed to be in the non-polar zone and is not shown.

Whether this instability is due to the added complexities of molecular orientation of the solvent layer imposed by the additional methyl group in the molecule is a matter for speculation, but is presumably open to testing experimentally. This instability produced by xylene offered, however, a convenient system to examine the nature of the problem. By adding either a long chain or a short chain alcohol to this system, it would be expected that the former may become associated with the detergent and the hydrocarbon layer, whereas the latter would become associated with the detergent and the aqueous layer. In the experiments to be described here, the effect of octan-1-ol on this system was examined systematically by replacing the detergent molecule in stepped amounts with the alcohol and measuring the efficiency and stability parameters of the resulting four-component systems. The results of the addition of 2% octan-1-ol are shown in Fig. 4. It can be seen from this figure that the region of less than 1% coefficient of variation has spread to cover a greater area of the phase diagram as a whole. Increasing this percentage to 5% increases the area still more (Fig. 5).

Concomitant with these changes are changes in the Merit Value Contours of the system. The two terms found most useful in working in this system are defined in Fig. 6.

The Instrument Corrected Merit Value (MĪV) should always be quoted rather than the merit value itself if comparison with other systems is to be made since the basic efficiency of the instrument itself can lend an apparent advantage to the system where no such advantage exists. The Instrument Stability Corrected Merit Value (MISQ) was devised[28] to try to include a term which would also stress the importance of the counting stability in any advantage gained. The MIV values for the system with Triton X-100 alone (i.e. without octanol) are shown in Fig. 7. It can be seen that similar to the coefficient of variation plot (Fig. 3) the disturbed molecular architecture is reflected in a similar pattern of contours, the maximum value observed in this subdivision being 1182, using tritiated water as β-emitter.

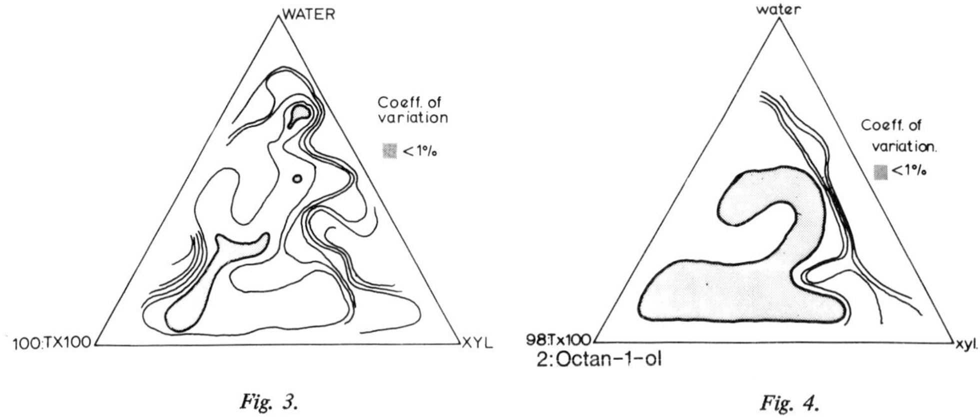

Fig. 3. *Fig. 4.*

Fig. 3. A triangular plot of the percentage coefficient of variation of counts over a 48-h period for the water:Triton X-100:xylene system. The shaded zone represents the region of less than 1%. The remainder of the contours are increasing in 1% increments from the shaded zone.

Fig. 4. As for Fig. 3, except that 2% (v/v) of the Triton X-100 component is replaced by 2-octan-1-ol.

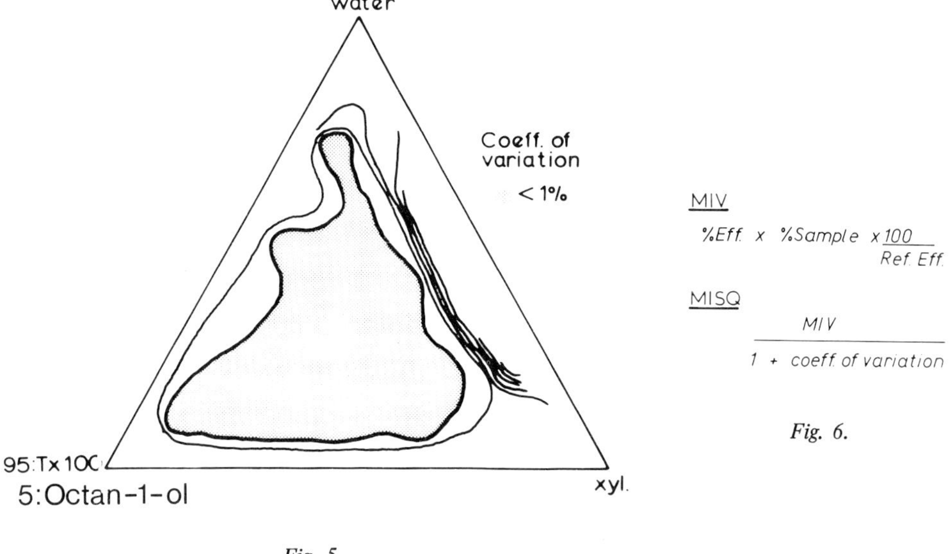

Fig. 5.

Fig. 6.

Fig. 5. As for Fig. 3, except that 5% (v/v) of the Triton X-100 component is replaced by 2-octan-1-ol.

Fig. 6. The Instrument Corrected Merit Value (MIV) uses the instrument reference efficiency under the conditions employed to attempt to relate results from difference sources. It is necessary to use suitable channel settings for both sample and reference measurements to avoid any cut-off, especially by the upper discriminator. The Instrument Stability Corrected Merit Value (MISQ) employs the percentage coefficient of variation.

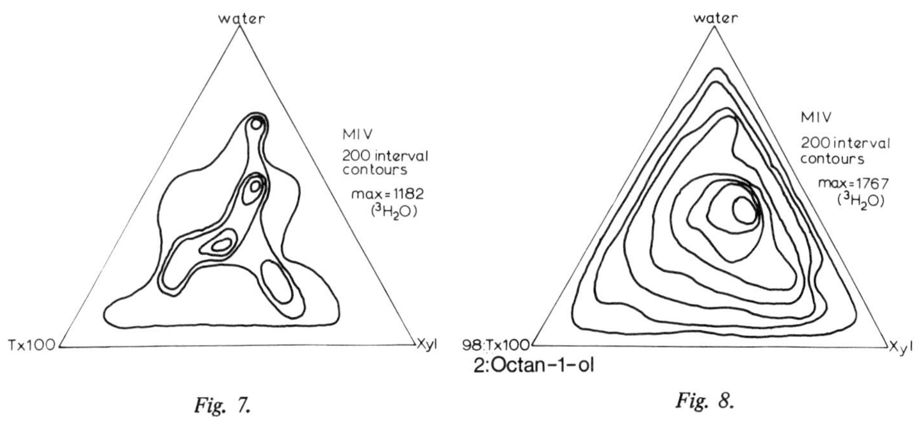

Fig. 7.

Fig. 8.

Fig. 7. A plot of the MIV values under the same conditions as in Fig. 3.

Fig. 8. A plot of the MIV values under the same conditions as in Fig. 4.

The addition of 2% octan-1-ol, however, spreads out the efficiency contours more evenly (Fig. 8), which suggests that there is a less discontinuous structure within the molecular architecture of the system from one combination to another. The maximum value is also significantly higher (1767), which suggests that a more suitable structure for efficient scintillation counting is present also. At the 5% level (Fig. 9) there is a similar continuity of the merit contours, except that the maximum is now lower. Thus, when 0 and 5% octan-1-ol is present the Triton X-100 layer clearly confers an advantage in terms of both the efficiency and stability by which water is assayed in this system. A more detailed bracketing of these concentrations, and extension to other alcohols would almost certainly lead to real advantages.

Finally, I would like to refer to some work which I am intermittently undertaking with Dr. C. W. Gilbert of our Institute. We have been examining the pulse height distributions throughout the toluene:Triton X-100:water system, using tritiated water and tritiated toluene as standards. When each composition at the 10% level is examined from the point of view of its efficiency in relation to the values of sample channels ratio and external standard ratio, all combinations with the exception of the 10% detergent level lie on the same sample channels ratio:efficiency curve. The contours drawn along sample channels ratio and external standards ratio should be parallel for areas in which they can be both used, but diverge in areas of high heterogeneity. Figure 10 shows preliminary data with regard to these measurements, and it can be seen that at relatively low detergent water levels such divergencies are very great. The efficiency plots closely parallel the sample channels ratio and not the external standard ratio. The degree of divergency, like the adsorption shift of Wigfield quoted earlier, could give a useful parameter by which to measure isotope concentration anywhere within the phase diagram, and overcome the ambiguity created by heterogeneity.

In conclusion, therefore, I think that the art of sample preparation for liquid scintillation counting is to be aware of the proportion of disintegrations that are being measured and to force this proportion to be as high and as constant as possible using the least effort to do so.

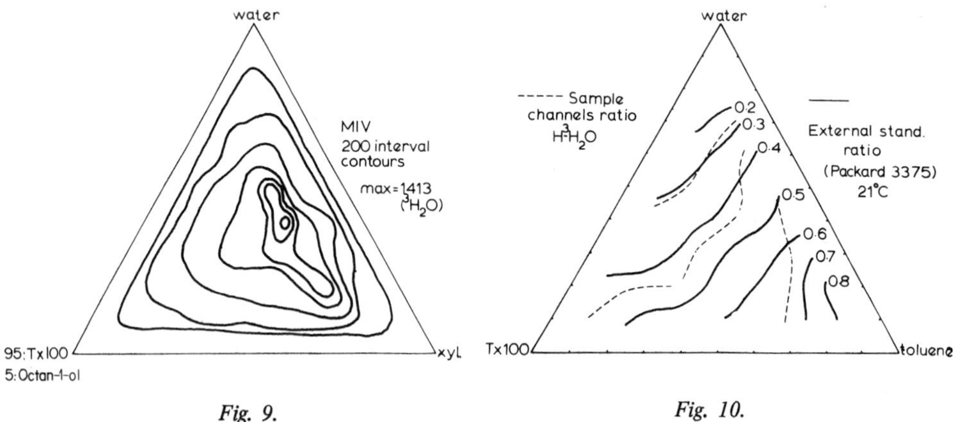

Fig. 9. *Fig. 10.*

Fig. 9. A plot of the MIV values under the same conditions as in Fig. 5.

Fig. 10. A plot of contours for external standard ratio and sample channels ratio for the toluene: Triton X-100:water system using tritiated water as standard. The mixture contained both PPO and POPOP.

ACKNOWLEDGEMENTS

I would like to thank Dr. C. W. Gilbert for helpful discussions and Mr. Robert Caffrey for excellent technical help. The work was supported by grants from the Medical Research Council and the Cancer Research Campaign.

REFERENCES

1. Elizabeth Bush Mueller, in *Liquid Scintillation Counting,* Vol. 3 (eds. M. A. Crook and P. Johnson), Heyden, London, 1974, pp. 47–64.
2. R. Steele, W. Bernstein and C. Bjerkness, *J. Appl. physiol.* **10**, 319 (1957).
3. G. B. Jones and N. F. Henschke, *Intern. J. Appl. Radn. Isotopes* **14**, 618 (1963).
4. J. Vikari, *Anal. Biochem.* **63**, 566 (1975).
5. L. Sankaran and B. M. Pogell, *Anal. Biochem.* **54**, 146 (1973).
6. D. R. White, *Proc. 1967 Beckmann Summer School,* Beckmann Instruments Ltd., 1967, 72 pp.
7. G. A. Bray, *Anal. Biochem.* **1**, 279 (1960).
8. S. Baba, U. Baba and J. Konishi, *Anal. Biochem.* **66**, 243 (1975).
9. A. A. Moghissi, E. W. Bretthauer, E. L. Whittaker and D. N. McNellis, *Intern. J. Appl. Radn. Isotopes* **26**, 339 (1975).
10. M. Laine-Boszormenyi and P. Fallot, *Intern. J. Appl. Radn. Isotopes.* **25**, 241 (1974).
11. A. Dunn, *Intern. J. Appl. Radn. Isotopes* **22**, 212 (1971).
12. K. Painter and M. J. Gezing, *Intern. J. Appl. Radn. Isotopes* **24**, 361 (1973).
13. L. R. Wetter and J. Dyck, *Intern. J. Appl. Radn. Isotopes* **24**, 430 (1973).
14. R. E. Johnsonbaugh, J. O. Kleiman and J. Sode, *Anal. Biochem.* **54**, 490 (1973).
15. P. A. Sandford and P. R. Watson, *Anal. Biochem.* **56**, 443 (1973).
16. R. M. McKenzie and R. K. Gholson, *Anal. Biochem.* **53**, 384 (1973).
17. F. G. Winder and G. R. Campbell, *Anal. Biochem.* **57**, 477 (1974).
18. P. O. Larsen, *Intern. J. Appl. Radn. Isotopes* **24**, 612 (1973).
19. D. C. Wigfield, *Anal. Biochem.* **59**, 11 (1974).
20. D. C. Wigfield and V. Srinavasan, *Intern. J. Appl. Radn. Isotopes* **25**, 473 (1974).
21. W. G. Crouthamel and K. Van Dyke, *Anal. Biochem.* **66**, 234 (1975).
22. P. Rauschenbach and H. Simon, *Z. Anal. Chem.* **256**, 119 (1971).
23. F. Gogan and P. Gogan, *Anal. Biochem.* **60**, 363 (1974).
24. D. L. Horrocks, *Intern. J. Appl. Radn. Isotopes* **26**, 243 (1975).
25. J. W. McBain, R. C. Merrill and J. R. Vinograd, *J. Amer. Chem. Soc.* **62**, 2880 (1940).
26. C. Neuberg, *Z. Elektrochem.* **42**, 289 (1936).
27. J. C. Turner, *Intern. J. Appl. Radn. Isotopes* **20**, 499 (1969).
28. B. W. Fox, *Intern. J. Appl. Radn. Isotopes* **25**, 209 (1974).

DISCUSSION

V. Tarkkanen: I wish to thank Dr. Fox for his contribution showing how important it is to solubilise a sample into a scintillator homogeneously. In your example showing addition of octanol into Triton-xylene the tests are only described for water. Adding the octanol showed an improvement in this case for pure water, but our experience is contradictory whenever true samples (i.e. those of buffers or any aqueous samples) are counted; addition of octanol destroys the stable micelle systems required for a correct sample. In practice the properties of most samples are not comparable with the properties of the water sample. Pure water samples are rare in liquid scintillation counting and are mostly found in hydrology or following sample combustion.

B. W. Fox: The system I described is simply designed in order to understand the basic principles involved at the level of the molecular architecture of optimal counting conditions. Water is used only as an example to study and it is not intended to suggest that this is the best method of counting, even water. Salts will introduce further changes in this architecture, but this must be the subject of further study.

S. Apelgot: I would like to say that liquid scintillators are able to extract lipids not only from liquids like plasma, but also from any material, even cells and tissue. With a liquid scintillator containing dioxane, it is possible to extract from such materials a lot of aqueous soluble compounds.

In the case of labelled tissue, it is possible to obtain results by using an enzyme, pronase, which are the same as those obtained by the combustion technique. These results were presented two years ago at the Sydney meeting.

I know from many years experience that accurate measurements are obtained in heterogeneous phases (solid–liquid) without elution. What happens in the case of ^{3}H on paper has nothing to do with elution. In my laboratory we were able to show that this is related to the structure of the paper, and never occurs with membranes or glass fibre papers. A paper on this topic will appear at the end of this year.

B. W. Fox: Substances adsorbed on to glass fibre discs often differ from their absorption on paper and membranes. In the latter cases, different substances, especially tritium-labelled, will count with different efficiencies depending on the extents to which they have penetrated the paper. In some cases, e.g. with nucleic acid precursors on paper discs, differential absorption of nucleosides and nucleotides has led to differences of counting efficiency which in turn gives erroneous data as to the extent of, say, an enzyme conversion of one to another. When absolute measurements are required, the only efficient way to count is to elute or combust. Glass fibre discs, of course, do not suffer from problems due to absorption within the fibre itself, and in this case the thickness of the deposited film, bearing in mind also that the counting is near 2π, will determine counting efficiency. In this case, higher counts which are reliably corrected for quenching can be obtained by elution, but for most practical purposes it is more convenient to assay the discs in non-polar scintillants following standardised disc preparation procedures.

N. G. L. Harding: At the last symposium we adduced structures for these micellar systems and were very pleased to have these confirmed by your diagrams of the micelles using polarised light. Since then we have produced further evidence strengthening the view that these are dynamic, rather than static, structures. This accounts for stabilising and destabilising effects observed with a variety of agents added to micellar systems. Have you explored the usefulness of such reagents for studying the solution structure?

B. W. Fox: In the present work we used octanol in an attempt to do this, but there is clearly considerable scope to extend to further alcohols to explore this system.

Chapter 8

Quenching Corrections for Emulsion Systems

H. S. Wagstaff

Fisons Scientific Apparatus, Laboratory Chemicals, Bishop Meadow Road, Loughborough, England

and

A. R. Ware*

Chemistry Department, Loughborough University of Technology, Loughborough, England

INTRODUCTION

The counting of aqueous solutions of radiolabelled material using emulsifiers of the alkyl phenol ethoxylate type is now a routine operation in many laboratories. Depending on the water content, various types of 'emulsion' can be obtained and these appear to fall into four general types. At low water concentrations a clear solution is obtained, at high concentrations a gel and at water concentrations between 15–25% by volume an unstable and/or a true emulsion. The concentration boundaries of these regions are affected by both temperature and the presence of dissolved electrolytes. In all systems, the water added and any other compounds present in solution or suspension may act as a quencher resulting in a loss in the light emitted by the scintillators. The effect of quench can be measured in liquid scintillation counting by the use of an internal standard, an external standard or the channels ratio method. All three methods have been tried with emulsion systems. The most convenient method of quench correction is to use an external standard, but this has given rise to contradictory statements in the literature,[1–10] in particular with regard to the measurement of tritium activity.

The suggestion that quench curves should be obtained for each possible quenching compound is clearly not always a practical possibility as the exact composition is not always known and may vary between samples. A general external standard quench correction curve would therefore be invaluable in routine analysis to obtain absolute activities.

In this work quench curves were obtained by use of an external standard and by the channels ratio method and the extent to which general quench curves may be applied will be discussed.

---

* Current address: Central Radiochemical Laboratory, Central Electricity Generating Board, Chalk Lane, Cockfosters, Hertfordshire, England.

EXPERIMENTAL

The emulsifier used was an iso-octyl phenol polyethoxyethanol containing 10 moles of ethylene oxide (Triton X-100). Previous work by Fox[10,11] has indicated that this emulsifier can be used with toluene to count aqueous solutions. A range of concentration ratios for emulsifier and toluene is possible. It was decided to use a concentration ratio of 1 part by volume emulsifier to 2 parts by volume toluene throughout the work. This value is one which is commonly used in practice and lies within the range suggested by Fox[11] to give a good merit stability quotient (MSQ).

All samples were prepared according to the method of Benson[12] which involves a warming and cooling cycle of the sample prior to counting. This method produced samples which gave reproducible counting rates for periods up to 10 days for a large range of water concentrations, thus allowing a series of samples to be counted several times on four different instruments for comparative purposes. The four instruments used were a Beckman LS100 operated at a controlled temperature of 12 °C and employing a ^{137}Cs external standard source, a Nuclear Chicago Unilux operated at a controlled temperature of 12 °C and employing a ^{133}Ba external standard source, a Packard 2420 operated at an ambient temperature of 23 °C and employing a ^{226}Ra external standard source and a Beckman operated at an ambient temperature of 26 °C and employing a ^{137}Cs external standard source.

All samples were counted in the preset ^{3}H channel of the instrument, the automatic external standard channels ratio (AESCR) as set on the instrument and a sample channels ratio (CR). The sample channels ratio was obtained by counting in the preset total ^{3}H channel (A) and a channel (B) set such that it contained the lower 20% of the ^{3}H count in A. The ratio CR was calculated after the counts had been corrected for background as follows:

$$\mathrm{CR} = \frac{(\text{count in channel A}) - (\text{count in channel B})}{\text{count in channel B}}$$

A concentrated scintillant solution of 10.5 g l^{-1} PPO and 0.15 g l^{-1} POPOP in toluene was prepared. Six millilitres of this solution was present in all samples ensuring that the total weight of scintillants was the same in all cases. Tritiated water (0.5 ml) containing 50,000 disintegrations min^{-1} was added to each sample; this enabled the sample channels ratio to be measured accurately. In practice the channels ratio is clearly not accurate for low active samples. Sufficient distilled water, toluene and Triton were then added to give a total volume of 15 ml and water concentrations ranging from 3.3–40% by volume.

The samples were then counted on the four counters described above. All samples were cycled and counted three times. The samples were counted long enough to give a 2σ error of 1.5% on all counts. Curves of percentage water concentration against counting efficiency are shown in Figs. 1 to 4. The AESCR and CR were also calculated and the curve of AESCR against counting efficiency is shown in Figs. 1 to 4. A similar smooth curve was obtained for CR against counting efficiency for water concentrations from 3.3–10% and from 26.7–40% by volume, but to avoid confusion it is not shown in the diagrams.

Three water concentrations were then chosen for further study of quench effects caused by chemical additives present in the system. The concentrations chosen were

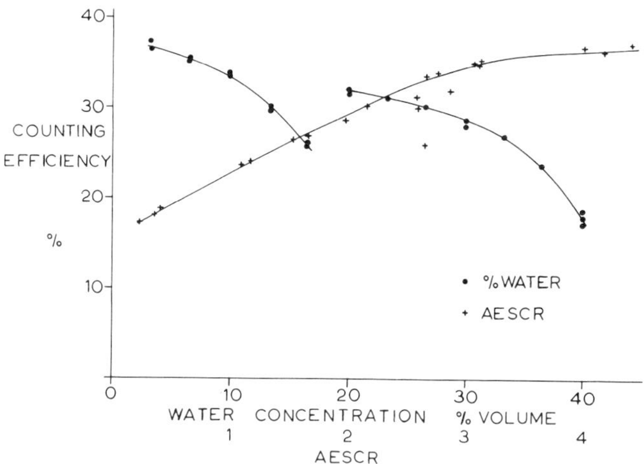

Fig. 1. The variation in counting efficiency with water concentration and AESCR for the Beckman LS100 controlled temperature counter (12 °C).

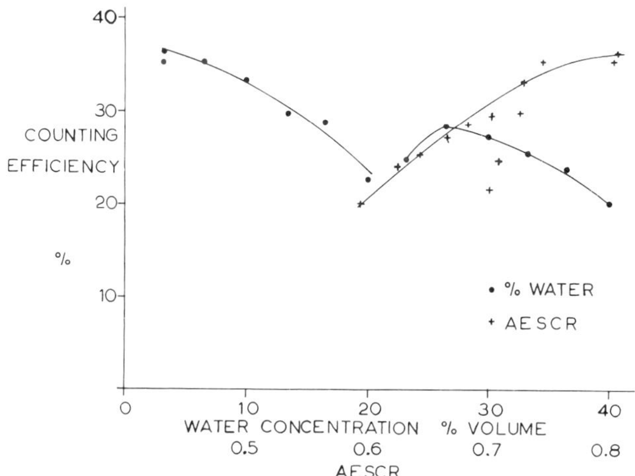

Fig. 2. The variation in counting efficiency with water concentration and AESCR for the Packard ambient temperature counter (23 °C).

10%, 26.7% and 40% water. Quenching agents investigated were carbon tetrachloride which should remain in the organic phase and the sodium salt of trichloracetic acid which should remain in the aqueous phase. A series of samples were made up at each of three water concentrations for each quench agent. The quench effect was measured by determining both the AESCR and the CR for all samples. The results obtained on the controlled temperature Beckman are shown in Figs. 5 and 6 for water concentrations of 10% and 26.7%. For comparison the solid line indicates the curve of AESCR and CR against counting efficiency obtained for the samples of variable water concentration.

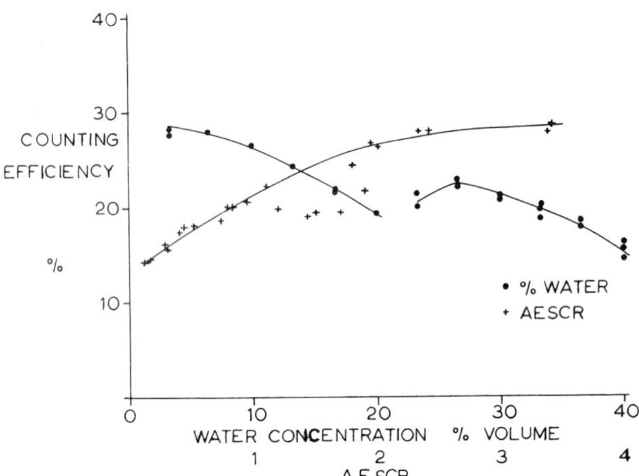

Fig. 3. The variation in counting efficiency with water concentration and AESCR for the LS100 ambient temperature counter (26 °C).

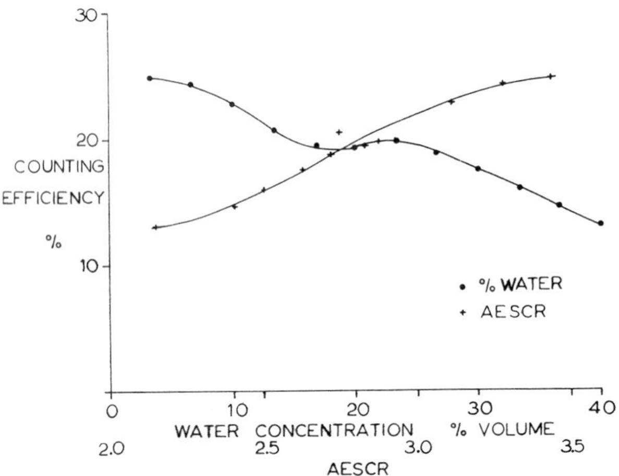

Fig. 4. The variation in counting efficiency with water concentration and AESCR for the Nuclear Chicago Unilux controlled temperature counter (12 °C).

Urine-quenched samples were also prepared and counted. The urine was decolourised to minimise colour quench effects. The results are also indicated in Figs. 5 and 6. No results were obtained for the 40% water samples as the quench had taken the spectrum below the range of the channels as set.

The results for the Packard instrument for all three water concentrations are shown in the same manner in Figs. 7 to 9.

The results for the Beckman LS100 operated at 26 °C are shown in Figs. 10 and 11. The results for the 26.7% water samples were very unreliable. This series of counts was

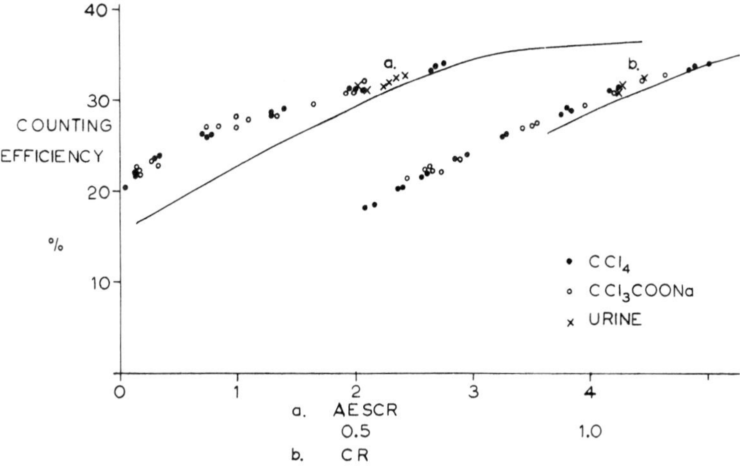

Fig. 5. Quench curves for 10% water samples showing AESCR and CR variations for the Beckman LS100 controlled temperature counter (12 °C).

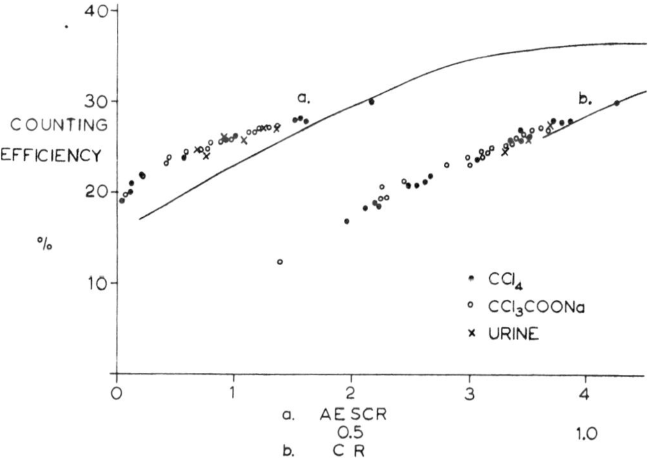

Fig. 6. Quench curves for 26.7% water samples showing AESCR and CR variations for the Beckman LS100 controlled temperature counter (12 °C).

performed at the highest temperature and as indicated by Tarkkanen[13] the unstable two-phase region for emulsions moves to higher water concentrations as the temperature is increased. The value of 26.7% water comes on the edge of the unstable region according to the AESCR against counting efficiency curve of Fig. 3 and a 1-2 °C rise in temperature would make the 26.7% water concentration samples become two-phase. This was in fact observed and the samples reverted to a single phase on cooling.

The quench correction curves obtained for the Nuclear Chicago Unilux are shown in Figs. 12 to 14.

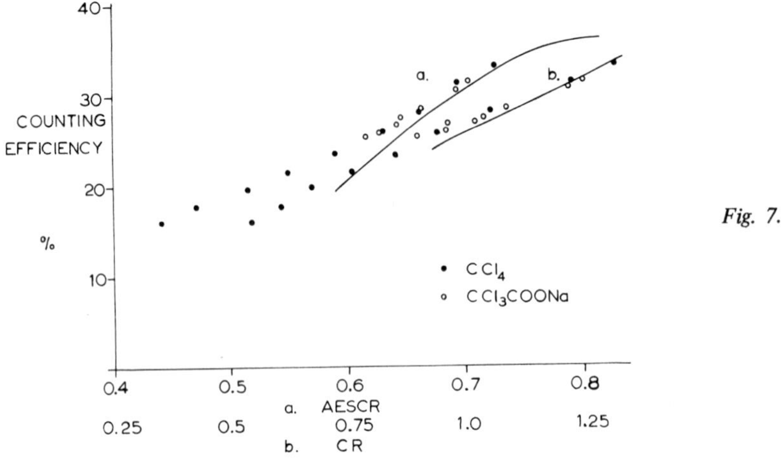

Fig. 7.

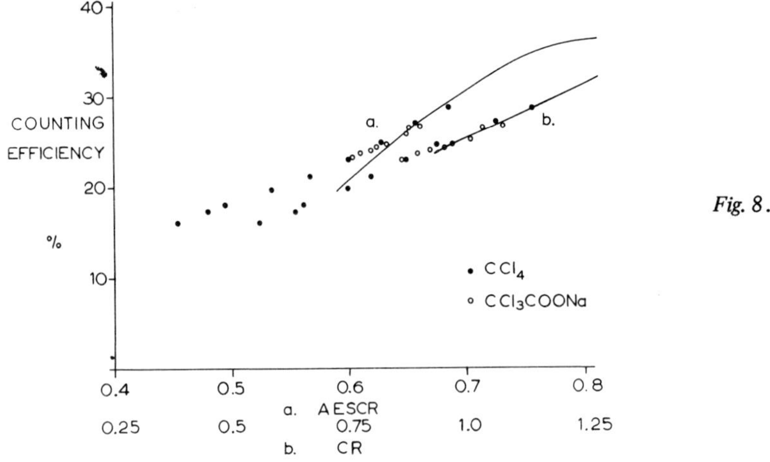

Fig. 8.

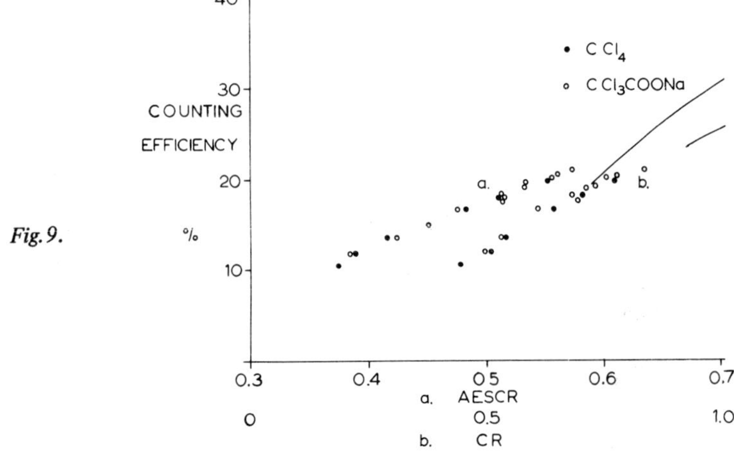

Fig. 9.

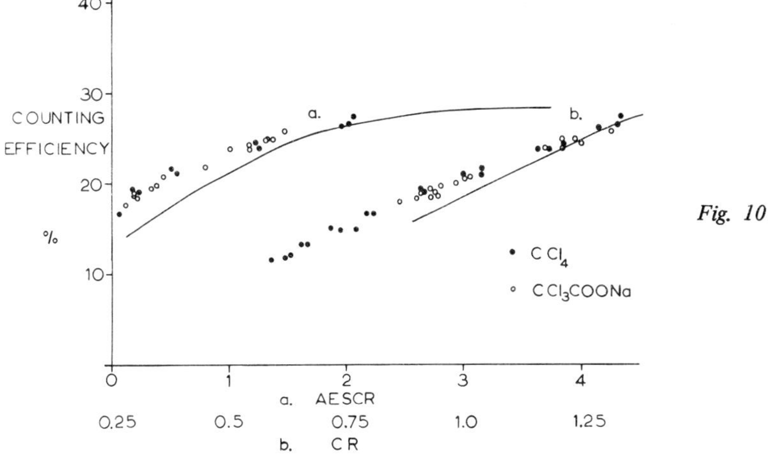

Fig. 10.

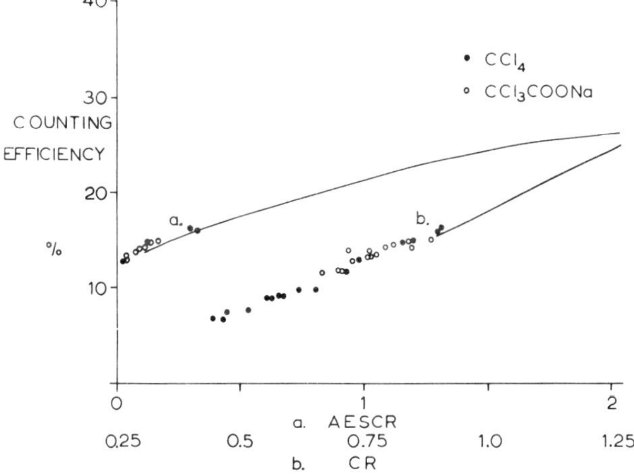

Fig. 11.

Fig. 7. Quench curves for 10% water samples showing AESCR and CR variations for the Packard ambient temperature counter (23 °C).

Fig. 8. Quench curves for 26.7% water samples showing AESCR and CR variations for the Packard ambient temperature counter (23 °C).

Fig. 9. Quench curves for 40% water samples showing AESCR and CR variations for the Packard ambient temperature counter (23 °C).

Fig. 10. Quench curves for 10% water samples showing AESCR and CR variations for the Beckman ambient temperature counter (26 °C).

Fig. 11. Quench curves for 40% water samples showing AESCR and CR variations for the Beckman ambient temperature counter (26 °C).

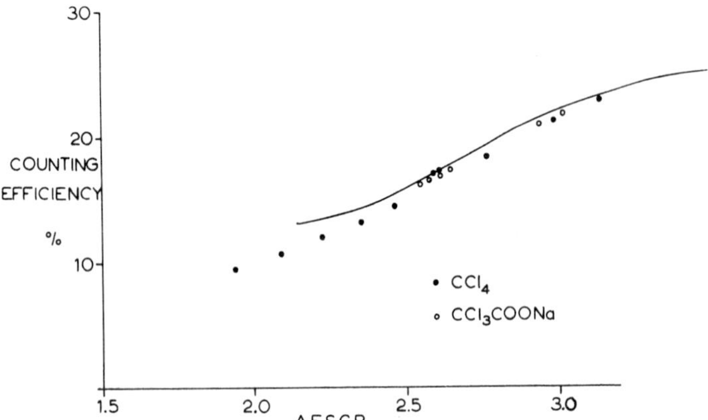

Fig. 12. Quench curves for 10% water samples showing AESCR variations for the Nuclear Chicago controlled temperature counter (12 °C).

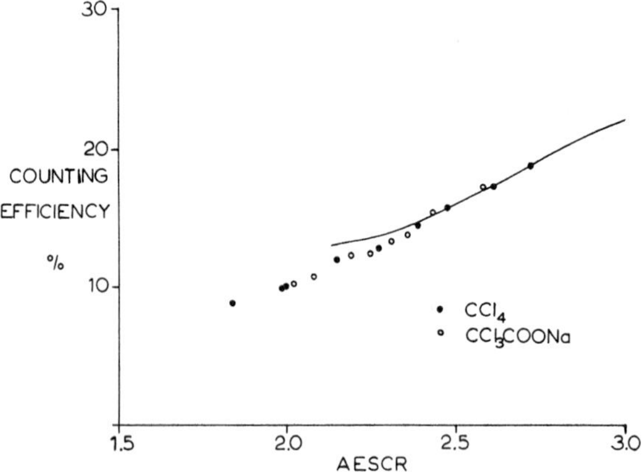

Fig. 13. Quench curves for 26.7% water samples showing AESCR variations for the Nuclear Chicago controlled temperature counter (12 °C).

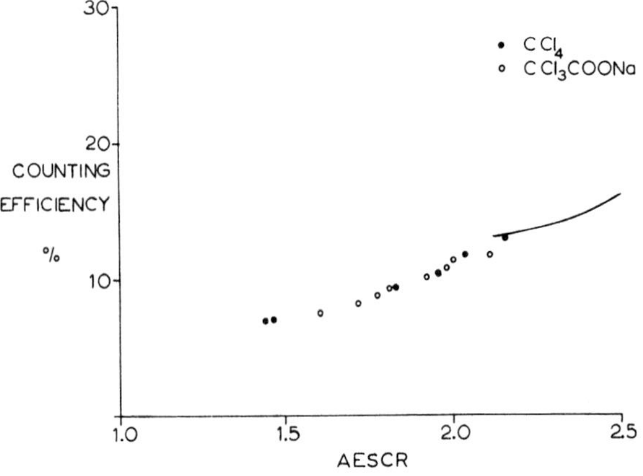

Fig. 14. Quench curves for 40% water samples showing AESCR variations for the Nuclear Chicago controlled temperature counter (12 °C).

DISCUSSION AND CONCLUSIONS

All the instruments gave the now familiar curve for water concentration against counting efficiency showing the region of two-phase separation leading to a discontinuity in the curve. Fox[11] has indicated that the two-phase region, although being physically unstable, can give consistently stable but low counting efficiency and this was confirmed. The AESCR against counting efficiency curves showed a smooth curve for water concentrations from 3.3% to 10% and 26.7% to 40%, but the results in the two-phase region did not fit the curve and recycled counts were not reproducible. As the counts had been shown to be stable this suggests that the AESCR was changing with time. This was found for both AESCR and CR measurements. This suggests that the use of any channels ratio technique in this two-phase region is unreliable. Quench curves for variable water concentration could be set up outside this region provided no other quenching agent was present.

The samples quenched with CCl_4 and CCl_3COONa at a fixed water concentration appear to give the same quench curve for any of the four counters even though the curves were not neccesarily the same between counters. All are apparently reliable and useful. The urine samples counted on the Beckman counter also fitted the quench curves. Within the errors of the method it would appear that the same quench curve may be obtained for many different quenching agents.

The effect appears to be similar no matter whether the agent is soluble in aqueous or organic solution. Further, the quench effect can be satisfactorily measured by the AESCR or the CR values. This is not to say that the methods can be universally applied to all counters and samples. As can be seen in all the diagrams, Figs. 5 to 14, the quench curve obtained for the variable water concentration samples is different from that obtained for the samples quenched with CCl_4, CCl_3COONa and urine. The quench mechanism would appear to be different for the two cases and quench curves are only reliable if either the water concentration or additives is kept constant. The four instruments employed three isotopes of widely differing γ-energies.

From the results, each of the isotopes gave suitable quench correction curves and consequently are all equally applicable. It was noticed that the manufacturers preset channels were not necessarily the optimum and more suitable channels might yield improved curves.

Provided samples are not counted in the two-phase region, which will vary with temperature, and that samples are made at a given water concentration, quench correction curves would appear to be a practical possibility for ^{3}H-labelled compounds. It is recommended that samples are put through a warming and cooling cycle during preparation and that the counter is operated at a controlled temperatrue.

ACKNOWLEDGEMENT

The authors wish to thank the School of Agriculture, Nottingham University, Sutton Bonnington, and Fisons Pharmaceuticals for loan of the Nuclear Chicago and Packard counters.

REFERENCES

1. L. A. Baillie, *Intern. J. Appl. Radn. Isotopes* 8, 1 (1960).
2. V. Tarkkanen, in *Liquid Scintillation Counting,* Vol.2 (eds. M. A. Crook, P. Johnson and B. Scales), Heyden, London, 1972, p. 177.

3. J. Van der Laarse, *Intern. J. Appl. Radn. Isotopes* **18**, 485 (1967).
4. J. C. Turner, *Intern. J. Appl. Radn. Isotopes* **20**, 499 (1969).
5. J. C. Turner, *Intern. J. Appl. Radn. Isotopes* **19**, 557 (1968).
6. P. Rauschenback and H. Simon, *Fresenius Z. Anal. Chem.* **256**, 119 (1971).
7. M. A. Friedman, A. McEvoy, G. Miller and S. Epstein, *Anal. Chem.* **43**, 780 (1971).
8. L. E. Anderson and W. O. McClure, *Anal. Biochem.* **51**, 173 (1973).
9. E. Rapkin, *Gel and Emulsion Counting of Aqueous Solutions*, Intertechnique Review.
10. B. W. Fox, in *Liquid Scintillation Counting*, Vol. 2 (eds. M. A. Crook, P. Johnson and B. Scales), Heyden, London, 1972, p. 189.
11. B. W. Fox, in *Liquid Scintillation Counting*, Vol. 3 (eds. M. A. Crook and P. Johnson), Heyden, London, 1973, p. 202.
12. R. H. Benson, *Anal. Chem.* **38**, 1353 (1966).
13. V. Tarkkanen, in *Liquid Scintillation Counting*, Vol. 3 (eds. M. A. Crook and P. Johnson), Heyden, London, 1973, p. 177.

Chapter 9

Sample Preparation Techniques for Tritium Counting in Biological Systems

P. Johnson, T. J. Rising and B. R. Twite

Drug Development, Hoechst Pharmaceutical Research Laboratories, Milton Keynes Buckinghamshire, England

INTRODUCTION

It was not until the late 1950's that the first commercially available automatic liquid scintillation counters appeared on the market, and the following decade saw considerable improvements in instrument design and in data presentation.[1] In the last few years, instruments designed for routine sample preparation have been introduced, particularly sample oxidisers.

A number of different types of sample oxidisers are currently available which provide samples for liquid scintillation counting from labelled biological material and which also separate ^{3}H and ^{14}C. The principle of operation is the same in all models, namely that the sample is combusted totally to water vapour and carbon dioxide. The water vapour is condensed and collected in a counting vial, and the carbon dioxide is absorbed in a suitable trapping agent. The vials are then automatically filled with scintillator.[2] Thus, biological samples containing either ^{14}C or ^{3}H or both can be processed in this way.

In some cases, it is insufficient simply to ascertain the total amount of radioactivity in a sample, as, for instance, following the administration of tritium-labelled compounds to animals and man. Biological exchange processes can lead to the incorporation of tritium into body water. This results in the half-life of the compound in the body being overestimated, misleading tissue retention data and incorrect kinetics. We have overcome this problem by using a Packard Tri-Carb 306 Oxidiser which has been modified to allow differentiation between tritium present as body water and tritium present as parent compound or metabolites.

THE MEASUREMENT OF TRITIATED WATER USING A MODIFIED SAMPLE OXIDISER

The oxidiser has been modified by the addition of an ignition switch which cuts the electrical supply to the heating coil, thereby preventing ignition of the sample in the combustion chamber (Fig. 1). Samples are held in a paper combustion cone. Tissue samples are cut into small pieces and pulped against the sides of the combustion cone using a spatula and liquid samples are absorbed onto fluted filter

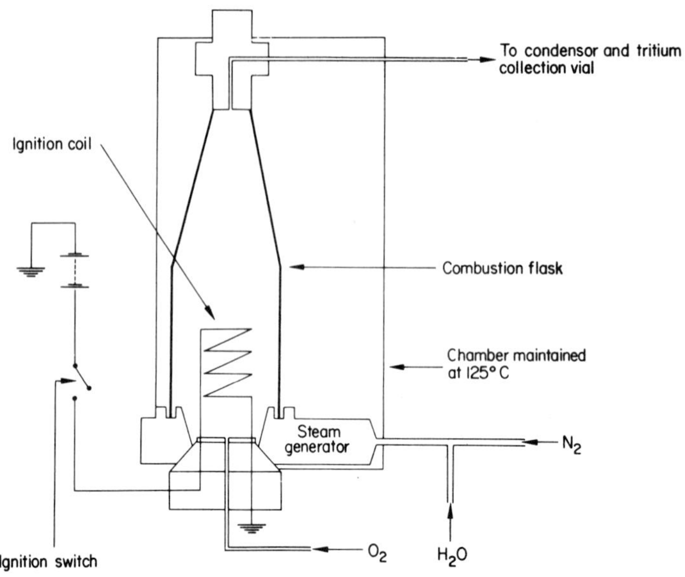

Fig. 1. Diagram of the combustion chamber of a Packard Model 306 Sample Oxidiser.

paper. When the programme is started, the sample is lifted into the combustion chamber, which is maintained at 125 °C. Provided a hole is made in the base of the combustion cone to allow the hot gases to circulate, a 200 mg sample will dry in 6 min, and the resultant water vapour is collected in the tritium collection vial. The ignition is then re-set and about 20 μl of water is added to the now dry sample to slow down the burning process. When the programme is restarted, the sample combusts in the normal way. The drying step gives a measure of tritiated water in the sample and the combustion step gives a measure of the tritium content of the compound and any metabolites.

VALIDATION OF METHOD

We have checked the efficiency of drying in two ways, using a wide range of biological samples. First, we have compared the weight loss after drying using the oxidiser with the weight loss after freeze-drying using a conventional Virtis freeze-dryer (Table 1). The weight loss figures, drying either with the sample oxidiser or with the freeze-dryer, give percentage weight losses that differ by no more than 3%. Second, we have measured residual radioactivity by combustion of samples previously spiked with tritiated water and dried with the oxidiser (Table 2). No more than 4% of the original radioactivity is found in the samples after the drying step. Therefore, the modified sample oxidiser allows very efficient drying of biological samples.

However, although the trapping efficiency is 98% when a sample containing tritium is combusted with the oxidiser, only 85% of the radioactivity is retained in the tritium collection vial when a sample is dried (Table 3). Three percent of the radioactivity remains in the sample and appears in the tritium collection vial on combustion of the sample. If two acetone/dry ice traps are put in series on the gas line, a small amount of radioactivity is retained, whilst 7–8% is collected in the effluent from the wash cycle, giving an overall recovery of 96–97%.

Table 1. Weight loss after drying samples (the results are expressed as the percentage loss in weight of the samples after drying ± standard error of three determinations).

Tissue	Sample oxidiser	Freeze-dryer	Difference
Liver	69.7 ± 1.7	69.0 ± 0.3	+ 0.7
Kidney	75.7 ± 2.1	75.7 ± 1.0	0.0
Heart	75.7 ± 0.7	75.3 ± 0.1	+ 0.4
Brain	75.4 ± 2.3	75.8 ± 2.2	- 0.4
Urine	94.3 ± 0.7	93.6 ± 0.1	+ 0.7
Faeces	62.2 ± 0.5	60.6 ± 1.3	+ 1.6
Blood	81.5 ± 1.7	82.3 ± 0.4	- 0.8
Plasma	89.6 ± 1.5	90.6 ± 0.6	- 1.0
Liver homogenate	79.2 ± 0.8	82.2 ± 1.2	- 3.0
Lung homogenate	92.7 ± 0.6	94.2 ± 0.1	- 1.5
Faecal homogenate	82.9 ± 0.7	85.1 ± 1.2	- 2.2
Water	100.7 ± 0.8	100.0 ± 0.0	+ 0.7

Table 2. Residual radioactivity after drying samples (the results are expressed as the percentage of the original radioactivity remaining in the samples after drying ± standard error).

Tissue	Number of determinations	Residual radioactivity
Kidney	4	3.1 ± 0.6
Heart	4	3.6 ± 0.4
Brain	4	2.0 ± 0.2
Urine	3	3.6 ± 0.8
Faeces	4	2.9 ± 0.5
Blood	4	4.0 ± 0.6
Plasma	4	3.4 ± 0.7
Liver homogenate	4	2.8 ± 0.5
Lung homogenate	3	2.9 ± 0.4
Faecal homogenate	3	2.9 ± 0.4
Water	5	3.3 ± 0.3

Table 3. Recovery of tritiated water from the sample oxidiser (the results are expressed as a percentage of the total sample radioactivity).

Drying step	Combustion step	Gas effluent	Wash effluent	Total recovery
85.8	3.3	0.5	7.4	97.0
85.2	2.6	$<$ 0.1	8.0	95.8
86.5	3.0	$<$ 0.1	6.8	96.4

The retention efficiency of 85% is remarkably consistent. Using a wide range of biological samples, and using three different Packard 306 sample oxidisers, the retention of radioactivity in the tritium vial varied from 83–87% of the total sample radioactivity (Table 4).

Thus two methods can be employed for the estimation of tritiated water in biological samples. As an approximate measurement, the tritiated water content can be obtained by applying the 85% retention efficiency figure to the radioactivity appearing in the tritium vial during the drying step. This procedure allows a simple and rapid check for tritium exchange and is particularly suitable where the amount of tritium exchange is small. A more accurate assessment is made by a lengthier process. A sample is combusted to give total radioactive content, and a second aliquot of the same sample is first dried and then combusted. The difference in radioactivity between the two combustions is a measure of the tritiated water in the sample.

Table 4. Retention of tritiated water in the tritium collection vial (the results are expressed as the percentage of the total sample radioactivity retained in the tritium vial on drying ± standard error).

Tissue	Number of determinations	Radioactivity retained
Liver	4	84.3 ± 2.5
Kidney	4	86.9 ± 1.4
Heart	4	85.9 ± 0.9
Brain	4	87.2 ± 2.8
Urine	3	83.7 ± 0.5
Faeces	4	85.7 ± 0.7
Blood	4	86.9 ± 1.2
Plasma	4	84.2 ± 0.8
Liver homogenate	4	87.3 ± 1.2
Lung homogenate	3	85.8 ± 0.9
Faecal homogenate	3	86.3 ± 1.2
Water	10	86.5 ± 1.6

PRACTICAL USE OF METHOD

We have recently carried out a study with a tritiated compound in laboratory animals and it was first necessary to determine the degree of biological exchange. The compound was an acridine derivative, Metifex, uniformly labelled with tritium in the aromatic ring (Fig. 2) with a radiochemical purity of at least 95%. As part of the study, the compound was administered orally to male beagle dogs at a dose level of 5 mg kg^{-1}, and urine, faeces and plasma samples obtained at suitable times after dosing. The radioactive content of samples was determined either by direct counting in a suitable scintillator for urine and plasma or by combustion of aliquots of faecal homogenates.

Table 5 shows the excretion of radioactivity in urine and faeces from one male dog. The excretion of radioactivity in the urine over 148 h was no more than 4% of

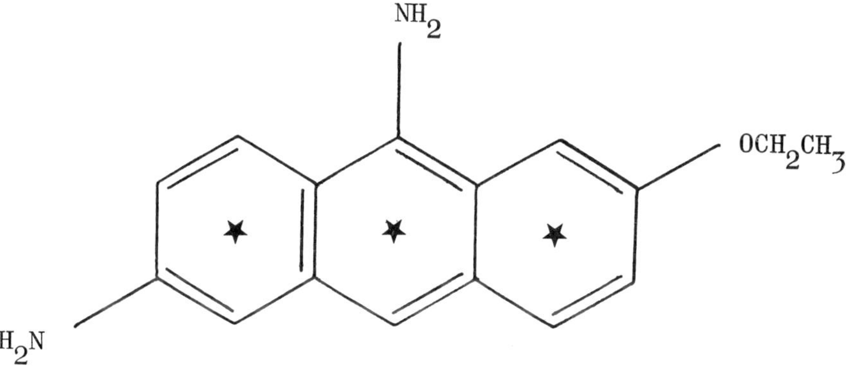

Fig. 2. 2-ethoxy-6,9-diaminoacridine(Metifex). (* denotes aromatic ring uniformly labelled with tritium.)

Table 5. Excretion of radioactivity in urine and faeces following oral administration of ^{3}H-Metifex to one male beagle dog (the results are expressed as the cumulative excretion of radioactivity – as a percentage of the administered dose).

Time after dosing (h)	Cumulative percentage of the dose	
	Urine	Faeces
2	0.6	ND
24	1.9	38.7
48	2.6	57.5
72	3.3	64.7
148	4.1	67.7

ND = not determined.

the dose. Faecal excretion was much greater, and after 148 h the cumulative urinary and faecal output was 72% of the dose, and the balance, including cage washings, was 76%. This poor recovery suggested that tritium exchange might be occurring, resulting in retention of tritiated water with a long biological half-life. Urine, plasma and faecal samples were therefore analysed for their tritiated water content. The results, using the 85% retention efficiency factor, are shown in Table 6. Little tritium exchange had occurred in faeces. However, urine showed an increasing amount of exchange with time from 17% at 24 h to 77% at 148 h after dosing. The majority of the radioactivity in the plasma was tritiated water. Pre-dose urine samples spiked with aliquots of the labelled dosing solution were shown to contain less than 0.3% tritiated water at zero time and after two days storage at 4 °C, indicating that spontaneous chemical exchange in urine was not occurring.

Table 6. Tritium exchange following oral administration of ^{3}H-Metifex to one male beagle dog (the results are expressed as the percentage of tritiated water in the excreted radioactivity).

Time after dosing (h)	Faeces	Urine	Plasma
2	ND	3.0	ND
4	ND	ND	67.5
6	ND	ND	73.5
12	ND	ND	81.5
24	0.7	18.6	85.0
48	1.4	26.0	92.6
72	2.2	56.9	94.5
148	1.2	77.1	93.9

ND = not determined.

Table 7. Three methods for the estimation of tritiated water in urine (the results are expressed as the percentage of tritiated water in the excreted radioactivity – means of two determinations).

Time after dosing (h)	Sample oxidiser		Freeze-dryer
	One-stage method	Two-stage method	
2	3.0	1.1	5.0
24	18.6	18.0	16.5
48	26.0	26.6	26.3
72	56.9	52.1	53.1
148	77.1	73.1	73.7

In order to compare the methods, tritium exchange in urine was also measured by the lengthier method of two combustions, as well as by freeze-drying. Table 7 shows the percentage of urinary radioactivity which is tritiated water. This was measured using

(i) the 85% retention efficiency factor as in Table 6;
(ii) the two-stage method, i.e. combustion of first fresh and then dried tissue and subtracting the result to give the tritiated water content;
(iii) freeze-drying using a conventional Virtis freeze-dryer.

The two-stage method is probably less accurate when the percentage of tritium exchange is small, but otherwise the three methods are in good agreement.

SUMMARY

Using a simple modification to a commercially available sample oxidiser, we have been able to measure reliably both the total tritium and tritiated water content of biological samples. The method is faster and less tedious than freeze-drying, and is an extension of a normal oxidation method for preparing biological samples for liquid scintillation counting.

REFERENCES

1. E. Rapkin, in *Liquid Scintillation Counting,* Vol. 2 (eds. M. A. Crook, P. Johnson and B. Scales), Heyden, London, 1971, pp. 61–100.
2. N. Kaartinen, *A New Oxidation Method for the Preparation of Liquid Scintillation Samples*, Packard Technical Bulletin No. 18, 1972.

DISCUSSION

M. J. Rance: I would like to know if you have compared the sensitivity of this technique with more standard methods (such as enclosed metabowl studies, distillation of urine etc.) for the determination of tritium exchange.

B. R. Twite: I have no experience of enclosed metabowl studies, but the method has been compared to freeze-drying and shows comparable sensitivity.

M. J. Rance: Do you feel that with your method it is possible to generate reliable drug distribution data routinely in the presence of biological lability?

B. R. Twite: In the example used to illustrate the method, 1% tritium exchange was detectable.

T. J. Rising: The method is applicable to the determination of small levels of tritium exchange in the order of 1%, provided the specific activity of the original dosing solution is high enough. I believe that results so obtained could be confidently submitted to regulatory authorities.

P. Johnson: Could Dr. Rance elaborate his first question because I am not clear as to the relevance of studies of expired air from metabolism cages to our technique of measuring tritiated water in tissue and body fluids. Are you referring to the amount of volatile tritium that might be exhaled? This is not measured by our technique.

M. J. Rance: In our laboratory, any tritiated compound showing biological lability would not be used in further studies. Nevertheless, distribution studies such as you describe could constitute a screen for exchange in the same way as we at present use total recovery studies, distillation of biological fluids, etc.

G. Ayrey: Is it true to say that your method assumes that the only volatile metabolic product is water? I would be interested to know if you had to check this and if there was any chance that the labelled drug was partially volatile or steam volatile under the conditions used.

B. R. Twite: It is, of course, absolutely essential to check the amount of tritium exchange in the dosing solution using the drying method described – indeed, this is best checked in the biological fluid in which the tritium exchange will be measured. Other volatile metabolites are evaporated from the sample, but are not necessarily retained in the tritium collection vial. The presence of other volatile metabolities will be shown by the 'two-stage' drying method described.

D. Case: What modifications, if any, have to be made to the Packard 306 in order to use the technique you describe?

B. R. Twite: A simple modification is required, involving the insertion of a switch in the circuitry.

Chapter 10

A New Method of Sample Handling and its Application to Liquid Scintillation Counting Analysis

Bohdan Bakay

Department of Pediatrics, School of Medicine, University of California, U.S.A.

The development of coincidence counters capable of reliable measurement of the weak β-emitting radioisotopes by liquid scintillation counting techniques made the use of radioactive tracers safe and simple. The first liquid scintillation counters were nothing more than replicas of one-channel γ-counters existing at that time. However, since their introduction 30 years ago, scintillation counters have benefited from many refinements in design which include sophisticated programmable features and computerised data processing.

In contrast, the original method of sample handling employing the counting vials was left unchanged and, as such, it remains the weakest link in the otherwise excellent analytical system. In fact, the use of counting vials has far more undesirable weak features than strong positive ones: (i) the fragility of glass counting vials continues to be responsible for the loss of samples and the main cause of costly decontaminations of laboratory tables, floors and instruments; (ii) glass (and plastic) vials have irregularities which usually cause optical distortions resulting in significant counting errors; (iii) low background counting vials must be made of high-quality, low in ^{40}K content glass – consequently, such vials are expensive items; (iv) in order to maintain the highest possible counting efficiency and the lowest error, counting vials must be filled with large volumes of costly scintillation fluids; (v) the processing of samples in the counting vials requires a great deal of labour. After deposition of the sample in the vials, the vials must be filled with scintillation fluid, closed with screw caps, wiped clean and counted. Next, the vials must be removed from the counter, opened and emptied. Used vials, caps and the liquid matter must be collected in separate containers and carried to specific burial grounds. With the rise of material and labour costs, petroleum-product shortage and ecological pressures, the counting vial is a luxury which for most researchers is difficult to support.

These disadvantages provided the impetus which prompted a search for new ways to present the sample to the liquid scintillation counter. Many years of experimentation have led to an efficient system of sample handling which eliminates the use of counting vials, generates no solid wastes, considerably decreases liquid wastes and reduces the cost of analysis by a factor of ten.

It is well established that various parameters of compounds separated by chromatographic columns can be measured quantitatively employing different flow analysers. Most of the modern high-pressure, high-speed analysers are equipped with highly sensitive detectors and while using very small volumes of fluids, they can analyse submicrogram quantities of chemical compounds at a greatly accelerated rate.[1,2] Similarly, autoanalysers used in clinical laboratories can perform many analyses on samples of biological fluids transported in a discrete manner between gas bubbles without involving human hands.

However, the quantitation of radioactive compounds in the effluent of high-speed chromatographic analysers, or in the series of discrete samples supplied by the autoanalyser or other similar instruments, is a more difficult problem. In most such systems, submicrogram quantities of compounds are rapidly eluted in very small volumes. This essentially precludes collecting fractions for conventional analysis in counting vials or analysis in flow cells packed with solid scintillators.[3-8] One possibility is to mix the effluent with scintillation fluid and monitor the activity of the mixture in a scintillation counter by means of a hollow-tube flow cell.[9-15] This method of analysis is notably more efficient than counting in the flow cell filled with solid scintillators and therefore it is especially suitable for measurement of compounds labelled with tritium. However, because of drag along the wall at high velocity, the fluids flow significantly faster in the centre of small-bore conduits than along the wall (Fig. 1). This causes spreading of the separated components. When followed closely by another sample, the part of the sample flowing along the wall is overtaken by the centrally flowing part of the subsequent sample resulting in remixing of the samples. Introduction of air bubbles into the stream of liquid unifies the flow and prevents mixing of samples. However, air bubbles do not wipe off all of the passing liquid from the wall of the conduit and therefore they only reduce but do not eliminate the sample carryover.

It was found that if small segments of a semi-solid resilient gel are introduced into the conduit at constant intervals, the gel acts as a series of pistons that, for all practical purposes, maintain uniform flow along the wall and in the centre of the conduit. Furthermore, semi-solid spacer wipes off the traces of passing sample from the wall of the conduit far more efficiently than air bubbles. Because of this distinct action, such gel spacers as polyacrylamide or agarose can maintain discrete

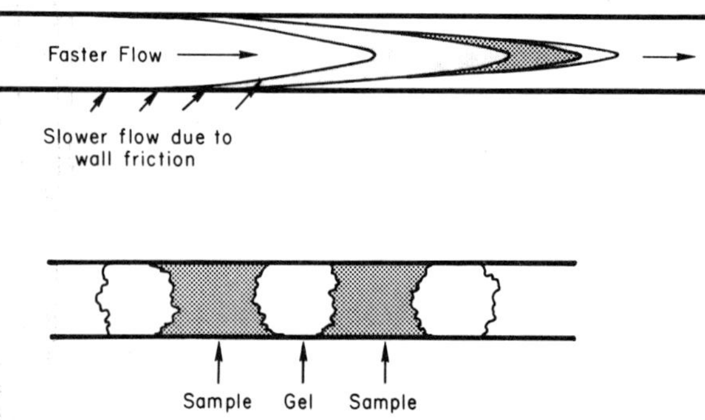

Fig. 1. Schematic illustration of wall effect and its elimination by the gel spacer.

transport of samples in long stretches of conduits. Also, because of the resilient nature, gel spacer glides freely through the restrictions, sharp turns and orifices in the apparatus and the interconnections without blocking them.

The basic apparatus used for the analysis of radioactivity in the effluent from a high-pressure amino acid analyser and of discrete samples is illustrated schematically in Fig. 2. Briefly, all interconnections were made using Cheminert tubing and connectors. The photometer of the amino acid analyser was connected with a piece of tubing to one port of a mixer. Another port of the mixer held a septum for manual injection of samples. The septum was connected to the spacer-gel pump, and to the piston 'a' of a Duplex Mini Pump. The third port of the mixer was connected to the piston 'b' of the Duplex Mini Pump. Both pistons of the Mini Pump were pumping multi-purpose scintillation fluid. The fourth port of the mixer was connected to the flow cell. The flow cell was mounted in a one- or two-channel liquid scintillation spectrometer which was equipped with a chart recorder and a printer. The outlet of the flow cell was connected to the waste reservoir. The details of this apparatus, materials and procedures are described elsewhere.[14]

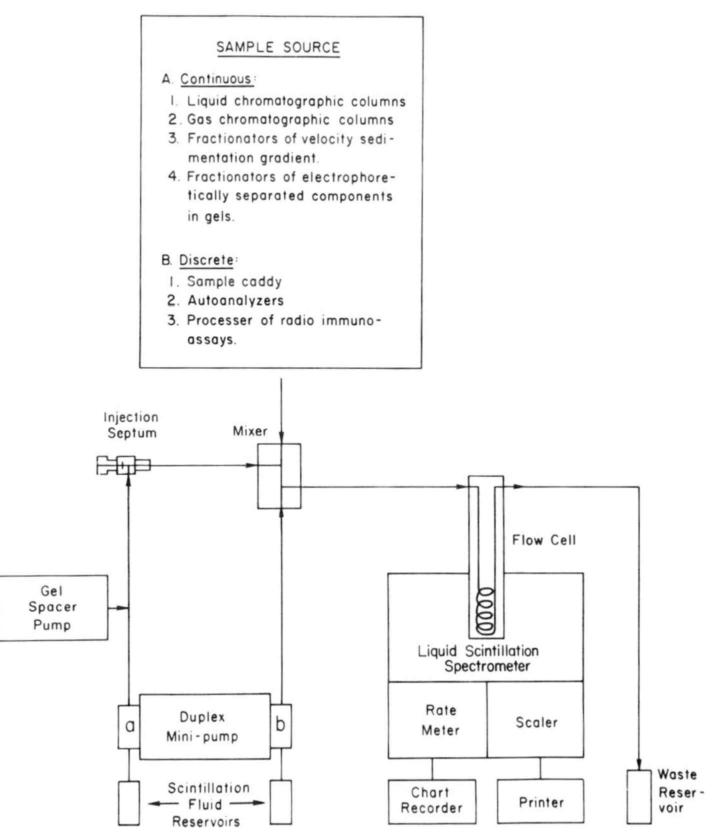

Fig. 2. Apparatus and connections used in analysis of discrete samples of effluent from amino acid analyser, and other sources.

The same apparatus was used for the analysis of individual samples, except that individual samples were introduced into the stream of flowing scintillation fluid and gel via a septum using a microsyringe, or samples were deposited with a microsyringe in a re-usable sample holder (Fig. 2). The sample holder was placed on a pneumatically operated sample caddy which, on command from a timer, moved the sample holder to the sample pick-up station. Once the sample holder was in place, the pumps were turned on and the sample was moved by the stream of scintillation fluid and gel spacer from the holder into the mixer. In the mixer, it was mixed with an equal volume of fresh scintillation fluid then directed into the flow cell and to the waste reservoir. Once in the flow cell, the sample was counted dynamically without stopping the pumps, or the pumps were stopped and the sample was counted for either a preset time interval or a preset error. Also, in the stop and count mode, the efficiency of counting was determined automatically with an external standard. The mode of counting can utilise essentially all features of all modern counters.

The effects of the spacer gel on the measurement of the radioactivity in the flowing stream of liquid are illustrated in Fig. 3. Twenty, 15, 10 and 5 μl aliquots of a solution containing 741.5 disintegrations min^{-1} for each microlitre of ^{14}C-leucine were injected at 45 s intervals with a microsyringe through the septum into the stream of scintillation fluid and analysed in a 1.06 ml flow cell at a rate of 3.45 ml min^{-1}, without adding polyacrylamide spacer gel. The same experiment was performed in the presence of gel spacers.

When the spacer gel was left out, each sample needed 48 s to clear the flow cell. Also, it generated a peak with a long trail. As a result, samples injected in succession produced a profile of fused peaks. This indicated that the front of the oncoming sample caught up with the preceding sample and caused mixing of the samples.

However, when the same samples were analysed in the presence of the spacer gel, the sample cleared the flow cell without trailing. This is seen especially clearly in the profile produced by samples injected in rapid succession. Each sample formed a sharp peak separated by a valley dropping down to the baseline.

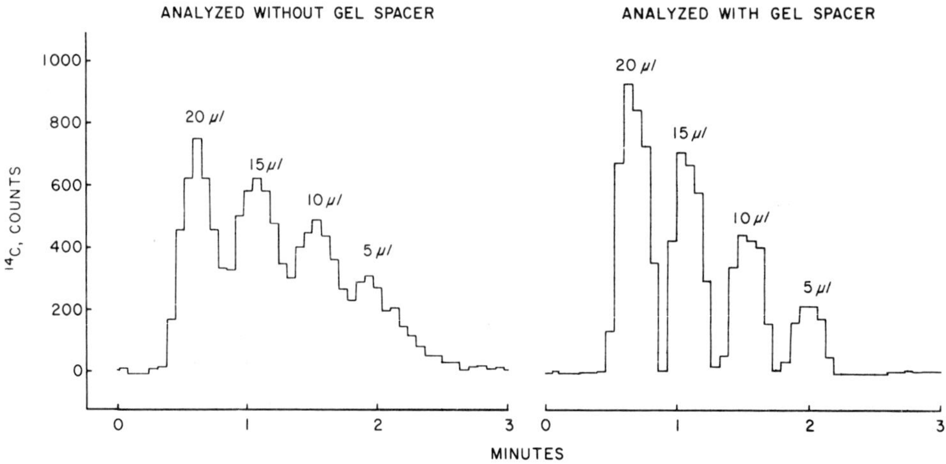

Fig. 3. Effect of spacer gel on the transport of samples in conduits of small diameter.

Measurement of radioactivity in the samples delivered by the sample caddy produced similar results (Fig. 4). Series of 5, 10, 20, 25, 30, etc. μl aliquots of ^{14}C-leucine standard, containing 741.5 disintegrations min^{-1} for each microlitre, were deposited with a microsyringe into the sample holder and placed on the sample caddy. The sample caddy was programmed to change the samples at 2 min intervals. The analysis was done in a 1.06 ml flow cell at 3.2 ml min^{-1} liquid flow. The activity was documented by the chart recorder and by the printer.

As shown in the top portion of Fig. 4, the size of the peaks were proportional to the size of the analysed aliquots. Also, each sample generated a sharp peak which dropped to the natural background without trailing. As shown on the bottom of Fig. 4, the plot of numerical values versus corresponding aliquots of standard produced a straight line. Data accumulated by repeated analysis of the same samples

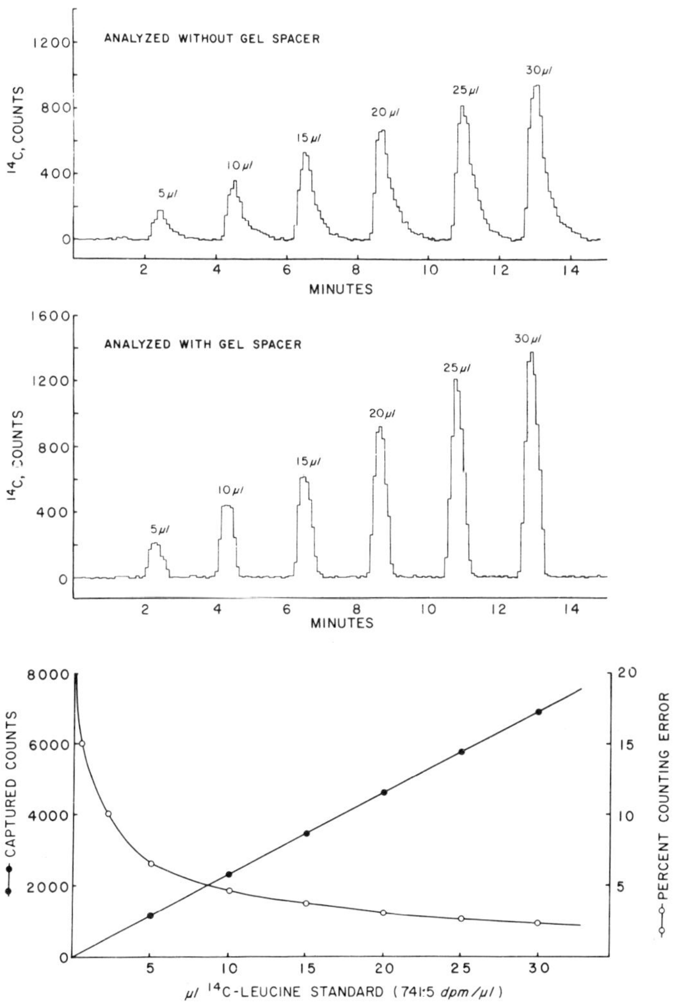

Fig. 4. Effect of spacer gel on the analysis of discrete samples delivered by an automatic sample caddy.

revealed that the counting error for the aliquot containing 500 disintegrations min^{-1} was ± 11%, for 1000 disintegrations min^{-1} was ± 7.3%, and that for 20,000 disintegrations min^{-1} was ± 1.6%. These figures are in complete agreement with values obtained in counting vials. Of course, greater counting error of less active samples is not due to the lack of precision of sampling or loss of the sample, but to uneven decay of the isotope.

The efficiency of counting of isotopes in the flow cell was calculated by using the equation which takes into account the rate of fluid flow and volume of the flow cell:[15]

$$E\,(\%) = \frac{(100\,\%) \times (\text{rate of flow, ml min}^{-1})}{(\text{disintegrations min}^{-1}\text{ per sample}) \times (\text{millilitres per flow cell})}$$

$$\times\,(\text{count per peak}) - (\text{background per peak})$$

When the data obtained for the standards were substituted in this equation, the efficiency of counting of ^{3}H-leucine and ^{14}C-leucine in a 1.06 ml flow cell in the above described experiment was 51% and 94.3%, respectively.

This equation shows that in continuous flow analysis, the rate of flow and volume of flow cell are critical parameters and require careful consideration for achieving optimal counting conditions.[15]

The results of measurements of radioactivity in the effluent of a high-pressure Durrum D-500 Amino Acid Analyser are shown in Fig. 5. In this experiment, the amino acid analyser was programmed to analyse protein hydrolysate. The column was charged with 20 μl of the amino acid calibration mixture containing 8 nmol (each) of 16 non-radioactive, and 7 to 70 pmol of 15 ^{14}C-labelled amino acids. The amount of radioactivity, contributed by individual amino acids, varied from 4800 to 7200 disintegrations min^{-1}. The analysis was carried out in a 0.53 ml flow cell at a total flow rate of 4.23 ml min^{-1}. The analysis was completed in about 55 min. As shown in Fig. 5, tracings of ninhydrin colour, recorded by the photometer of the D-500 Analyser, and of radioactivity, recorded by the scintillation spectrometer, showed essentially the same degree of resolution. They differed only in respect to the number and height of peaks. As expected in the tracings produced by the photometer, the height of the peaks was proportional to the amount of each amino acid; in the tracings of the scintillation spectrometer, they were proportional to the amount of radioactivity present in each amino acid. Furthermore, the scintillation spectrometer detected presence of proline, while the D-500 Analyser did not because the photometer, operating at 560 nm, could not detect colour produced by ninhydrin reacting with proline. In contrast, the D-500 Analyser detected cystine, methionine and ammonia which were non-radioactive. The separation of amino acids emerging within 30 s of each other, e.g. serine–threonine of isoleucine-leucine, was the same in both tracings. The profiles differed mainly in that the photometer tracings consisted of a smooth line while that of the scintillation counter, recorded at 1 mV sensitivity, consisted of a characteristic jagged line reflecting random isotope decay. The efficiency of counting ^{14}C in this experiment was about 68%. It should be noted that complete analysis consumed 220 ml of scintillation fluid. To the best of my knowledge, no other system can produce such results.

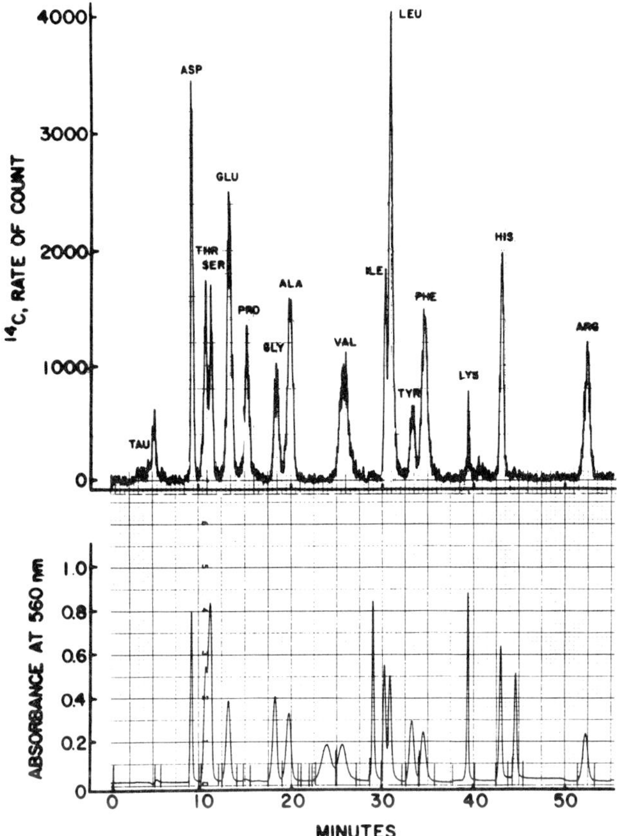

Fig. 5. Tracings of radioactivity and ninhydrin colour in the effluent from the Durrum D-500 Amino Acid Analyser.

In addition to analysis of individual samples or column effluents, this system is well suited for analysis of samples presented by a wide variety of analytical devices. Thus far, this system has been used to measure radioactivity of protein and nucleic acids separated by electrophoresis in gels,[16] and in velocity sedimentation gradients.[17]

Promising results were also obtained with gas chromatographic and nucleic acid analysers.[18] As the function of the gel spacer is similar to that of the air bubbles, the system can be adapted for clinical colorimetric assays – performed by Technicon's Autoanalyzer – and radio-immuno assays. In fact, gel spacer is not subject to compression, it does not lose its effectiveness with change of pressure or temperature, and therefore it may have a wider range of application that the air bubbles.

The composition of the spacer is especially important because it governs the effectiveness of separation of the samples and cleaning of the inner walls of the conduits. The spacer must be compatible with the liquid scintillation counting process; it must be inert enough not to absorb or interact with either the sample or the scintillators; and its physical properties should not change when it is mixed with the scintillation fluid or the sample. Of the number of semi-solid resilient materials tested, polyacrylamide gel fulfilled these requirements the closest. It is an effective spacer in concentrations between 3 to 7 g per 100 ml. However, it is most effective in concentrations between 5 and 7 g per 100 ml. It is also most effective when introduced in small amounts at 2 to 3 s intervals. The amount of spacer gel needed to maintain discrete transport depends on the rate of flow and diameter of the conduit. In the experiments described above, 40 to 60 μl of 7 g per 100 ml gel introduced in 31 segments per minute maintained discrete flow quite adequately.

The other critical factor is the composition of scintillation fluid. However, a number of commercially available multi-purpose scintillation fluids are suitable for flow analysis. Many of these products are formulated in such a way that they are equally suitable for the aqueous and non-aqueous samples.

As is evident from these experiments, this system of sample transport can satisfy the requirements of essentially all types of isotope measurement and thus it eliminates the need for counting vials. Furthermore, it can be adapted for various other analytical purposes.

ACKNOWLEDGEMENT

I am greatly indebted to the Bar Fran Corporation of San Diego for the manufacturing of specialised equipment, and the help from their skilled machinist in the construction of new pieces of apparatus.

REFERENCES

1. S. Moore, in *Chemistry and Biology of Peptides,* Proc. 3rd Amer. Peptide Symp. (ed. J. Meienhofer), Ann Arbor Science Publishers, Inc., Ann Arbor, Michigan, 1972, p. 629.
2. H. K. Knox, *Lab. Practice* **22**, 52 (1973).
3. E. Schram and R. Lombaert, *Anal. Chim. Acta* **17**, 417 (1957).
4. D. Steinberg, *Nature (London)* **182**, 740 (1958).
5. D. Steinberg, *Nature (London)* **183**, 1253 (1959).
6. B. L. Funt and A. Hetherington, *Science* **129**, 1429 (1959).
7. E. Schram and R. Lombaert, *Anal. Biochem.* **3**, 68 (1962).
8. K. Piez, *Anal. Biochem.* **4**, 444 (1962).
9. K. H. Clifford, J. W. Hewett and G. Popjak, *J. Chromatog.* **40**, 377 (1969).
10. H. W. Scharpensel and K. H. Menke, in *Tritium in Physical and Biological Sciences,* Vol. I, International Atomic Energy Agency, Vienna, Austria, 1962.
11. J. A. Hunt, *Anal. Biochem.* **23**, 289 (1968).
12. L. Schutte, *J. Chromatog.* **72**, 303 (1972).
13. D. R. Eyre, *Anal. Biochem.* **54**, 619 (1973).
14. B. Bakay, *Anal. Biochem.* **63**, 87 (1975).
15. B. Bakay, *Clin. Chem.* **21**, 1212 (1975).
16. B. Bakay, *Anal. Biochem.* **40**, 429 (1971).
17. B. Bakay, unpublished.
18. W. Weyler, L. Sweetman and B. Bakay, unpublished.

DISCUSSION

P. Johnson: First I must congratulate you on your fascinating technique. Could I ask whether your method has any limitations with regard to the types of solvent or the pressures that may be used?

B. Bakay: We have used all types of aqueous systems and non-aqueous organic solvents with no problem. The full range of pressures normally used in high pressure liquid chromatography is acceptable.

P. Tothill: I should like to congratulate Dr. Bakay on an extremely elegant and novel development and ask whether the flow rate through the counter has to be compatible with the output of the analyser. Is it possible to have a longer counting time than the interval between the effluent samples?

B. Bakay: Yes. When the pumps are stopped there is no movement or mixing in the system.

B. Fox: I did not fully appreciate from your excellent paper whether the gel spacer itself contained scintillant or not.

B. Bakay: No it did not. We have considered it and can do it, but do not think that there would be any advantage.

Chapter 11

Technique for Sequential Čerenkov and Liquid Scintillation Counting after Concentration of Emitters on Submilligram Amounts of Carrier

T. H. Bates

British Nuclear Fuels Ltd., Windscale & Calder Works, Sellafield, Seascale, Cumbria, England

INTRODUCTION

Since the inception of liquid scintillation counting, there have been descriptions, too numerous to list, of procedures for radioassay of ^{14}C, ^{35}S and ^{3}H in tracer studies and/or biological and environmental monitoring analyses. A review of preparative methods used up to 1967 is available.[1] Electron capture nuclides and low-energy β-emitters for which this mode of assay has also been employed include ^{51}Cr, ^{54}Mn, ^{65}Zn, ^{88}Yt, ^{125}I,[2] ^{129}I,[3] ^{55}Fe and ^{59}Fe,[2,4,5] ^{45}Ca,[6,7] ^{63}Ni.[8] Preparative methods usually require specific procedures to (i) eliminate the presence of any other interfering radionuclides and (ii) preclude or reduce quenching arising from both the sample matrix and indigenous or added carrier (usually tens of milligrams).

Assay of α-emitters by liquid scintillation counting has been reported.[9-17] In the case of plutonium, a frequently used technique is to mix a solution of the carrier-free actinide in specific organic extractant with scintillation cocktail.[9,13,16] A technique of dispersing plutonium as a colourless ferriphosphate in a gel suspension system has been employed[18] to assay plutonium α (^{238}Pu + ^{239}Pu + ^{240}Pu) and β (^{241}Pu, E_{max} 21 keV) activity simultaneously. Given suitable counting parameters for these isotopes, resolution is sufficient to preclude cross-interference between counting channels.[19,20] A simple technique for concentrating radiochemically separated plutonium on submilligram quantities of yttrium or ferric hydroxides, collecting the precipitate on a membrane filter, and quantitatively transferring it to scintillant is presented in this paper; the final preparation is homogeneous, and therefore susceptible to quench correction.

Assay of high-energy β-emitters, e.g. ^{89}Sr and ^{90}Yt,[21] or β/γ-emitters[22] by liquid scintillation counting has received little attention, but assay of high-energy β-emitters by Čerenkov counting in aqueous media using a liquid scintillation counter is becoming an accepted alternative to other forms of β-counting. A wide range of high-energy ($E_{max} >$ 500 keV) β-emitters has been studied,[7,23-29] and in particular the technique is being applied to the determination of ^{90}Sr in a variety of biological and environmental samples by following grow-in of ^{90}Yt by repetitive counting, and in monitoring of low-level activity in drainage systems.[28]

It is known that some fission products can be co-precipitated on ferric or yttrium hydroxide, and the above-mentioned preparation technique developed for plutonium was applied to three different types of fission product (and daughter isotopes), namely (i) ^{90}Yt, a pure β-emitter, E_{max} = 2.27 MeV, (ii) ^{144}Ce in equilibrium with its daughter ^{144}Pr, both of which have complex β- and γ-decay schemes, and (iii) ^{95}Zr in equilibrium with its daughter ^{95}Nb, which are β (E_{max} = 890 keV (2%), 924 keV (0.1%), rest $<$ 400 keV) and γ-emitters. The membrane and precipitate were first dispersed in pure solvent (dioxan) to study the Čerenkov gain spectrum, followed by addition of dioxan-based scintillant to study the scintillation spectrum. The aim was to assess how specific the spectra characteristics would be with regard to determination of purity of a separated isotope, and by the same token to what extent the dual assay might also permit resolution in a mixture of emitters as compared with either assay alone. Advantages and potential applications of the technique are discussed.

METHODS AND RESULTS

Assay of plutonium

A plutonium residue, separated from a sample by a previously described technique[30] using ion-exchange resin, is dissolved in a small beaker in about 5 ml of 3M nitric acid containing one drop of hydrogen peroxide (100 vol.). Ferric iron or yttrium (100 μg) as nitrate in 3M nitric acid solution is added, and the pH adjusted to 10 using phenolphthalein indicator by adding sodium hydroxide solution (50% w/v followed by 1% w/v near the colour change). The solution is warmed to 60–80°C for 10 min, allowed to cool and the precipitate is filtered off on a 20 mm diameter, 0.2 μm pore size, cellulose nitrate membrane filter ('Sartorius' is suitable) supported on a conventional filter stick fabricated with a porosity 3 glass sinter whose effective diameter is not more than 19 mm. The filtration procedure involves mounting the membrane on top of the sinter, wetting it with distilled water and applying suction from a water pump. The filter stick is then inverted and the membrane immersed in the solution in the beaker which is carefully agitated as the supernate passes through the membrane; two additions to the beaker of distilled water (about 5 ml) complete transfer of the precipitate and washing. The filter stick is re-inverted, suction removed and the membrane and precipitate are transferred to a desiccator containing silica gel drying agent. After 15 min, the membrane is transferred to a polyethylene counting vial (20 ml capacity), source upwards, and 100 μl of 1M nitric acid is pipetted on to it. After 5 min, when the source appears to have dissolved, 1 ml of a dioxan-based scintillant (NE220 is suitable) is added; after a further 5 min, to ensure complete dissolution of precipitate, further scintillant (11 ml) is added and the vial carefully agitated to achieve an homogeneous solution of source and membrane. The vial is transferred to a liquid scintillation spectrometer (a Packard Tri-Carb (three-channel) Model 3320 was employed in this work), and counted after a dark adaption time of 1 h, against standards and blanks prepared in the same way.

Three counting channels are employed, and settings and typical counting characteristics are shown in Table 1.

At low levels of activity (less than 5 pCi α), a precise α-determination is not possible because of the high background obtained from a counter set up optimally for tritium. For α-determination only, the background may be reduced to about 2 counts min^{-1} [16] by adjustment of EHT on photomultipliers, but it is possible by the technique described here to carry out a conventional α-assay, in a low background counter (0.3 counts h^{-1}), of the source mounted on the membrane prior to introduction to scintillant. The activity ratio of ^{241}Pu to Pu-α is not usually less than 5, but at low levels of activity it is not possible to obtain precise internal channels ratios. This can be overcome by carrying out

Table 1. Counting characteristics of plutonium-α and ^{241}Pu.

Assay	Spectrometer channel	Amplifier gain (%)	Window	Background (blank cpm)	Counting efficiency (%) (100 x cpm/dpm)
α	1	0.5	50–210	10	100
β	2	50	50–1000	20	37
Channels ratio	3	50	290–1000	10	–

internal standardisation by transferring 1 ml from the counting vial containing standard to the counted sample vial; counting efficiencies are insensitive to volume changes of ± 1 ml. When using the internal channels ratio method a quench curve may be set up by preparing standards with varying amounts of iron or more simply by the conventional technique of adding increments of acetone to a standard. A quench curve for iron is shown in Fig. 1(a). The line was obtained by least squares analysis of the results. Within the limits of experimental error the quench curve for acetone fell on this line. At very low energies, there is little difference between quench curves for chemical and colour quenchers.[31] In the range investigated, no quenching of α-pulses was observed. Addition of a second membrane filter to a vial did not increase quenching, so it may be concluded that any observed quenching arises from iron and nitric acid. Figure 1(b) indicates the relationship between iron concentration and degree of quench. (N.B. Herein lies a potential means for determining low concentrations of iron.)

Pure yttrium did not show quenching in a range from 40 to 200 μg as is evident from the results shown in Table 2. In practice, plutonium separated from a sample and precipitated on yttrium contains a few micrograms of iron arising from reagents and slight quenching is observed.

The optimum amount of 'carrier' iron or yttrium to be employed routinely was chosen as 100 μg to ensure quantitative recovery; serious losses occur when less than 20 μg are used, ascribed to mechanical loss and possibly peptisation.

Table 2. ^{241}Pu counting efficiency as a function of mass of yttrium employed.

Yttrium (μg)	Internal channels ratio	^{241}Pu counting efficiency (%)
40	2.02	36.7
80	2.00	37.1
120	2.01	35.8
160	2.00	38.4
200	2.04	37.7
240	1.99	34.5

Mean of results (1–5) = 37.1 ± 0.4 (1S.D.).

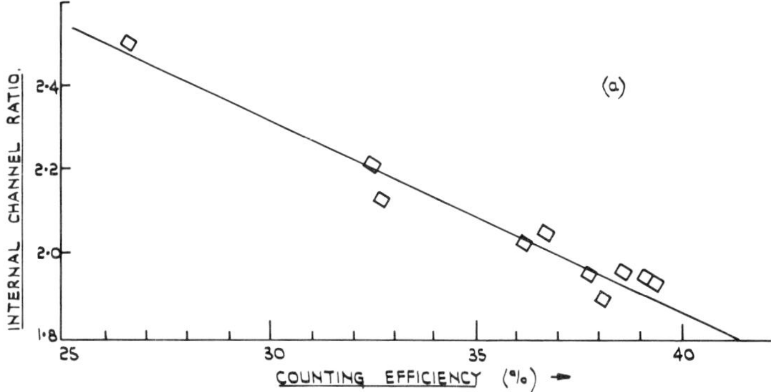

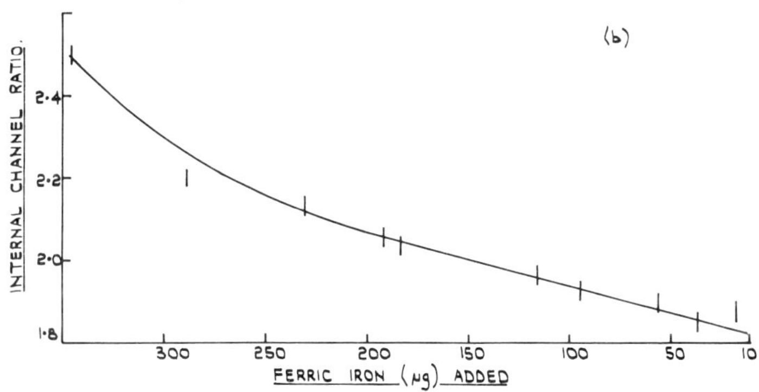

Fig. 1. Liquid scintillation counting of ^{241}Pu using ferric iron. Plot of internal channels ratio versus (a) counting efficiency, (b) mass of iron used.

YTTRIUM-90

Aliquots of ^{90}Yt standard solution (TRC Amersham, 2 x 10^4 disintegrations min^{-1} per aliquot) were separately subjected to precipitation on 100 μg of ferric iron and 100 μg of yttrium and the precipitates mounted on membranes as described above. After dissolution of a source in 100 μl of 1M nitric acid, 6 ml of dioxan (scintillation grade) was added, and the Čerenkov gain spectrum scanned. Six millilitres of dioxan-based scintillant (NE220) were then added, mixed with the solvent, and the liquid scintillation gain spectrum scanned (window = 50 to 1000 = maximum). The spectra using iron and yttrium are shown in Figs. 2 and 3 respectively, employing a logarithmic gain scale to cover the range; counting efficiency (%) is 100 x (count rate/disintegration rate). The maximum Čerenkov response in the presence of yttrium carrier is enhanced by a factor of 1.14 as compared with the response in 1M acid solution by virtue of the lower β-energy threshold in a solvent of higher refractive index ($n^{20°}{}_D$(water) = 1.333, $n^{20°}{}_D$(dioxan) = 1.422). The Čerenkov photons are attenuated by colour quenching in

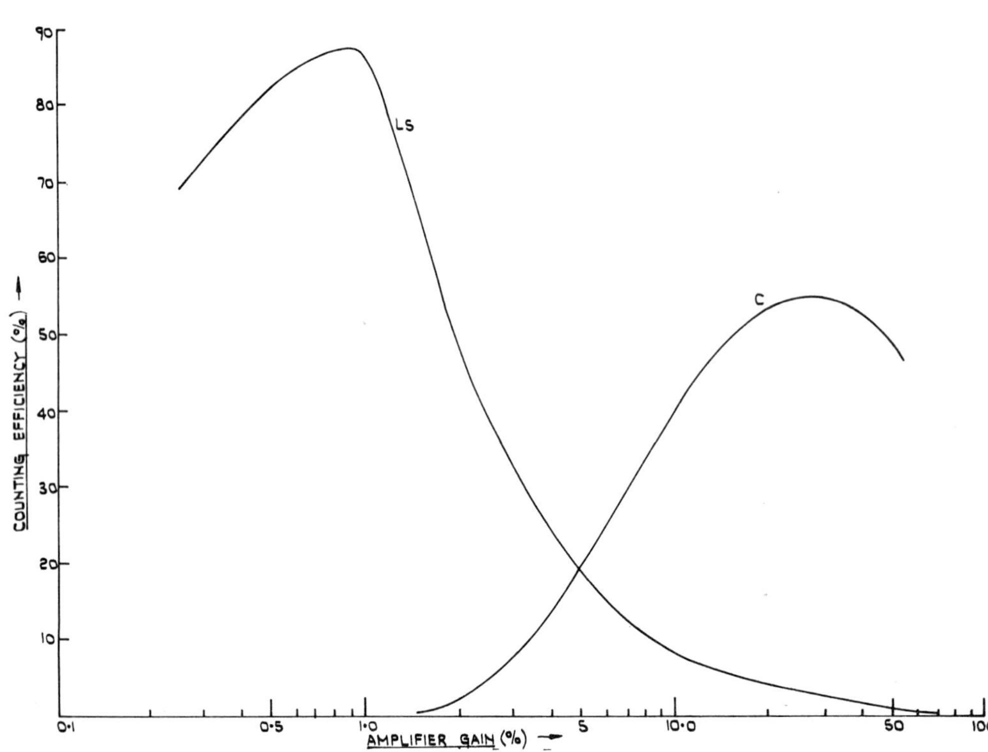

Fig. 2. Gain spectra of ^{90}Yt using ferric iron (100 μg). C = Čerenkov, LS = scintillation.

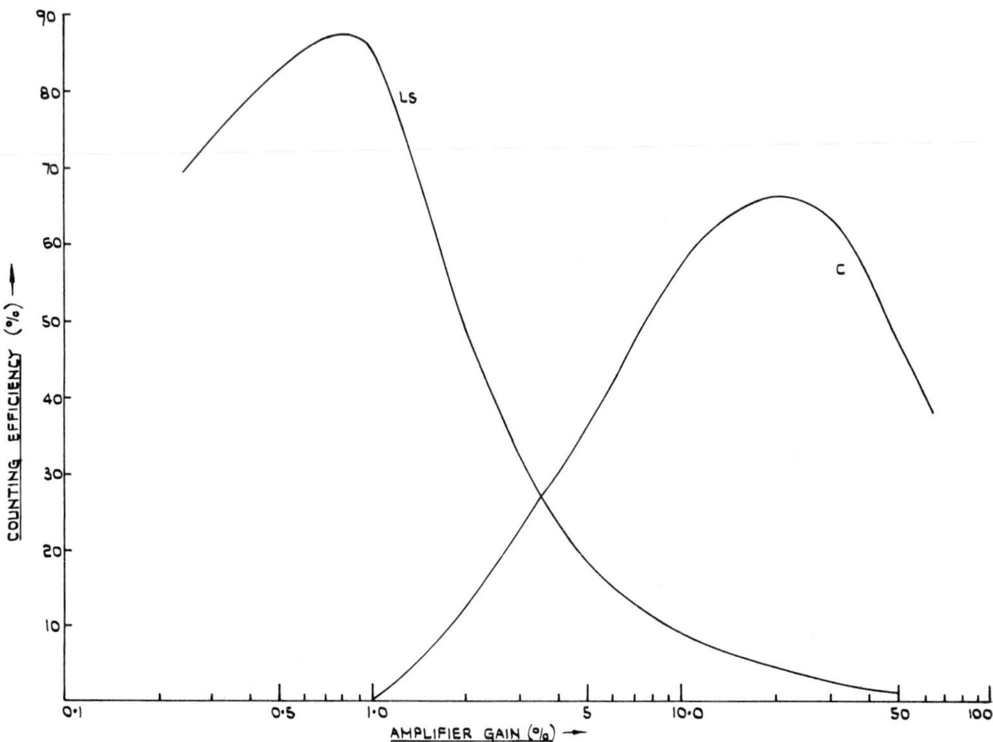

Fig. 3. Gain spectra of ^{90}Yt using yttrium (100 μg). C = Čerenkov, LS = scintillation.

the presence of ferric iron (chemical quenching cannot occur), and at the concentration employed the response is 0.96 of that in a colourless aqueous system. Quench correction in Čerenkov counting is possible.[32] On addition of scintillant, the response is at low gain and at the optimum Čerenkov gain setting is reduced by a factor of 16.4, which presumably arises from radiative transfer of photon energy to the solutes in the scintillant; the net result is an enhancement of the 'β'-spectrum produced by energy transfer from ionising particles to the scintillation cocktail.

CERIUM-144 AND PRASEODYMIUM-144

Aliquots of ^{144}Ce (10^4 disintegrations $\min^{-1}$) standard solution (carrier-free) in equilibrium with its daughter ($t_{1/2}$ = 17.3 min) were separately subjected to precipitation on 100 μg of ferric iron and 100 μg of yttrium, and the procedure adopted for ^{90}Yt carried out. Recovery was quantitative as determined by γ-assay of the dioxan solution against 6 ml of the standard solution in the same geometry. The gain spectra (Figs. 4 and 5) must be examined in relation to the decay schemes. Cerium-144 emits three types of β-particle, namely E_{max} = 309 keV (70%), 240 keV (ca.5%) and 180 keV (25%) with associated γ-emissions ranging from 34 keV to 134 keV; ^{144}Pr emits β-particles of E_{max} = 2.996 MeV (97.8%), 2.30 MeV (1.2%) and 0.81 MeV (1.0%) with associated γ-emissions ranging from 0.697 to 2.186 MeV. The Čerenkov spectrum may therefore be attributed mainly to the high-energy β-particles of the daughter isotope, which by analogy with ^{90}Yt produce responses in scintillation cocktail in the region of 1% amplifier gain. The responses in the rest of the spectrum can be ascribed to ^{144}Ce β-particles and superimposed pulses arising from γ-photon interactions in the medium. The discernible peaks appearing in the 2 to 8% gain region are tentatively ascribed to the electrons arising from the non-coincident groups of γ-photons of widely different energy range and internal conversions. A theoretical calculation of mean Compton energies has been reported.[22]

ZIRCONIUM-95 AND NIOBIUM-95 (IN EQUILIBRIUM)

Quantitative recovery of parent and daughter in equilibrium (3 x 10^4 disintegrations $\min^{-1}$) was possible using ferric iron (100 μg) and ammonium hydroxide solution as precipitant. In all other respects, the procedures were as described for ^{144}Ce plus ^{144}Pr. Zirconium-95 emits β-particles of E_{max} = 890 keV (ca.2%), 396 keV (55%) and 360 keV (45%) with associated γ-energies of 722 keV and 754 keV. Niobium-95 emits β-particles of E_{max} = 160 keV (100%) and a γ-photon of energy 765 keV. Consequently, very little Čerenkov response can be seen (Fig. 6) since few electrons exceed the energy threshold. The smooth liquid scintillation gain spectrum arises from the β-particles (N.B. for ^{35}S, E_{max} = 167 keV, the optimum gain in NE220 is 5.5%), and the electrons produced by interaction of the almost mono-energetic gammas with the medium.

CONCLUSIONS

The technique is presented as a potentially simple means of source preparation whenever quantitative collection of emitter(s) on a small amount of precipitate, amenable to later dissolution, is possible. The prepared solid source may be submitted to other types of radioassay, e.g. low background α-counting, γ-spectrometry, prior to quantitative transfer to a liquid scintillation counting vial.

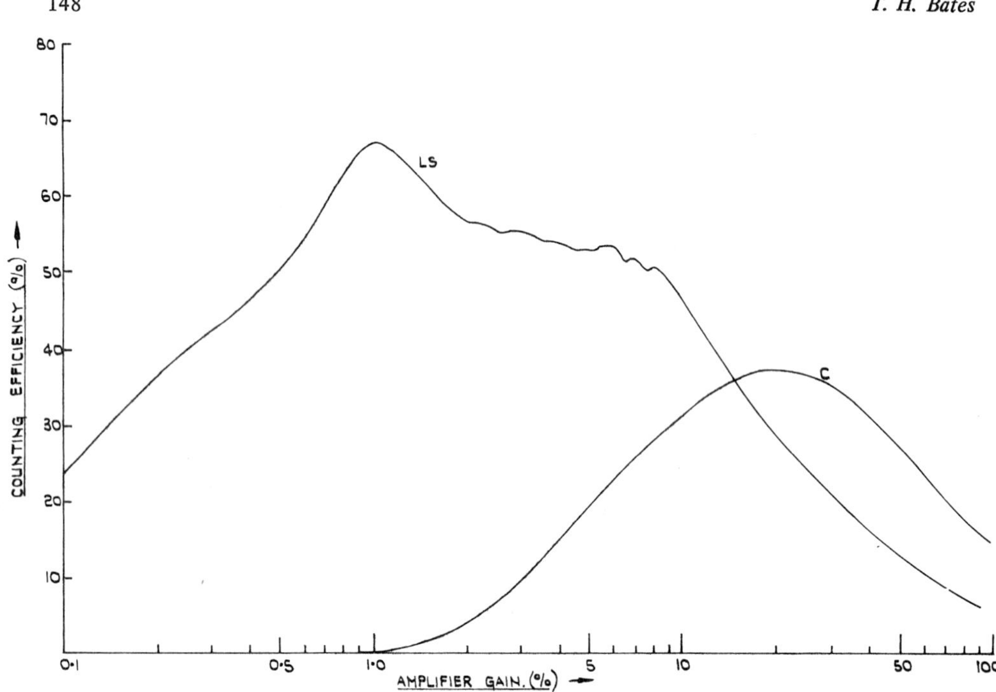

Fig. 4. Gain spectra of ^{144}Ce and ^{144}Pr using ferric iron (100 μg). C = Čerenkov, LS = scintillation.

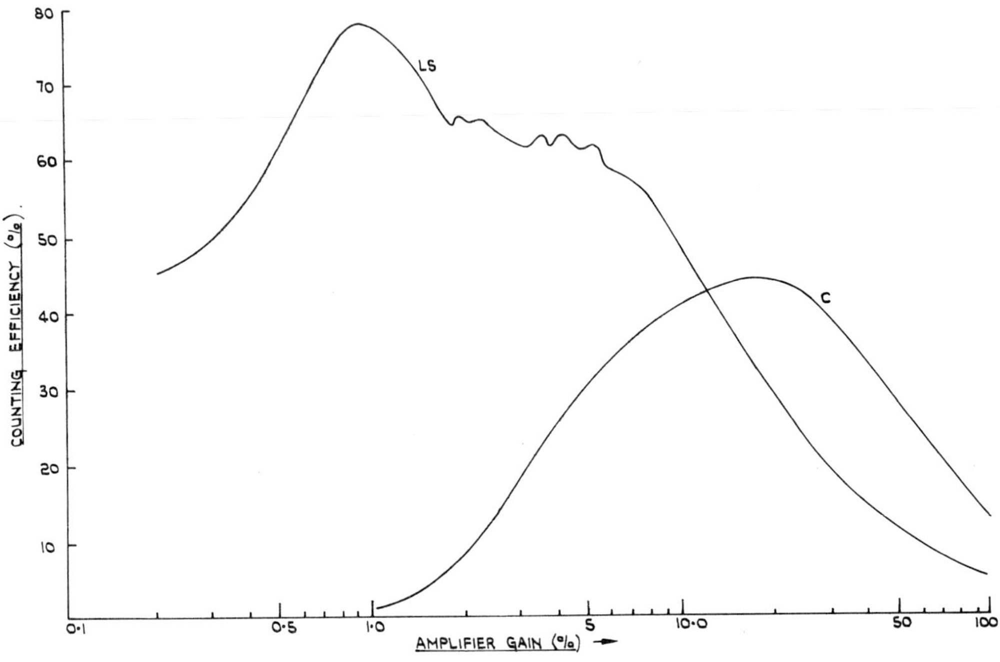

Fig. 5. Gain spectra of ^{144}Ce and ^{144}Pr using yttrium (100 μg). C = Čerenkov, LS = scintillation.

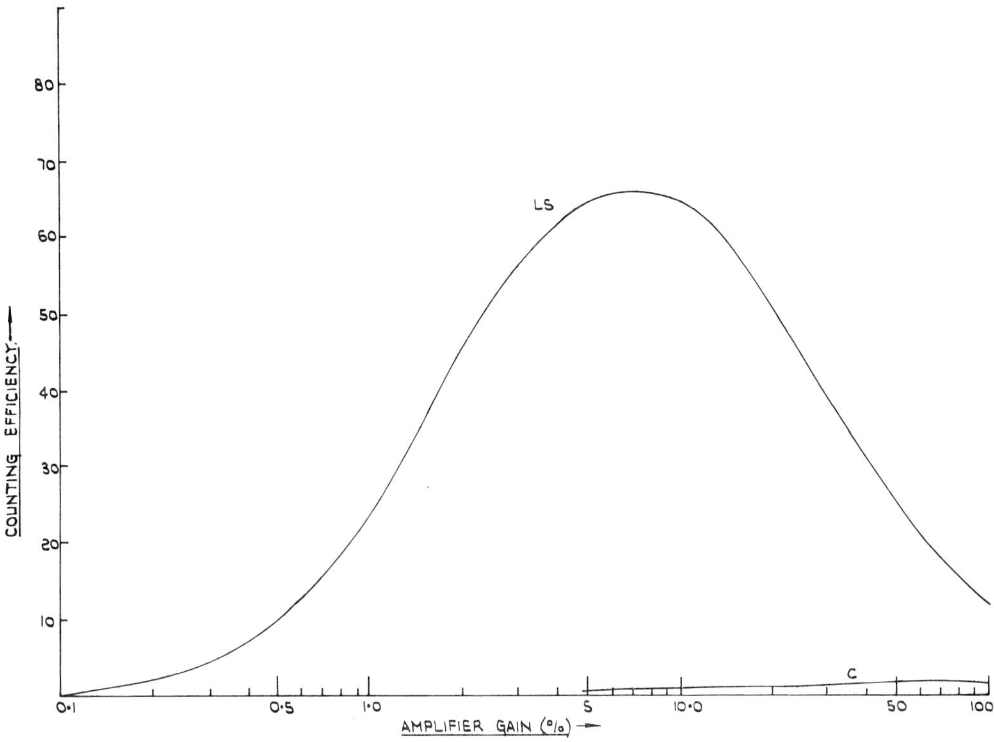

Fig. 6. Gain spectra of ^{95}Zr and ^{95}Nb (at equilibrium) using ferric iron (100 μg). C = Čerenkov, LS = scintillation.

Specific applications of the technique will be described in future publications, but it is worth noting that in addition to use for assay of indigenous ^{241}Pu in a sample, the technique has been used successfully employing pure ^{241}Pu as an added internal spike for recovery determination of plutonium in biological samples as an alternative to the use of spikes of an α-emitting isotope (^{236}Pu or ^{242}Pu) in conjunction with α-spectrometry. Assay of other actinides has not been thoroughly tested, but recent work[33] indicates that quantitative precipitation of actinides on phosphate and assay under similar conditions is possible.

The pairs of gain spectra for the β- and β/γ-emitters give extra information to characterise a radiochemically separated radionuclide in terms of purity as compared with that to be obtained from a single spectrum; the procedures involve little extra operational effort especially if scanning can be carried out with a multi-channel analyser. Alternatively, assay at selected gains, e.g. at maxima and at the intersection points of the spectra, might be useful. Pulse shape discrimination techniques, not available in this work, would permit significantly increased resolution between emissions inducing different ionisation/excitation densities which differently affect the fast and slow components of a pulse,[34] e.g. between α-pulses and β-pulses[17] in a mixture.

Determination of emitters in an unknown mixture could never be possible because of inevitable limits in resolution, but in routine control measurements of, say, low levels of activity in water, for which the technique is attractive, any changes in level or spectra characteristics would prompt closer investigation. The use of the membrane material for obtaining smear samples[35] in health physics monitoring is worth considering.

ACKNOWLEDGEMENTS

The author is indebted to the collaboration of Mr. J. B. Fardon who made substantial contributions to the preparative technique for plutonium, and to Mr. J. Middlemas for contributions to the radioassay and for useful discussions.

REFERENCES

1. J. C. Turner, *Sample Preparation for Liquid Scintillation Counting,* Review 6, The Radiochemical Centre, Amersham, 1967.
2. A. Tada and T. Radoszewski, in *Liquid Scintillation Counting,* Vol. 1 (ed. A. Dyer), Heyden, London, 1971, p. 49.
3. J. J. Gabay, C. J. Paperiello, S. Goodyear, J. C. Daly and J. M. Matuszek, *Health Phys.* **26**, 89 (1974).
4. J. D. Eakins and D. A. Brown, *Intern. J. Appl. Radn. Isotopes* **17**, 391 (1966).
5. A. A. Moghissi, E. L. Whittaker, D. N. McNelis and R. Lieberman, *Anal. Chem.* **46** (9), 1355 (1974).
6. J. Nolan, in *Liquid Scintillation Counting*, Vol. 2 (eds. M. A. Crook, P. Johnson and B. Scales), Heyden, London, 1972, p. 147.
7. C. W. Baker, G. A. Sutton and J. W. R. Dutton, Symposium on the Determination of Radionuclides in Environmental and Biological Materials, CEGB London, AED CONF. 73-085 04, 1973.
8. B. R. Harvey and G. A. Sutton, *Intern. J. Appl. Radn. Isotopes* **21**, 519 (1970).
9. D. L. Horrocks and M. H. Studier, *Anal. Chem.* **30**, 1747 (1958).
10. D. L. Horrocks, *Rev. Sci. Instrum.* **34**, 1035 (1963); *Intern. J. Appl. Radn. Isotopes* **17**, 441 (1966).
11. T. Y. Toribara, C. Predmore and P. A. Hargrove, *Talanta* **10**, 205 (1963).
12. A. Lindenbaum and C. J. Lund, *Radiation Res.* **37**, 131 (1969).
13. R. F. Keough and G. J. Powers, *Anal. Chem.* **42**, 419 (1970).
14. A. Seidel and V. Volf, *Intern. J. Appl. Radn. Isotopes* **23**, 1 (1972).
15. A. Delle Site, G. Santori and S. Bazzari, *CNEN RT/PROT* (74), 23 (1974).
16. D. L. Bokowski, *J. Amer. Ind. Hyg. Assoc.* 333 (June 1974).
17. J. W. McKlveen and W. Reed Johnson, *Health Phys.* **28**, 5 (1975).
18. J. D. Eakins and A. E. Lally, in *Liquid Scintillation Counting,* Vol. 2 (eds. M. A. Crook, P. Johnson and B. Scales), Heyden, London, 1972, p. 155.
19. T. H. Bates, see Discussion in above symposium, p. 165
20. T. H. Bates, T. H. Boyd and J. Middlemas, Internal Report, 1971.
21. H. V. Piltingsrud and J. R. Stencel, *Health Phys.* **23**, 121 (1972).
22. H. Prochazka and R. Jilek, IAEA Symposium on Rapid Methods for Measuring Radioactivity in the Environment, Neuherberg, July 1971, IAEA-SM-148/52.
23. S. E. H. Rippon, *Nucl. Instrum. Methods* **21**, 185 (1963).
24. A. De Volpi and K. G. A. Porges, *Intern. J. Appl. Radn. Isotopes* **16** (8), 496 (1965).
25. K. Haberer and W. Kolle, *Atompraxis* **11**, 664 (1965).
26. R. H. Elrick and R. P. Parker, *Intern. J. Appl. Radn. Isotopes* **17**, 361 (1966); **19**, 263 (1968).
27. A. Lauchli, *Intern. J. Appl. Radn. Isotopes* **20**, 265 (1969).
28. T. H. Bates, Symposium on the Determination of Radionuclides in Environmental and Biological Materials, CEGB London, AED CONF. 73-085 14, 1973.
29. A. Dyer, in *Liquid Scintillation Counting,* Vol. 2 (eds. M. A. Crook, P. Johnson and B. Scales), Heyden, London, 1972, p. 121.
30. T. H. Bates, T. H. Boyd and J. P. Clarke, BNFL Report 1(W), 1971.
31. R. J. Herberg, Packard Technical Bulletin No. 15, 1965.
32. A. J. Kemp and F. A. Blanchard, *Anal. Biochem.* **44**, 369 (1971).

33. T. H. Bates and J. B. Fardon, publication in preparation.
34. G. W. McBeth, R. A. Winyard and J. E. Lutkin, *Pulse Shape Discrimination with Organic Scintillators,* Koch-Light Laboratories Ltd., 1971.
35. M. J. Slobodien and R. W. Granlund, *Health Phys.* **27**, 128 (1974).

DISCUSSION

A. Dyer: You observed that your quench correction curve might be used to estimate concentrations of ions. We have carried out some preliminary investigations into the determination of Ti by developing the pertitanyl ion and observing quench correction and also for Fe as the orthophenanthrolene complex. In both cases encouraging results are observed over wide ranges of concentration down to at least 10^{-10} M.

SECTION III

THE USE OF THE SCINTILLATION COUNTER IN RADIOIMMUNOASSAY

Chapter 12

The Use of the Scintillation Counter in Radioimmunoassay

P. Tothill

Department of Medical Physics, Royal Infirmary, Edinburgh, Scotland

INTRODUCTION

The inclusion of radioimmunoassay and related procedures in the topics for this symposium is particularly timely, as the field has grown enormously in the past few years and still has considerable further potential. The procedures probably account for something like 90% of the use of γ scintillation counters and a substantial proportion of liquid scintillation counting at the present time. The inclusion represents something of an innovation, as topics other than liquid scintillation counting must be discussed. Radioimmunoassay (RIA) was introduced by Yalow and Berson in 1959[1] with an assay for insulin. At about the same time, Ekins[2] independently introduced a saturation analysis for thyroxine, based on similar principles.

RIA is one example of a collection of procedures for which there is not yet a universally accepted collective name. Perhaps the best suggestion to date is that of Landon[3] that they should be designated as 'binding assays'. Other examples in this category are competitive protein binding, immunoradiometric, radioenzymatic and radiomicrobiological assays. However, the same considerations apply regarding the eventual measurement of radioactivity of samples and these will form the main consideration of this presentation.

Binding assays require a substance with a limited binding capacity, which is an antibody for RIA and perhaps a natural plasma protein for competitive protein binding assays. Also required is a labelled substance similar to that to be assayed (and preferably identical with it) which would be in the form of an antigen for RIA. The tracer antigen is mixed with the sample and incubated with the antibody. The binding sites are then occupied by the labelled and unlabelled molecules in proportion to their concentration in the mixture, so that the degree of binding of radioactivity is dependent on the amount present of the substance to be assayed. Other requirements are a method of separating the bound from the free fraction of the agent and a means of determining the radioactivity in one or other of the separated components. The chief power of the method is its sensitivity – for many substances picogram quantities can be determined in 1 ml of plasma or other biological fluid. Ideally and usually another property of the system is its specificity. RIA was applied first to the determination of protein hormones which are now assayed in very large numbers. The other big class of compounds assayed comprises the steroids in plasma and urine. More recently the method has been applied to the assay of drugs such as digoxin and morphine.

The raising of antisera is by exposure of an experimental animal to the substance to be assayed over a period and is beyond the scope of this presentation. Proteins and other large molecules may be naturally immunogenic. Smaller molecules such as steroids and drugs have to be conjugated with a protein, e.g. bovine serum albumin, before they will be recognised by a host as immunogenic. When this happens, antibody is raised to the 'hapten', i.e. the drug or steroid which is attached to the albumin, as well as to the albumin itself. Complete immunoassay kits are available but expensive. If it is desired to initiate a new assay procedure it will usually be necessary for the experimenter to raise his own antisera and possibly also to label the molecule concerned.

RADIOACTIVE LABELS

A discussion of the available labels is of importance to our considerations as they will affect counting procedures. Desirable properties are that they should be easy to substitute or attach, easy to count, cheap, have a convenient half-life and a high specific activity. The last two considerations are related and conflicting. The best labels are carrier-free but even then one cannot get away from the fact that the longer the half-life the lower the specific activity. Table 1 outlines the main labels which have been used in RIA procedures. The specific activities quoted are for carrier-free material and relate to the elements listed. Compounds may have more than one atom of radionuclide per molecule, particularly with tritium. However, although it is theoretically possible to add more than one atom of iodine to a protein, it is usually undesirable as there will be resulting damage to the labelled compound. Carbon-14 has been used as a label, but is really included in the table to indicate why it is not suitable, owing to the low specific activity. Carbon-14 compounds also tend to be more expensive than those labelled with ^{3}H. Selenium-75 has been used as a γ-emitting label for steroids;[4] at the moment, carrier-free material is not available and the specific activities attained in practice are far

Table 1. Labels used in binding assays.

Radionuclide	$T_{1/2}$	Emission	Theoretical specific activity (Ci mg-atom^{-1})
^{3}H	12y	β	29
^{14}C	5700y	β	0.06
^{125}I	60d	γ +X	2160
^{131}I	8d	β +γ	16200
^{75}Se	121d	γ +X	1070
^{57}Co	270d	γ +X	480

below the theoretical maximum. Because γ-counting is so much easier and cheaper liquid scintillation counting, considerable efforts are being made to develop the γ-labelling of steroids and drugs. In general, they cannot be labelled directly with the isotopes of iodine, but labelling of the conjugated pair is possible. By far the most common radionuclides used for RIA are ^{125}I and ^{3}H. Although ^{131}I has some advantages as far as theoretical specific activity is concerned, these advantages are not fully realisable as ^{131}I is not available at above about 30% abundance, while ^{125}I is available very close to carrier-free. In addition, the efficiency of counting for ^{125}I is usually higher, and its longer half-life makes it much more convenient.

It is worth remarking in passing that tracers other than radioactivity have been proposed for immunoassay, such as enzymes, viruses, phage and electron spin. However, these alternatives cannot at the moment match the sensitivity of RIA, and there is more likelihood of interference upon detection.

SEPARATION OF BOUND FROM FREE AGENT

Once again the importance of the discussion under this heading relates to the way it may affect counting procedures. Table 2 lists the chief methods that have been adopted. Electrophoresis is principally of historical importance, as it was the technique adopted by Berson and Yalow in their original experiments. It is not suitable for large numbers or automation, and compared with other techniques the counting is somewhat difficult, particularly of ^{3}H-labelled compounds. In gel filtration, use is made of separation by molecular size, the higher molecular weight bound portion coming through the column first. The eluate provides liquid samples so there is no special problem of counting. However, the method is slow and tedious and is suitable only for small numbers of samples. Of much more widespread use are the techniques which result in a precipitation of either the free or the bound fraction. The double antibody technique extends the use of immunology. Another antibody is raised in a different species to the immunoglobulin used for the first antibody. When this is added to the mixture, precipitation of the bound fraction occurs. The technique is of wide application and suitable for automation. Chemical precipitation achieves the same end; commonly ammonium sulphate or ethanol are used in well-defined concentrations in order to precipitate the bound fraction. The adsorption techniques use an agent to which the free antigen is attached after incubation. The most used material is charcoal coated with dextran or albumin. Alternative materials that have been used include talc, powdered glass, Fuller's earth and ion exchange resin. The mixture is then centifruged and the supernatant decanted or sampled for counting.

Table 2. Some methods of separating antibody-bound from free antigen.

Electrophoresis	Adsorption
Chromatography	Solid phase systems:
Gel filtration	Antibody-coated tubes
Immunoprecipitation	Polymerised antibodies
Chemical precipitation	

'Solid phase' systems refer to the attachment of the antibody. In the coated tube method the antibody is merely applied to the inside of a plastic vial. The technique is undoubtedly simple, but in many people's experience lacks reproducibility. Better precision is obtained when the antibody is covalently linked to a polymer. One advantage over precipitation techniques is that centrifugation is much quicker.

During the initial validation of an assay it is often advisable to use more than one separation method. Techniques which do not lend themselves to large-scale routine use may thus be valuable at the development stage.

MEASUREMENT OF RADIOACTIVITY

It is worthwhile examining the radiations from the most commonly used label, ^{125}I. The decay scheme, electron capture followed by the emission of a single γ-ray, is deceptively simple. However, most of the γ-radiations are internally converted and subsequent re-arrangements in the atom give rise to the emission of X-rays and particulate radiation. The electron capture process itself also gives rise to similar radiations. The principal radiations are summarised in groups in Table 3. The photon radiations in the first group are the only ones with sufficient penetrating power to be measured by sodium iodide scintillation detectors. As the latter are much simpler to use, requiring no sample preparation, quench correction or continuing expense, they are almost universally preferred. The counting efficiency commonly obtained in many standard well counters is around 50%, but if the latter are designed specially for ^{125}I,

Table 3. Principal radiations from ^{125}I.

Radiation	Energy (keV)	% Abundance
γ	35	7
Kα X	27	112
Kβ X	31	24
		143
K internal conversion electron	3.7	75
L X	3.8	22
LMM Auger electron	2.9	149
		246
L internal conversion electron	31	11
M internal conversion electron	35	8
KLL Auger electron	23	14
KLX Auger electron	26	6
KXY Auger electron	30	1
		40

an efficiency of about 80% may be obtained. Nor need such a system be expensive, as thin sodium iodide suffices to absorb the low-energy radiation and consequently a minimal amount of lead shielding is required.

Some of the radiations from ^{125}I are emitted in coincidence, and if both components are intercepted by the detector, the pulse heights will be added together, giving a signal corresponding to the sum of the two individual energies. The pulse height distribution will then show a photopeak from the X- and γ-rays at around 30 keV (the resolution of a NaI crystal is not good enough to separate the different energies in this group), together with a smaller peak at about 60 keV. The sum peak only appears if the detector geometry is good, as with a sample at the bottom of a well crystal, as the probability of interception of both constituents by the crystal varies as the square of the solid angle subtended by the detector. Whether the sum peak should be included within the channel width set on the analyser depends on the background characteristics of the system. A channel set wide enough to include both peaks will undoubtedly give higher efficiency counting but the background will also increase. If a special thin crystal is chosen, background from higher energy radiation is minimised and it may not be necessary to provide an upper threshold to the analyser, a simple discriminator sufficing. In this case the sum peak will be automatically included. It must be remembered that the photon radiations from ^{125}I are relatively easily absorbed and the use of glass tubes with appreciable heavy metal content can sometimes lead to variations of efficiency between samples.

Although a well scintillation counter provides excellent counting geometry, an alternative that comes close to it in efficiency is the use of two opposed thin flat crystals. This arrangement is adopted in at least one automatic counting system.

The chief radiation from ^{131}I is of a much higher energy than that from ^{125}I, so that separation in double isotope studies is quite easy. Some 'crossover' from the higher energy radiation is inevitably included in the ^{125}I photo-peak setting, and it is interesting to note that there is a similar peak in the ^{131}I spectrum, arising from internal conversion processes, as well as Compton scatter continuum. Selenium-75 has principal γ-radiations not too dissimilar from those of ^{131}I and so could also be readily detected and separated from ^{125}I.

An examination of Table 3 shows that, although there are no β-radiations from ^{125}I, there is an abundance of low-energy radiation which could be detected by liquid scintillation counting. Indeed, the latter gives a higher efficiency of counting than the use of the X- and γ-rays alone with a NaI crystal. However, the main interest in the liquid scintillation counting of ^{125}I is in its influence on double labelled experiments which also make use of ^{3}H. There are to be papers dealing with this subject later in the session but perhaps I can introduce it in general terms. Figure 1(a) shows the pulse height distribution from the liquid scintillation counting of ^{125}I. Two broad peaks are noticeable, corresponding to the groups with energy about 3 and 30 keV. Most of the latter come from internal conversion electrons, but some from X- and γ-ray absorption or Compton scatter in the scintillator. That the latter is a relatively small proportion can be seen from Fig. 1(b), which was obtained by placing a ^{125}I source in an inner tube suspended in the centre of a vial of scintillator. All the particulate radiation was absorbed, and the remaining photon radiation contributed 18% of the counting rate.

The similarity between the spectra of ^{125}I and ^{3}H is illustrated in Fig. 1. The pulse height distributions are not identical, and it would be theoretically possible to perform double label assays by liquid scintillation counting alone using two

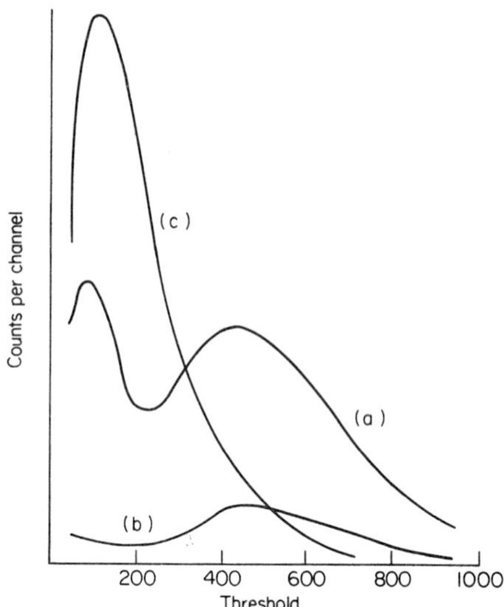

Fig. 1. Pulse height distributions from the liquid scintillation counting of ^{125}I and ^{3}H: (a) ^{125}I all radiations; (b) ^{125}I X + γ-rays only; (c) ^{3}H.

channels. However, it is much better to use γ-counting for ^{125}I and then make a correction for its contribution to the combined liquid scintillation counts.

The consideration of the liquid scintillation counting of ^{3}H alone in connection with RIA is primarily concerned with sample preparation. The subject does not warrant detailed consideration here as the techniques used are in general no different from those already discussed in this and previous symposia. If the separation procedure results in an aqueous solution, such as the decanted bound fraction from a charcoal precipitation or the free fraction from a double antibody separation, there is no difference from other aqueous sample counting. Standard procedures that have been used include dioxane-based scintillator, solubiliser, emulsion or gel systems. It is worth remembering that if only a small aqueous volume is required, a correspondingly small volume of scintillator suffices and the mixture can be placed in a disposable plastic tube held in a standard vial with no loss in efficiency but a gain in economy.

Another useful technique, introduced by Abraham *et al.*,[5] and applicable to non-polar substances such as progesterone, is simply to mix the aqueous solution with a toluene-based scintillator and allow the mixture to equilibrate. The activity is extracted into the scintillator and can be counted reproducibly in the two-phase system. It is even possible to rely on partition into liquid scintillant without removal of an ammonium sulphate precipitate.[6]

If it is desired to assay a precipitate, such as the bound fraction from a double-antibody or chemical precipitation, two courses are open. The precipitate can be suspended in a gel scintillator. This places reliance on the reproducibility of self-absorption of the β-rays which is a problem with heterogeneous systems, as no method of quench correction can allow for variations. An alternative is to extract the label from the precipitate and treat it as a liquid sample. Once again, small volumes often suffice.

It is not usual to apply quench corrections when ^3H is used for RIA, as conditions are so standard that the degree of quenching is considered the same for each sample. However, it has been shown that in some circumstances quenching is somewhat variable, and that correction is worthwhile. This factor should always be investigated in any new technique.

If separation has been by electrophoresis or chromatography, the simplest way of measuring ^3H is to cut up the paper or cellulose acetate and suspend the strips in liquid scintillator. The best efficiency and reproducibility are obtained if this process results in the complete elution of the active material into the scintillator, but reasonable counting can be performed if there is no removal from the medium.

AUTOMATION

The rapid growth and wide applicability of RIA, the large number of samples generated and the relatively complex data processing required all call for automation. Scintillation counters themselves have already led the way, but there is also a need for the automation of dispensing, dilution, addition of reagents and separation. Equipment such as the Analmatic or Micromedic systems lend themselves well to some of the procedures, but there is still some transfer of trays or samples to and from centifruge and counter required. The traditional chemical auto-analyser system using continuous flow is not well suited to processes requiring finite and possibly lengthy incubation or counting, although efforts are being made to develop such a technique for RIA. The fact that counting time may place a limit on the number of samples processed has been recognised in some commercial systems which provide multiple detectors.

The most fully automatic system dedicated to RIA is that developed by Bagshawe[7] and now commercially available. Special mechanical modules handle the dispensing, dilution and addition, but an important feature is that separation of the precipitate is by filtration. The glass fibre filter pads are mounted on a plastic tape, simplifying passage through a detector module containing five opposed pairs of counters. Another feature is the inclusion of a computer, which controls all the modules, stores the sample information, acts as a scaler for counting and performs the calculations and curve fitting necessary to provide a direct output of agent concentration.

Automated equipment is expensive and in considering its justification attention must be paid to the work load and to the costs of alternative methods. In any case, it is a salutory exercise to assess the component costs of any assay procedure. Any estimates must be dependent on local circumstances, and the following are offered just as an example.

Antiserum is cheap enough not to be a significant factor in considering costs. A labelled tracer might cost up to 1p per assay and materials for separation another 1p. These items can be compared with the cost of commercial kits, which range from about 10–100p per assay. Of course, one is then paying for development, quality control and convenience.

Costs of radioactivity counting must include a figure for the capital equipment, based on write-off over a suitable period, costs of servicing and number of assays performed. It is likely that the figure will be 1-2p per assay. A vial might cost 1-10p and, for ^3H-assays, scintillator another 1-10p. The overall total of about 4–25p does not include any salaries of technical staff. These might account for 5–20p per assay, so a substantial investment in automation can readily be justified if there is sufficient work load. Nor are other laboratory overheads included, and these must obviously be borne in mind when comparing any estimates with the cost of having assays performed commercially, currently in the range of £2-£10 per assay.

ACKNOWLEDGEMENT

It is a pleasure to acknowledge my indebtedness to Dr. W. M. Hunter, head of the M. R. C. Radioimmunoassay Team in Edinburgh, for many helpful discussions.

BIBLIOGRAPHY

Radioimmunoassay Methods, European Workshop (eds. K. E. Kirkham and W. M. Hunter), Churchill Livingstone, Edinburgh, 1971.
Steroid Immunoassay, Fifth Tenovus Workshop (eds. E. H. D. Cameron, S. G. Hillier and K. Griffiths), Alpha Omega Publishing, Cardiff, 1975.
Radioimmunoassay and saturation analysis, *British Medical Bulletin* **30**, No. 1 (1974).

REFERENCES

1. R. S. Yalow and S. A. Berson, *Nature (London)* **184**, 1648 (1959).
2. R. P. Ekins, *Clin. Chim. Acta* **5**, 453 (1960).
3. J. Landon, Report of 26th Expert Committee on Biological Standardisation, WHO/BS/75. 1095, World Health Organisation, Geneva, 1975.
4. V. E. M. Chambers, J. S. Glover and R. Tudos, in *Steroid Immunoassay* (eds. E. H. D. Cameron, S. G. Hillier and K. Griffiths), Alpha Omega Publishing, Cardiff, 1975, p. 177.
5. G. E. Abraham, R. Swerdloff, D. Tulchinsky and W. D. Odell, *J. Clin. Endocrin. Metab.* **32**, 619 (1971).
6. C. R. W. Edwards, A. A. Taylor, C. K. Baum and A. B. Kurtz, in *Steroid Immunoassay* (eds. E. H. D. Cameron, S. G. Hillier and K. Griffiths), Alpha Omega Publishing, Cardiff, 1975, p. 229.
7. K. D. Bagshawe, *Acta Endocrin.* Supplement 177, 384 (1973).

DISCUSSION

R. P. Parker: In view of the wide variations in sample volume, tube size and detector size that are catered for in commercial equipment, would you comment on the relative merits and accuracy of using the various amounts of sample, etc?

P. Tothill: The efficiency of a well counter varies considerably with height and therefore with volume. Thus it is important to standardise the volume of the samples. If this is small, as is common, then it is best for a narrow sample container to be used and a small diameter well. Indeed, the sample volume of either precipitate or supernatant can be kept to no more than 1 ml in the great majority of RIA procedures and this could lead to simplification and economy of γ-counters designed specially for RIA work, and to savings on scintillator and vials in liquid scintillation counting.

R. Evans: Although it is true that in a well scintillation counter the portion of the sample that is within the well will be counted with higher efficiency than any portion of the sample outside the well, nevertheless one can improve sensitivity of measurement by using a larger sample extending outside the well. A further possible benefit in this case is that small variations in sample volume will affect the counting result much less critically than will the same volume variation in a sample entirely within the well.

P. Tothill: I agree that the plot of count rate against volume reaches a plateau where changes of volume are not critical, but I doubt whether such large volumes are available from RIA procedures.

N. Harding: For some of our experiments we would like absolute activity estimates of ^{125}I. For this purpose both standards and methods of estimating both liquid and crystal counting efficiencies are required. Would Dr. Tothill comment on the best methods available for estimating the efficiency of counting ^{125}I?

P. Tothill: We have only measured efficiencies using manufacturers' quoted activities on ordinary commercial ^{125}I and I would refer Dr. Harding to the Radiochemical Centre for ^{125}I standards. One can use the relationship between the areas under the sum peak and the singles photopeak as an absolute determination of ^{125}I in a well counter, and it appears to give reasonable results. There was a paper in *Nucleonics* some years ago.

A. McNair: Iodine-125 can be accurately standardised by a double crystal coincidence method measuring coincidences between the X-rays. Standards are readily available, those from the Radiochemical Centre being guaranteed to have an accuracy of better than ± 2% overall uncertainty. In order to standardise a liquid scintillation counter it would probably be necessary to use a ^{125}I standard and derive some quench correction curves.

D. W. T. Lee: Does not the peak occurring in the ^{125}I region of the ^{131}I γ-spectrum partially consist of γ-rays from ^{131}Xe^{m} (half-life 12d) which results from the decay of the ^{131}I? This xenon can build up in the solution so that for dual labelling experiments the crossover of ^{131}I into the ^{125}I channel can vary with time.

P. Tothill: I do not know the decay scheme of ^{131}Xe^{m} and whether it includes radiation within the counting window for ^{125}I. Certainly some of the ^{131}I γ-rays are internally converted and the resulting X-rays account for most of the peak in the 30 keV region.

[**Editor:** 0.6% of the ^{131}I disintegrations result in ^{131}Xe^{m} with a γ-ray emission at 160 keV; the remainder result in stable ^{131}Xe. The contribution to the spectrum in the 30 keV region is therefore negligible.]

F. E. L. ten Haaf: I should like to make a comment regarding the counting efficiency of large samples that extend beyond the well of a well-type γ-counter. As the counting efficiency is high within the well and low outside, with large samples one is measuring the specific activity of the sample rather than the total activity. If the samples are of unequal volume (and height), this will lead to errors in the determination of the sample activity.

P. Tothill: I can only agree and emphasise that the volume used must be known and either kept fixed or corrections applied for variations.

G. W. Pearson: I should like to point out that in view of the considerable amounts of radioactive iodine that are involved in labelling procedures, there can be radiation protection problems.

P. Tothill: This is a very important point, and particular care must be taken owing to the volatile nature of iodine. Provided the correct precautions are used, including operation in a fume cupboard, there should not, however, be any undue hazard in using radioactive iodine for protein labelling. Monitoring is important and any portable scintillation counter held against the neck will indicate a significant thyroid accumulation.

R. P. Parker: We have found whole body counting a useful check on the possibility of contamination or ingestion of radioactive iodine by personnel. Iodine-131 in particular can be detected at amounts considerably below the maximum permissible body burden. Routine checks by this method have been of particular value in training new staff.

W. R. Carr: In view of the criticism of plastic vials for liquid scintillation counting, has Dr. Tothill any comment to make concerning their use in RIA procedures?

P. Tothill: In our experience, modern vials appear to behave quite satisfactorily, although earlier versions showed loss of toluene and deformation.

Chapter 13

The Determination of Tritium in the Presence of Iodine-125

G. Ayrey, K. L. Evans, D. Exley and B. J. Woodhams

Queen Elizabeth College, University of London, England

INTRODUCTION

For some time there has been a need for a reliable assay method for samples labelled doubly with ^{3}H and ^{125}I. A procedure utilising both a γ-counter and a liquid scintillation counter was proposed during discussion following a contribution to an earlier meeting in this series.[1] Nevertheless, our own investigations and those of others[2] have revealed certain difficulties and anomalies associated with this isotope pair and it was felt that a systematic study of the problem was justified. We have considered three distinct approaches, two of which are based on well-known techniques in radiochemical assay. The third is a new technique and has the advantage of requiring only a liquid scintillation counter. It relies neither on energy discrimination nor on decay of the relatively short-lived ^{125}I, but as the work is still at an exploratory stage this paper must be regarded as a preliminary progress report.

ASSAY METHODS

Method 1. Double instrument

A liquid scintillation counter may be used to detect and count the decay events of both ^{125}I and ^{3}H; the former with high efficiency, the latter with moderate efficiency. Unfortunately, a considerable overlap of the pulse height spectra of the two isotopes (Fig. 1) renders energy discrimination an unreliable basis for an assay method. However, since ^{125}I emits weak γ-photons and X-rays in addition to energetic electrons, it is possible to determine the iodine activity using a γ-counter. If the γ counting efficiency for ^{125}I is *assumed to be constant,* then the following simple method should be applicable:

$$dpm_{\text{I-125}} = cpm_{\gamma}/e_{\text{I-125}(\gamma)}$$

$$dpm_{\text{H-3}} = (cpm_{\beta} - dpm_{\text{I-125}}\, e_{\text{I-125}(\beta)})/e_{\text{H-3}(\beta)}$$

where *cpm* = net count rate (counts min^{-1}); *dpm* = activity (disintegrations min^{-1}); e = efficiency; γ = γ-counter; β = liquid scintillation counter.

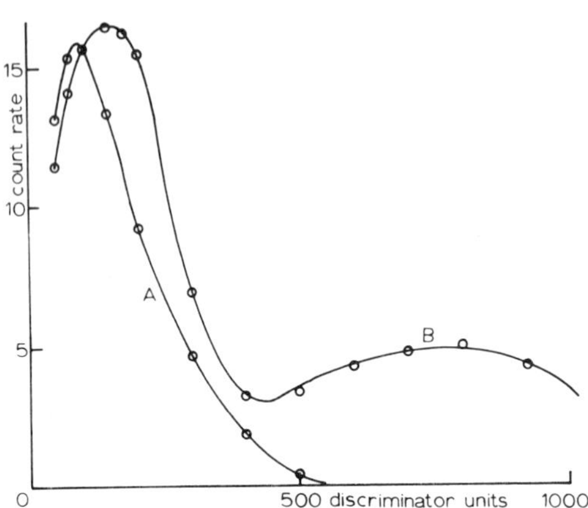

Fig. 1. Pulse height spectra of unquenched samples of ^3H (curve A) and ^{125}I (curve B) (0.8% butyl-PBD in toluene).

The variable liquid scintillation counting efficiencies are readily monitored by automatic external standard channels ratio.

In a variety of conditions, such a method yields results of high accuracy as typified by those for duplicate standards recorded in Table 1. The standards were counted once in the γ-counter and three times for each of four attenuation levels in the liquid scintillation counter. Further investigation showed that sample composition was important and very poor results were to be expected in certain circumstances. These included samples containing high levels of iodine activity and samples containing molecules incorporating heavy atoms.

The effect of heavy atoms cast doubt on the assumption that the γ counting efficiency was constant because it is well known that the absorption of γ-photons is a function of the atomic numbers comprising the absorbing material. These doubts were justified by showing that the γ counting efficiencies of sodium [^{125}I] iodide and [^{125}I] iodobenzene are reduced by successive additions of carrier (Figs. 2 and 3).

If the γ counting efficiency is reduced by a significant presence of high atomic number atoms, then the ^3H activity will be overestimated on two counts. First, the calculated iodine contribution to the β count rate will be low, thus exaggerating the ^3H contribution. Second, the γ-energy absorbed by the heavy atoms will be dissipated largely as photoelectric or Compton scattered electrons leading to an increase in the β count rate.

The principal failing of this method then is that the γ counting efficiency is not constant and, unlike liquid scintillation counting, automatic monitoring of counting efficiency is not available. For this reason we turned our attention to methods relying solely on liquid scintillation counting.

Table 1. Results obtained using a γ-counter and a liquid scintillation counter (Method 1).

Observed ^{125}I activity (dpm)	Recovery (%)	Attenuator setting	Observed ^{3}H activity (dpm)	Recovery (%)
16178	99.4	0.5	109384	99.8
			111639	101.9
			109522	100.0
16380	100.6		108889	99.4
			111133	101.4
			109022	99.5
16178	99.4	1.0	107607	98.2
			107603	98.2
			107479	98.1
16380	100.6		106156	96.9
			107010	97.7
			106879	97.6
16178	99.4	1.5	108134	98.7
			111353	101.0
			108885	99.4
16380	100.6		107449	98.1
			110659	101.0
			108196	98.8
16178	99.4	2.0	107678	98.3
			111324	101.6
			108480	99.0
16380	100.6		106922	97.6
			110558	100.9
			107719	98.3

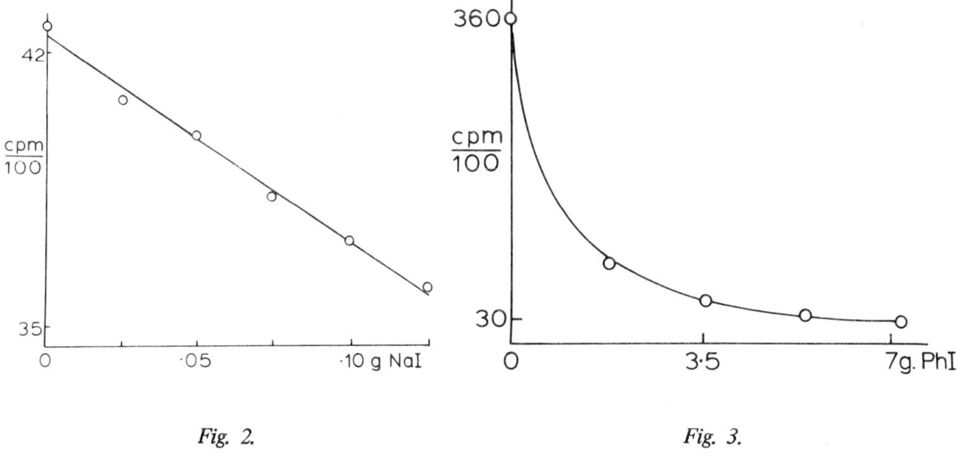

Fig. 2.   *Fig. 3.*

Fig. 2. The effect of adding sodium iodide carrier to an aqueous sample of ^{125}I counted in a polyethylene vial in the γ-counter.

Fig. 3. The effect of adding iodobenzene carrier to a toluene solution of [^{125}I] iodobenzene counted in a polyethylene vial in the γ-counter.

Method 2. Single instrument; iodine decay

Methods exploiting isotope decay are usually unattractive where many samples are to be routinely assayed. Such procedures necessitate storage of samples for a suitable period which in itself can be a problem where space is limited but more seriously leads to delayed results. This is at best frustrating and in certain laboratories, notably those undertaking clinical tests, intolerable. Nevertheless, we have looked at this approach for the sake of completeness.

The count rate of a double-labelled sample at day zero (cpm_0) is equal to the sum of the count rates due to the two isotopes:

$$cpm_0 = cpm_{\mathrm{H}} + cpm_{\mathrm{I}(0)} \tag{1}$$

Some few (t) days later the ^{3}H decay will be negligible while that of the ^{125}I will be significant. The reduced count rate of the sample at day t will be given by

$$cpm_t = cpm_{\mathrm{H}} + cpm_{\mathrm{I}(t)} \tag{2}$$

The difference between the sample count rates at day t and at day zero is equal to the difference between the iodine count rates:

$$cpm_0 - cpm_t = cpm_{\mathrm{I}(0)} - cpm_{\mathrm{I}(t)} \tag{3}$$

By normal decay kinetics the following relationship exists:

$$cpm_{\mathrm{I}(t)} = cpm_{\mathrm{I}(0)}\mathrm{e}^{-\lambda t} \tag{4}$$

where λ is the radioactive decay constant (days^{-1}) for ^{125}I. On changing the sign on both sides of Eqn. (4) and adding $cpm_{\mathrm{I}(0)}$ to both, we have, on rearrangement,

$$cpm_{\mathrm{I}(0)} - cpm_{\mathrm{I}(t)} = cpm_{\mathrm{I}(0)}(1-\mathrm{e}^{-\lambda t}) \tag{5}$$

Substituting from Eqn. (3) and rearranging gives

$$cpm_{\mathrm{I}(0)} = (cpm_0 - cpm_t)/(1-\mathrm{e}^{-\lambda t}) \tag{6}$$

Thus, the count rate due to ^{125}I at day zero is obtained by dividing the difference between the two sample count rates by $1-e^{-\lambda t}$. In practice the latter term may be tabulated for convenience. The ^{3}H count rate is then readily deduced from Eqn. (1). Table 2 records the results for a typical standard counted 2.67 and 5.67 days after the original count. Even after the longer period the results are unacceptable. Greater accuracy may be expected if several counts are taken during the decay period but this imposes even greater burdens on the routine analyst. An important point is that the decay period required to achieve a certain accuracy depends on the relative abundance of the two isotopes. Figure 4 illustrates plots of expected change in count rate (Δ *cpm*) divided by the maximum error (*E*), predicted from the 90% confidence level, against time. The figure shows that any given accuracy is achieved after a shorter period for a sample rich in iodine than an equally active sample rich in ^{3}H.

Both from the point of view of accuracy and labour this method had little to recommend it.

Table 2. Results obtained using a liquid scintillation counter and ^{125}I decay (Method 2).

Period between counts (days)	Attenuator setting	Observed ^{125}I activity (dpm)	Recovery (%)	Observed ^{3}H activity (dpm)	Recovery (%)
2.67	0.5	17927	110.9	105742	96.6
	1.0	34708	213.2	53573	48.9
	1.5	27110	166.5	71826	65.6
	2.0	29178	179.2	59908	54.7
5.67	0.5	13723	84.3	118193	107.9
	1.0	16444	101.1	109438	99.9
	1.5	18489	113.6	102658	93.8
	2.0	12563	77.2	122570	111.9

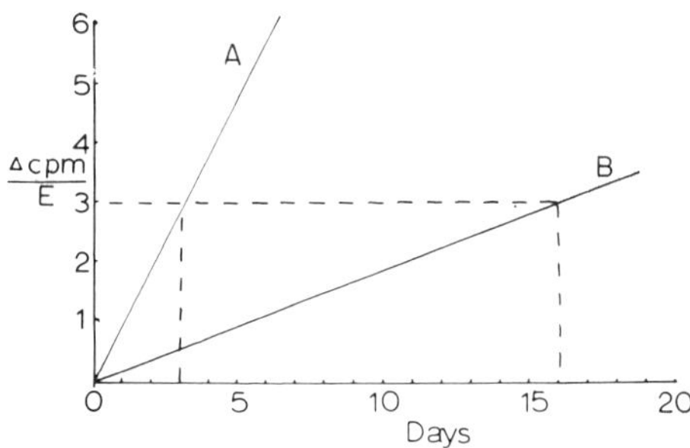

Fig. 4. Error curves for ^{125}I-decay in double-labelled samples. (A)^{3}H:^{125}I activity ratio of 1:8. (B)^{3}H:^{125}I activity ratio of 8:1.

Method 3. Single instrument; efficiency variation

This method exploits the effect of heavy atoms on the electron counting efficiency of ^{125}I. The addition of compounds containing such atoms should show two opposing effects on the count rate. First, there would be the normal quenching effect leading to a reduction in counting efficiency. Second, the energy of the γ-photons absorbed by the additive would be dissipated largely by electron emission which would lead to an increase in counting efficiency. By careful control of the nature and quantity of the additive, the latter effect might be made predominant. The effect of such additives on ^{3}H would be to reduce the counting efficiency by quenching. The following scheme indicates how we may exploit these effects in an assay method. The count rate of a double-labelled sample may be expressed as follows:

$$cpm = e_I a_I + e_H a_H \qquad (7)$$

where e = efficiency; a = activity; I = iodine-125; H = tritium.

Addition of a normal quenching agent would reduce the counting efficiency of each isotope. This leads to a second count rate cpm':

$$cpm' = e'_I a_I + e'_H a_H \qquad (8)$$

where $e'_I < e_I$ and $e'_H < e_H$.

Equations (7) and (8) would have an infinity of solutions in a_I and a_H since the quenching effects would be very similar for two isotopes with such similar energies. If, however, we are able to enhance the counting efficiency of the ^{125}I to e''_I, where $e''_I > e_I$, then a third equation may be written which can be uniquely solved with Eqn. (7) to yield a_H and a_I (Eqns. 10 and 11):

$$cpm' = e''_I a_I + e'_H a_H \qquad (9)$$

In practice the accuracy of the solution will depend not only on the precision of our measurements but also on the relative magnitudes of the efficiency changes:

$$a_H = (e''_I\, cpm - e_I cpm')/(e''_I e_H - e_I e'_H) \qquad (10)$$

$$a_I = (cpm - e_H a_H)/e_I \qquad (11)$$

Provided a suitable additive can be found it is only necessary to count the sample and then recount immediately after adding the reagent. Then by reference to four calibration curves of efficiency against automatic external standard channels ratio for each isotope in the presence and absence of additive, four efficiencies are obtained and hence the activity of each isotope from Eqns. (10) and (11). The alternative to recounting the sample is to prepare two samples, one with additive, the other without.

In our laboratories, certain organotin compounds were readily available. This availability, coupled with the fact that the counting of ^{125}I in the presence of such compounds has already been studied,[3] encouraged us to select them as our first additives. Initially we examined tetraphenyl tin, triphenyl tin acetate, tetramethyl tin and tetrabutyl tin. The first two were rejected because of poor solubility and our study was restricted to tetramethyl tin and tetrabutyl tin. The first stage was to determine the effectiveness of the reagents as γ- and X-ray absorbers by counting samples of ^{125}I iodobenzene containing increasing volumes of additive on the γ-counter. In terms of volume, tetramethyl tin is a more effective absorber than tetrabutyl tin (Fig. 5), but in terms of moles the behaviour of each additive is identical (Fig. 6).

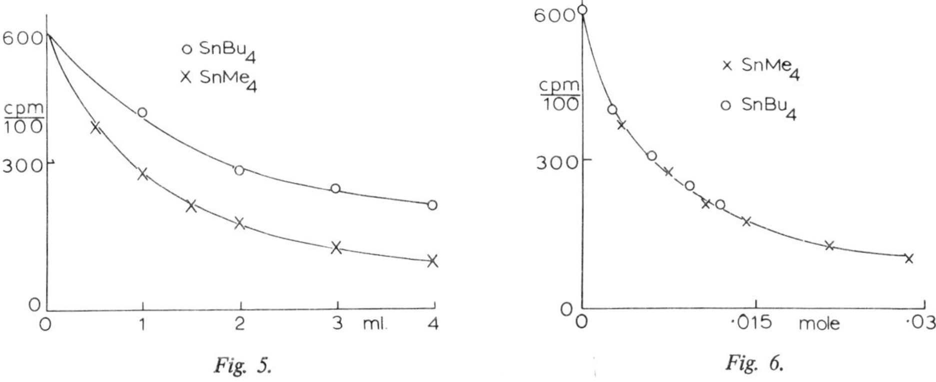

Fig. 5.

Fig. 6.

Fig. 5. The volume effect of adding organotin compounds to a toluene/butyl-PBD solution of [^{125}I] iodobenzene counted in glass vials in the γ-counter.

Fig. 6. The molar effect of adding organotin compounds to a toluene/butyl-PBD solution of [^{125}I] iodobenzene counted in glass vials in the γ counter.

Absorption of the energy of the γ-photons and X-rays within the sample will lead to electron emission and this should be evident from changes in pulse height spectra from the liquid scintillation counter (Figs. 7 and 8). The unquenched, untreated spectrum exhibits a fairly sharp, intense low-energy peak. This may be attributed to Auger electrons while the broad, low-intensity, higher energy band is associated with internal conversion electrons and a few electrons arising from photoelecrtic and Compton scattering interactions between γ-photons and X-rays and the counting medium. The presence of additive nicely demonstrates both a quenching effect and an electron enhancement effect. The low-energy peak, which arises exclusively from electrons produced within the decaying atom, is clearly quenched since the peak intensity is considerably reduced. The higher energy region associated partially with X-ray and γ-photon interactions exhibits the expected increased intensity because of the increased electron density of the medium.

For quenched samples, the effects are similar, although there is an apparent increased intensity of the higher energy band due to the normal pulse height shift with impurity quenching.

The pulse height spectra for ^{3}H (Fig. 9) indicate that the effect of the additive is merely to reduce the counting efficiency in the normal way.

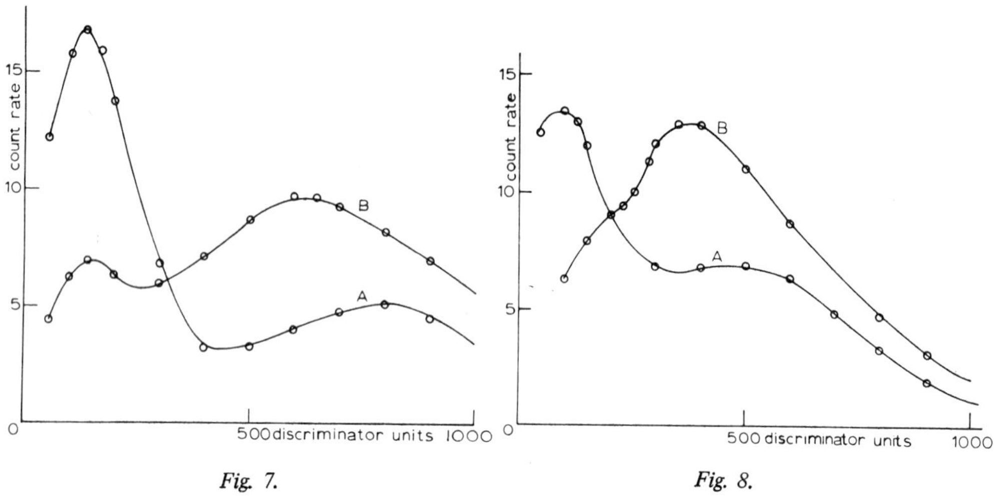

Fig. 7. *Fig. 8.*

Fig. 7. Pulse height spectra for unquenched samples of ^{125}I in toluene/butyl-PBD in the absence (curve A) and presence (curve B) of 1 ml $Sn(CH_3)_4$.

Fig. 8. Pulse height spectra for quenched samples of ^{125}I in toluene/butyl-PBD in the absence (curve A) and presence (curve B) of 1 ml $Sn(CH_3)_4$.

From the reduction in counting efficiency on the γ-counter for tetramethyl tin (Fig. 5), it should be possible to predict the increase in counting efficiency on the β-counter. The fact that the observed change in count rate with increase in additive deviates from that predicted (Fig. 10) is explained by the fact that the predicted value

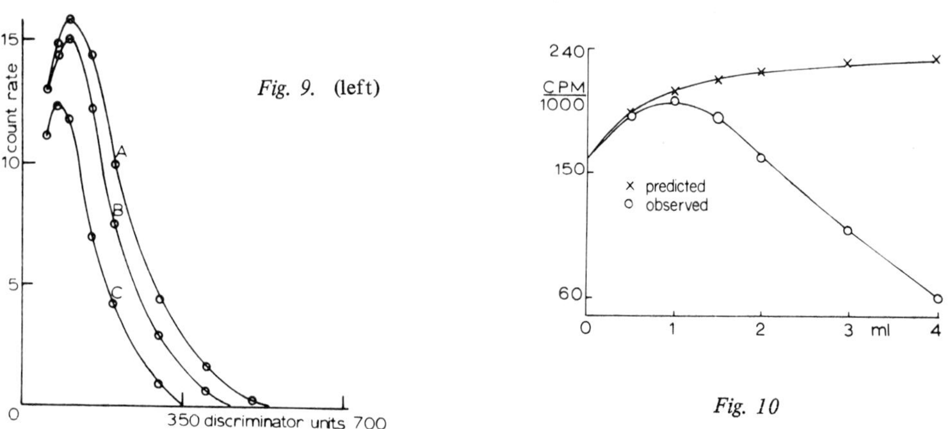

Fig. 9. (left) *Fig. 10*

Fig. 9. Pulse height spectra for samples of ^{3}H in toluene/butyl-PBD. (A) Unquenched sample. (B) Sample A + 1 ml of $Sn(CH_3)_4$. (C) Sample B + 0.45 ml of acetone.

Fig. 10. Plot of ^{125}I counting rate against volume of added $Sn(CH_3)_4$: 0 – observed curve for the liquid scintillation counter; X – curve predicted from γ-counting data assuming no impurity quench effect.

takes no account of the impurity quenching effect. Up to 1 ml of additive quenching is negligible, above this volume it becomes increasingly important. It would appear then that 1 ml is the optimum volume for tetramethyl tin.

Figure 11 compares the volume effect of both additives. Tetrabutyl tin has neither such a dramatic enhancement nor quenching effect as tetramethyl tin, although in terms of moles the differences are less pronounced (Fig. 12).

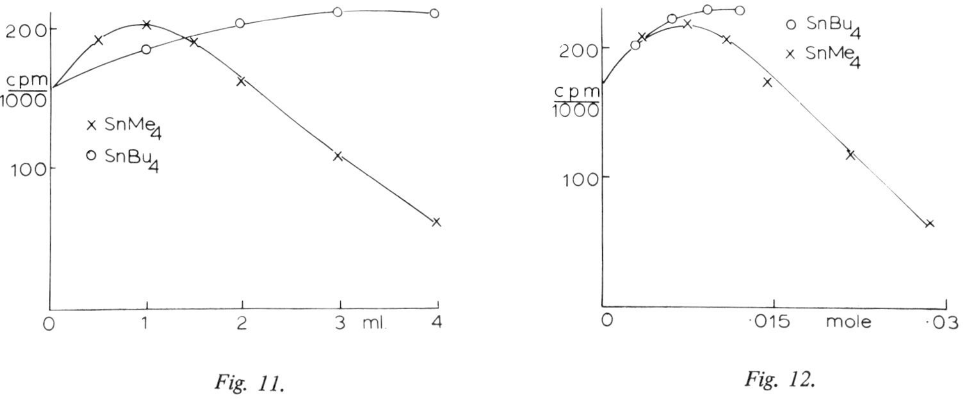

Fig. 11. Plot of ^{125}I counting rate against volume of added organotin compound.

Fig. 12. Plot of ^{125}I counting rate against moles of organotin compounds.

Figure 13 illustrates the volume effect of both additives on ^{3}H and ^{125}I samples both having an initial count rate of 150,000 counts min^{-1}. The optimum increase in count rate of ^{125}I requires 1 ml of tetramethyl tin and 2 ml of tetrabutyl tin. The same volumes result in ^{3}H count rates of 50,000 counts min^{-1} and 90,000 counts min^{-1} respectively. Bearing in mind that we wish to achieve the greatest possible relative change in counting efficiency, it is clear that 1 ml of tetramethyl tin represents the optimum volume of the more suitable reagent.

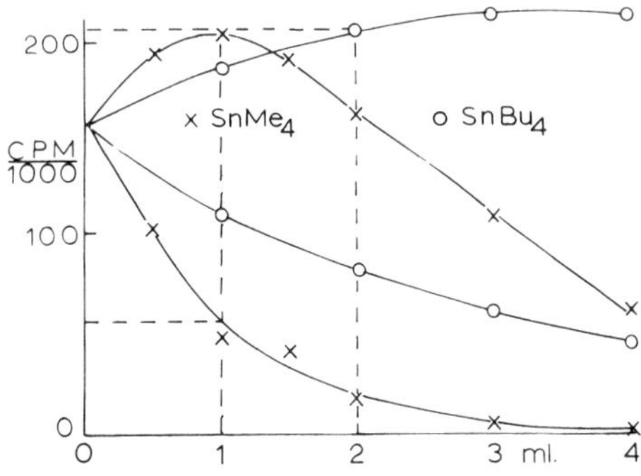

Fig. 13. Summary of the effects of added volumes of organotin compounds on the counting rates of ^{3}H and ^{125}I samples in toluene/butyl-PBD.

Figure 14 illustrates the four quench correction curves and the efficiencies of a typical sample. The effect of adding 1 ml of tetramethyl tin is to increase the ^{125}I efficiency from 68% to 85% and to reduce the ^{3}H efficiency from 31% to 24%. Typical results are shown in Table 3. The results are reasonable. They are better than those obtained using Method 2 and, although not as good as the best results obtained by Method 1, are vastly superior when Method 1 is subject to significant variations in γ counting efficiencies.

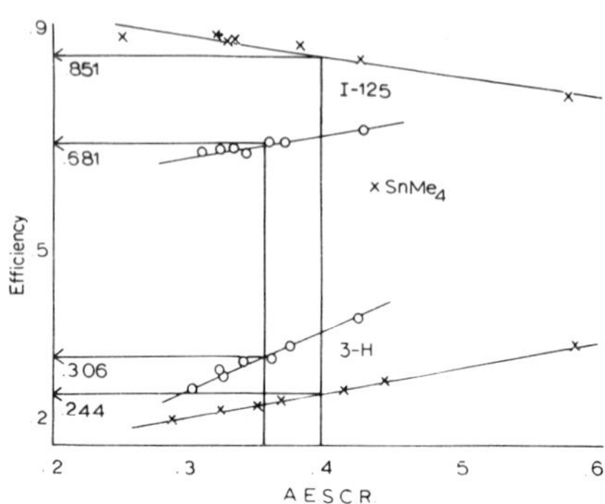

Fig. 14. Quench correction curves for ^{3}H and ^{125}I samples: 0 – in the absence of organotin compound; X – in the presence of 1 ml of $Sn(CH_3)_4$.

Table 3. Results obtained using a liquid scintillation counter and efficiency variation (Method 3).

Observed ^{125}I activity (dpm)	Recovery (%)	Attenuator setting	Observed ^{3}H activity (dpm)	Recovery (%)
17582	108.0	0.5	106512	97.2
17248	105.9		108057	98.6
17161	105.4		106971	97.6
17023	104.6	1.0	107154	97.9
17526	107.6		106635	96.8
17407	106.9		104792	95.7
16948	104.1	1.5	106286	97.0
17156	105.3		106470	97.2
15822	97.1		105188	96.0
16612	102.0	2.0	106599	102.0
16743	102.8		107480	98.1
16332	100.2		107724	98.3

EXPERIMENTAL

Sodium [^{125}I] iodide was supplied by the Radiochemical Centre, Amersham, England. [^{125}I] Iodobenzene was prepared from sodium [^{125}I] iodide and aniline by the method of diazotisation.[4] Counting equipment used was a Packard Model 3314 liquid scintillation counter, a Philips PW4510 liquid scintillation analyser and a Packard Model 578 γ-counter.

CONCLUSIONS

Three methods for the assay of doubly labelled samples containing ^{125}I and ^{3}H have been investigated.

Method 1 under ideal conditions provides results of the greatest accuracy, but is sensitive to γ-photon absorption which cannot easily be monitored. Method 2 is both time-consuming and inaccurate and is therefore of little value. The results obtained by Method 3 are reasonably accurate. It is confidently expected that improved results will be obtained using this novel method when other additives have been investigated. The method described uses only one window of a liquid scintillation counter. It may well be that, although energy discrimination alone does not provide a reliable basis for an assay method, a double channel approach with two-way spillover will improve the results obtained from Method 3. This also requires to be investigated.

REFERENCES

1. W. R. Greig, in *Liquid Scintillation Counting,* Vol. 1 (Ed. A. Dyer), Heyden, London, 1970, p.120.
2. W. R. Greig, private communication.
3. J. Ashcroft, *Anal. Biochem.* **37**, 268 (1970).
4. H. J. Lucas and E. R. Kennedy, *Org. Synth. Coll.* **2**, 351 (1943).

DISCUSSION

V. Tarkkanen: Your problem seems to arise from the simultaneous quenching of ^{125}I and ^{3}H in a liquid scintillation counting spectrometer. Have you tried the mentioned simultaneous quenching of ^{3}H and efficiency improvement of ^{125}I by an inorganic density increaser like zinc chloride in water, combined with a colloidal scintillator? The organic tin compounds are expensive and toxic. Also the nature of biological aqueous samples requires a colloidal scintillator. We have found better results with an aqueous zinc chloride solution combined with a colloidal scintillator (1:1 water solution: Insta-Gel) than with organic tin derivates.

K. L. Evans: Sodium iodide, for instance, gives a lower efficiency than the tin compounds. We have not tried zinc chloride merely because in this preliminary investigation we preferred a homogeneous counting sample. Indeed the cost of tetramethyl tin is considerable – about £1 per ml. Your proposal seems to be very interesting indeed.

Chapter 14

Correct Liquid Scintillation Counting of Steroids and Glycosides in RIA Samples: a Comparison of Xylene-based, Dioxane-based and Colloidal Counting Systems

H. Spolders

Packard-Becker B.V. Chemical Operations, Groningen, The Netherlands

INTRODUCTION

The last major step in the radioimmunoassay (RIA) technique (before calculation of the results) is the measurement of the radioactivity. Steroids, heart glycosides, prostaglandins and other small labelled molecules are commonly ^{3}H-labelled and counted by the liquid scintillation (LSC) technique. The correct choice of a counting solution is important to the reliability of the assay.

RIA sample types counted in LSC

After separation of free and antibody-bound antigens by various means, one of them is counted. The fraction to be counted contains either a free labelled compound or the antigen–antibody complex in a buffer or plasma solution.

Type of scintillator solution

Some investigators use ingenious methods to extract the free labelled antigens directly in the counting fluid. Steroids have been extracted in toluene,[1] dioxane[2] has been used for buffer solutions and some investigators have used mixtures of Triton X-100 and toluene for aqueous solutions[3] and plasma.[4]

CRITERIA OF A LIQUID SCINTILLATION SYSTEM FOR CORRECT MEASUREMENT OF RIA SAMPLES

In RIA, the following parameters are important for accurate liquid scintillation counting.

Absence of chemiluminescence assures a counting of the samples soon after preparation and eliminates the need to count the samples for a second cycle.

Stability of count rate. The quench level in samples prepared with the same amount of buffer in scintillator solution is often the same, which makes it possible to compare samples without quench correction. This, of course, requires a stable count rate.

Dissolving properties for the sample. For samples with varying colours, a quench correction must be applied. For any type of accurate quench correction, a homogeneous sample is necessary. This can be obtained if proteins and the buffer can be dissolved completely in the scintillator solution.

Comparison of scintillator solutions

In this paper, the above-mentioned criteria are compared in xylene-based, dioxane-based and colloidal scintillation solutions for either bound or free antigens of different polarity.

Using colloidal scintillators with plasma and buffer samples, phasing or sedimentation of salt or proteins sometimes occurs. The influence of sedimentation or phasing on count rate stability and correct quench correction is illustrated by varying the ratio between the scintillator solution and a RIA sample containing a semi-polar steroid aldosterone.

EXPERIMENTAL

Materials

Insta-Fluor (xylene-based), Permafluor II (a dioxane-based modified Bray solution), Insta-Gel and Triton/toluene (33% Triton X-100 in toluene with 9 g Permablend III/L) were chosen as scintillation solutions (all from Packard Instrument Co.).

(1, 2, 6, 7 (*n*) - ^{3}H) Testosterone (Amersham), a non-polar steroid, (1, 2, 6, 7 (*n*) - ^{3}H) aldosterone (Amersham), a semi-polar steroid, and (G - ^{3}H) digoxine (Amersham), a polar glycoside, were chosen as examples of compounds of different polarity.

Methods

To obtain bound antigens, the labelled compounds of known activity were bound to antibodies (CEA, IRI, Sorin) in PBS buffer (0.100M phosphate-buffered saline with pH 7.4). The dilution of the antiserum was chosen in such a way that at least 80–90% of the antigens and antibody was bound. RIA plasma samples were prepared by adding 10% antigen–antibody complex in PBS buffer to normal plasma. Various amounts of these antigen–antibody complex solutions were pipetted into counting vials and 10 ml of the scintillation solutions were added. To obtain free antigens, identical samples were prepared without the antibody.

Liquid scintillation counting

Immediately after preparation, the samples were placed in a Model 3385 Tri-Carb with optimum ^{3}H setting and operating at 12 °C. The samples were counted a total of six times with 10000 counts accumulated each time during a 24-h period. The count rate stability and sample appearance was noted and quench correction was performed with the automatic external standard ratio (ESR) technique.

Phase diagrams of Insta-Gel with the PBS buffer and various other buffers applied in the RIA technique were defined at various temperatures. This was done by adding 0–10 ml of buffer solution to 10 ml of Insta-Gel and recording the appearance after 16 h of stabilisation in water baths at 5, 7, 12, 21 and 25 °C.

RESULTS

Free antigen

Free testosterone (non-polar) was extracted out of the aqueous sample completely by xylene and dioxane (Table 1). Free aldosterone (semi-polar) was extracted partially by xylene and totally by dioxane. Free digoxine (polar) was counted in dioxane with a yield of 90% but could not be counted in lipophilic xylene. In colloidal scintillators, the aqueous sample was dispersed homogeneously and all types of free antigens could be counted with a 100% yield.

Table 1. Recovery of free antigens with various polarity[a] in the xylene-based, dioxane-based and colloidal scintillation solution.

Scintillator solution	Free antigen recovered (%)		
	Testosterone	Aldosterone	Digoxine
Xylene (Insta-Fluor)	102	62	0
Dioxane (Permafluor II)	100	100	90
Colloid (Insta-Gel)	100	100	100

[a] 0.5 ml buffered solution (PBS) in 10 ml scintillator solution.

Antigen–antibody complex

Non-polar testosterone. Figure 1 shows the behaviour of testosterone in different scintillators for a sample load of 5% buffer (PBS). In the dioxane cocktail, testosterone in buffer was measured for 100%, based on ESR quench correction. The count rate was variable until testosterone was separated from its antibody and the salt from the buffer was sedimented out completely. In the two-phase solution of xylene (Insta-Fluor), the count rate was changing until the testosterone was separated and completely extracted. In the homogeneous colloidal scintillation solution (Insta-Gel), the radioactivity was measured at 100% recovery immediately and the count rate was constant.

Semi-polar aldosterone. Figure 2 shows the difference between the scintillators for aldosterone in a 5% aqueous buffer (PBS). In the dioxane cocktail, aldosterone was measured for 70–100%, depending on the time of measurement. In the xylene-based scintillation solution, it was counted for 24–62%, the recovery increasing with time. In the homogeneous colloidal scintillation solution (Insta-Gel), aldosterone was measured for 95%. If sedimentation was present in the colloidal scintillation solution or phasing occurred, results were different. The count rate changed with time in dioxane and in xylene, as shown in Fig. 3.

Polar digoxine. Figure 4 shows the performance of digoxine with 0.5 ml of buffer and 20% of plasma in various scintillation solutions. In dioxane, digoxine was counted for 80–90%. Digoxine could not be counted in xylene, but was measured for 100% in the homogeneous colloidal scintillation solution. A stable count rate was obtained for Insta-Gel within 30 min and for Permafluor II (dioxane) in 1 h.

Influence of phasing of colloidal scintillators on liquid scintillation counting of RIA samples. Various amounts of buffer (from 0.2 to 2.0 ml) in 10 ml of Insta-Gel and in 10 ml of Triton X-100/toluene resulted in samples of different appearances. The influence of this parameter on LSC was studied with buffer and plasma samples containing the aldosterone-antibody complex.

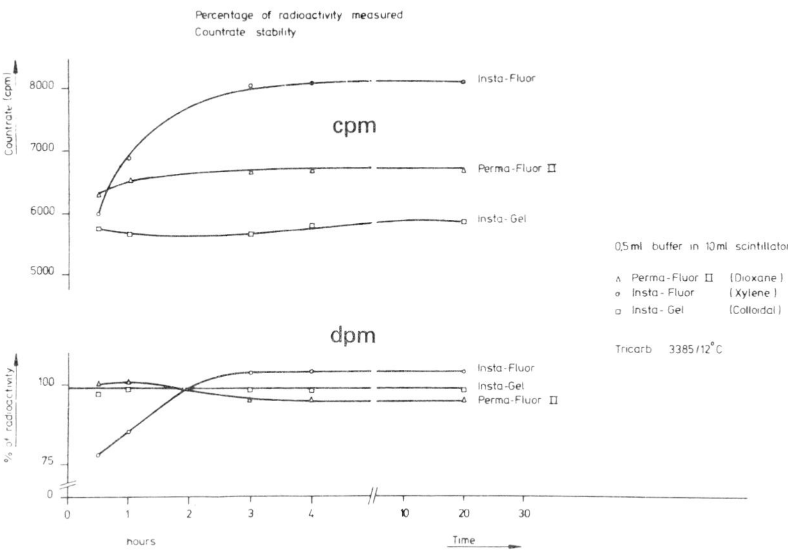

Fig. 1. Count rate stability and recovery of disintegrations $\min^{-1}$ of ^{3}H-testosterone–antibody complex in various scintillation solutions.

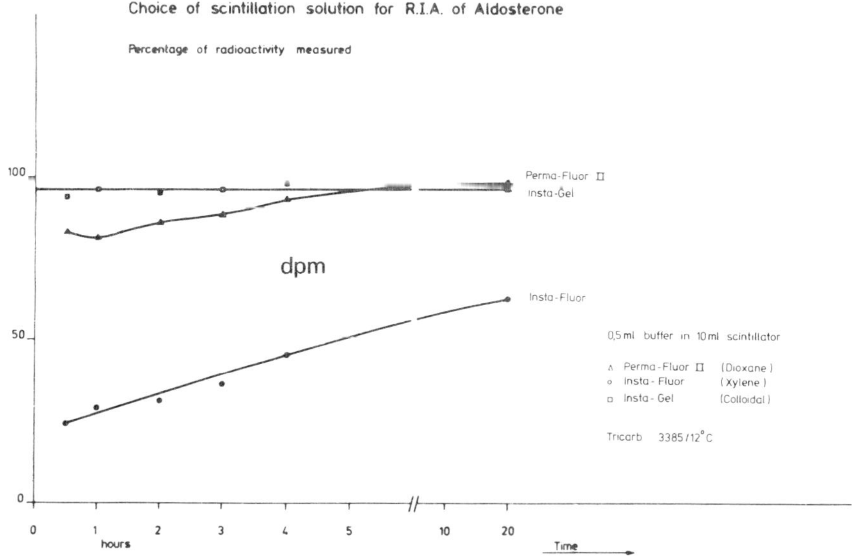

Fig. 2. Recovery of disintegrations $\min^{-1}$ of ^{3}H-aldosterone–antibody complex in various scintillation solutions.

Figure 5 presents the count rate stability and the percentage of the activity measured in Insta-Gel with various volumes of a buffer solution containing aldosterone bound to its antibody. The sample with sedimentation (0.2 ml) showed a 21% increase in the count rate in 20 h, and this was not correctable with the ESR quench correction method. Single-phase samples (0.5 and 1.0 ml) had a stable count rate which was

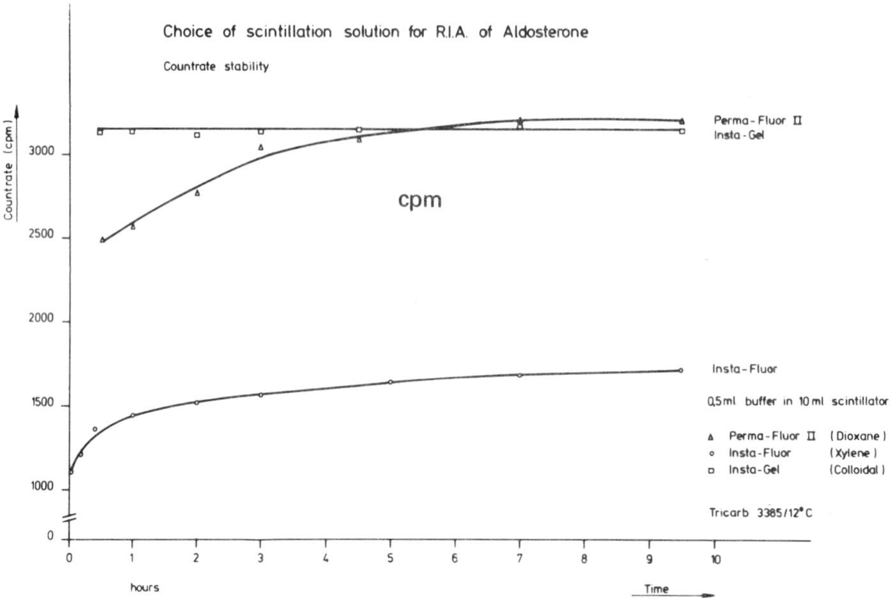

Fig. 3. Count rate stability of ^{3}H-aldosterone-antibody complex in various scintillation solutions.

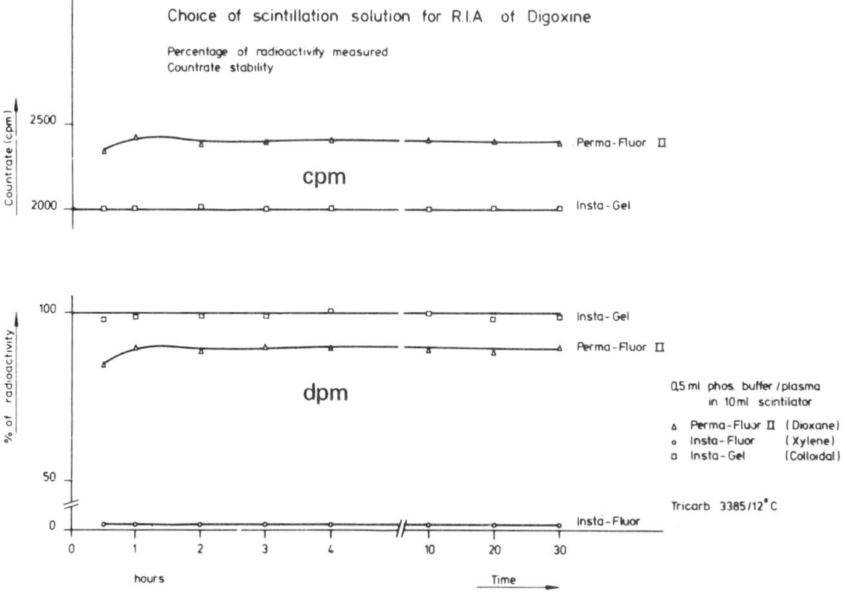

Fig. 4. Count rate stability and recovery of disintegrations min^{-1} of ^{3}H-digoxine in various scintillation solutions.

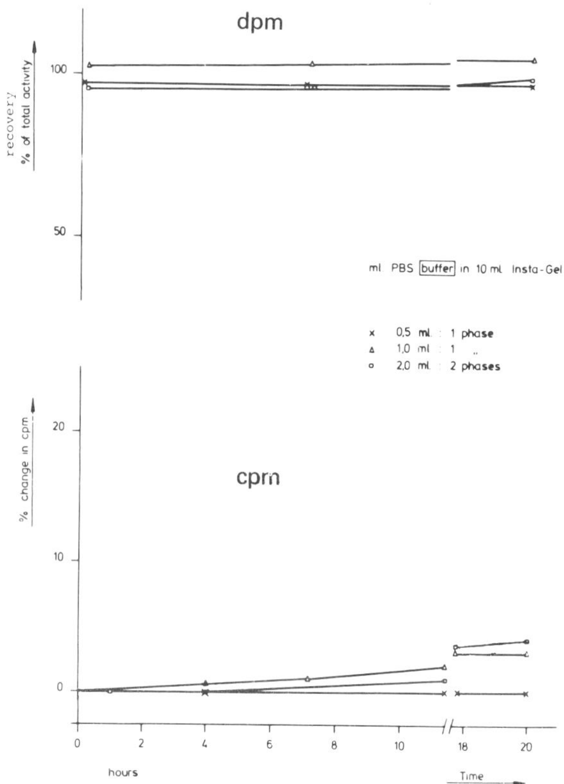

Fig. 5. Count rate stability and recovery of disintegrations min^{-1} in Insta-Gel with various volumes of buffer (PBS) solution containing ^3H-aldosterone-antibody complex.

correctable with the ESR method. The phasing samples (2.0 ml) showed a slow increase in the count rate, due to a slow extraction of free aldosterone out of the hydrophilic layer into the lipophilic layer. Figure 6 shows the same parameters for the Triton/toluene scintillator with various amounts of a buffer solution containing the aldosterone–antibody complex. The sample containing sedimentation (0.2 ml) had an unstable count rate. Calculation of disintegrations min^{-1} of this sample provided incorrect values. The single-phase sample maintained a constant count rate and its activity could be calculated correctly with the ESR method.

Plasma solutions with the aldosterone–antibody complex in Insta-Gel showed an increasing count rate with time (as shown in Fig. 7). This change in count rate was, in all single-phase samples, completely corrected by the ESR quench correction method.

The results showed that the sample appearance in colloidal scintillation solutions determines the accuracy of the liquid scintillation counting of RIA samples. In homogeneous samples, the count rate was stable for buffer solutions and quench correction with the ESR method could be applied.

The sample appearances of Insta-Gel with commonly used RIA buffers are presented as phase diagrams in Fig. 8. The filled areas indicate sample percentages

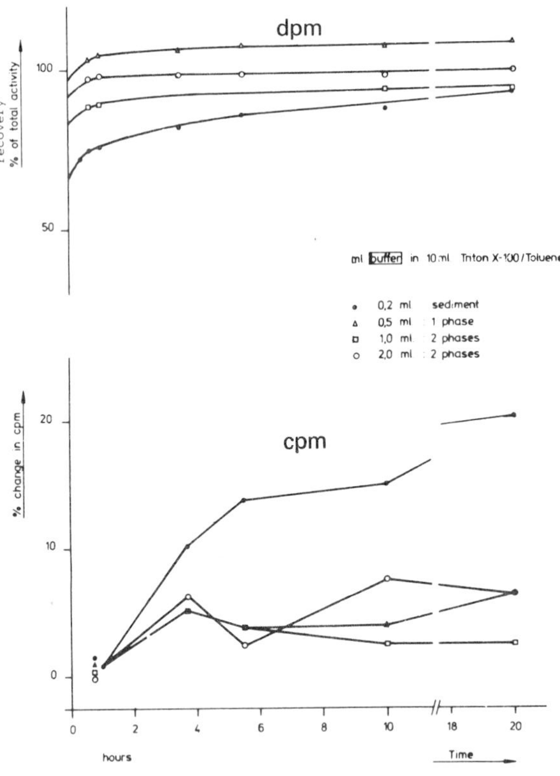

Fig. 6. Count rate stability and recovery of disintegrations min^{-1} in Triton X-100/toluene with various volumes of buffer (PBS) solutions containing ^3H–aldosterone–antibody complex.

providing homogeneous counting solutions; 5 to 10% sample load of buffer for all investigated buffers in Insta-Gel results in homogeneous counting solutions.

COMMENTS AND CONCLUSIONS

In this study, the commonly used scintillation solutions were compared for counting RIA samples. These samples had either free labelled antigens or antigens bound to an antibody in an aqueous buffer or plasma solutions. In toluene scintillator (Insta-Fluor), only non-bound, non-polar antigens could be counted. Non-polar, bound antigens required a long time for dissociation of the antigen–antibody complex before they could be counted. The toluene probably extracted the lipophilic antigen out of the water layer of the phasing liquid scintillation sample.

In Bray solutions (dioxane), non-bound polar and non-polar antigens provided reliable results. The antigen–antibody complex dissociated in dioxane,[5] generating a free antigen to be counted after a stabilisation time varying from 1 to 3 h. However, in Bray solutions the sample load was limited to 9–10% and proteins were poorly soluble.

Colloidal scintillators (Insta-Gel) were generally applicable. Samples could be measured within a short time after preparation. An antibody–antigen complex in

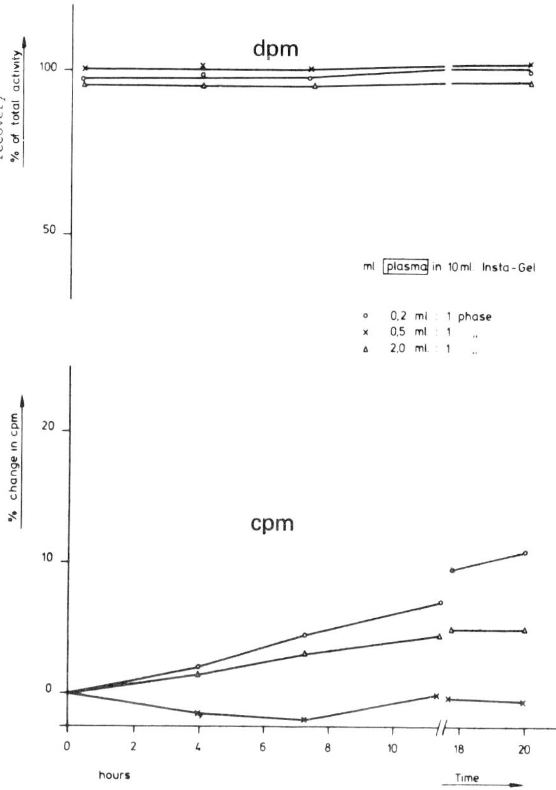

Fig. 7. Count rate stability and recovery of disintegrations min^{-1} in Insta-Gel with various volumes of plasma containing ^{3}H-aldosterone–antibody complex.

buffer exhibited a constant count rate and existing variations in the count rate for plasma samples could be corrected with the ESR quench correction method. Colloidal scintillators had a higher sample capacity than Bray solutions, thus requiring less scintillator. Because two-phase colloidal samples gave erroneous results, the sample volume had to be adjusted to produce homogeneous samples. In determining this volume, phase diagrams appeared very useful. Phase diagrams of colloidal scintillators varied for different RIA buffers, and even for Triton X-100 there could be variations from batch to batch for the same buffer.[6] Chemiluminescence was not evident in any of the investigated samples.

In our laboratory, colloidal scintillators have been developed for RIA buffers or plasma without phasing or sedimentation. Their sample holding capacity is higher than the commonly used scintillators, thus allowing the use of less scintillator solution. Comparative tests with other scintillator systems will be reported elsewhere.

SUMMARY

In this study the commonly used scintillation solutions were compared for liquid scintillation counting of RIA samples containing free or bound antigens of

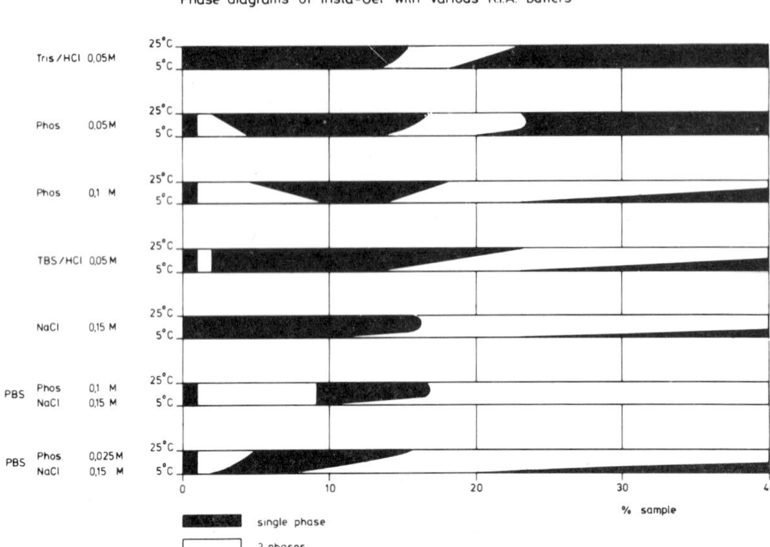

Fig. 8. Phase diagrams of Insta-Gel with various buffer solutions at temperatures of 5°C to 25°C.

different polarity.

Absence of chemiluminescence, stable count rate and good dissolving properties of the labelled compound were required for counting RIA samples. Xylene or toluene scintillators met these requirements only for non-bound, non-polar antigens.

Dioxane-based scintillation solutions could be applied for free antigens of each polarity. The antibody–antigen complex was counted after a stabilisation time needed for dissociation of the complex. The sample capacity of dioxane was limited and proteins were poorly soluble in dioxane.

Colloidal scintillators could be used for either bound or non-bound antigens in plasma or buffer solution. Variations in count rate in plasma samples could be corrected by the external standard ratio quench correction method, provided the solution was homogeneous.

Two-phase colloidal scintillators could provide varying count rate and erroneous quench correction. Bound antigens could be counted in dioxane with limited sample capacity after a stabilisation time for dissociation varying from 1 to 3 h.

REFERENCES

1. M. Castenier and R. Scholler, *C. R. Acad. Sci. (Paris)* **271**, 1787 (1970).
2. G. E. Abraham, in *Modern Methods in Steroid Analysis* (ed. E. Heftmann), Academic Press, New York, 1973, p. 451.
3. H. S. Patterson and R. C. Green, *Anal. Chem.* **37**, 854 (1965).
4. P. I. Chapman and J. Mancroft, *Intern. J. Appl. Radn. Isotopes* **22**, 371 (1971).
5. G. E. Abraham, *Acta Endocrin.* **75**(S 183), 7 (1974).
6. H. Spolders, unpublished data.

DISCUSSION

D. I. Chapman: One of the problems in counting large numbers of samples is the prohibitive cost of commercial scintillators. To reduce the volume of scintillant one either has to reduce the volume of sample, which means discarding counts, or find a scintillator which will accept large volumes of aqeuous sample. Figure 8 of your paper indicates that Insta-Gel will accept up to 40% of certain aqueous buffers. Are these mixtures really homogeneous solutions or are they heterogeneous? Are they fluids or are they gels?

H. Spolders: The mixtures are really homogeneous for liquid scintillation counting. At ambient temperatures, the samples with aqueous buffer are viscous fluids, but at sub-ambient temperatures they are gels.

N. Lucas: With regard to Triton X-100 systems, preparation of a single phase is difficult, but if a batch of Triton X-100 is assayed, results can be stabilised. Normally, Triton X-100/toluene can accommodate a 10% sample. Samples should be counted at 6–8°C and allowed to equilibrate for 20–30 min before counting.

D. I. Chapman: May I disagree with Dr. Lucas? My experience is that batch to batch variation of Triton X-100 does occur, and on standing a precipitate may settle out. Acceptable material can be obtained by decanting clear material from the top of the drum, the residue being discarded or used as a washing-up detergent.

N. Lucas: The precipitate can be removed by warming – it apparently dissolves.

H. Spolders: If precipitation occurs in drums with large volumes of Triton X-100, warming can cause chemical changes which are responsible for chemiluminescence. Triton X-100 changes not only from batch to batch but also large quantities are inhomogeneous. To obtain reproducible results from Triton X-100 in large drums, they should be shaken well before use, which is rather difficult to do and does not change the final poor performance.

V. Tarkkanen: Heating does not actually remove the reason for the poor and variable sample holding characteristics of Triton X-100 based scintillator systems. This is based on the fact that the polymerisation process involved in production of Triton does not produce an equal product and the higher molecular weight homologs which do interfere cannot be removed by heating. The same applies to the peroxides present in varying amounts in Triton causing alterations in background.

SECTION IV

LOW LEVEL COUNTING

Chapter 15

Considerations for Achieving Low Level Radioactivity Measurements with Liquid Scintillation Counters

John E. Noakes

Geochronology Laboratory, University of Georgia, Athens, Georgia, U.S.A.

INTRODUCTION

The liquid scintillation (LS) counter as we know it today is a product of 25 years of continuous development. Its evolution has been made possible by the joint efforts of the scientific community in seeking a superior research tool and industry's desire for a commercial product. The early creation of the LS counter grew out of the need to detect and measure efficiently low energy β-radioactivity. Today's counter is a direct result of this successful endeavour and owes much of its popularity to the extreme versatility to which it has been applied in counting a wide spectrum of samples.

Today's commercial LS counters boast high E^2/B values with ^{14}C and ^{3}H counting efficiencies of 98% and 67% with equally low backgrounds of 10–20 and 20–30 counts min^{-1} respectively. These performance figures have been obtainable through the use of and continuous improvement in such components as pulse height electronics, high quantum yield phototubes, fast coincidence circuitry and massive shielding. Other features of the newer instruments which have attracted the user are the capability of simultaneous single, dual or triple isotope counting, sequential multisample handling and automatic absolute activity correction of highly quenched samples. As a result, even with the widespread usage of LS instruments with non-optimised settings and highly quenched samples, the present-day user can in most cases expect to obtain valid counting statistics.

From what has been stated previously, it would appear the average LS user has many reasons to be well satisfied with the performance of his LS counter. There is, however, a small group, which we will collectively call the 'low level group', which is, from all outward appearances, continuously discontent, always striving to ask more of the LS counter than it was originally designed to produce. While the average user deals with milli- or microcurie sample activity levels and frequently neglects to subtract background activities from this data, the low level counter often deals in sample count rates equal to that of his instrument's background. In this region of signal-to-noise ratio approaching 1, a highly optimised instrument capable of long-term stable performance and minimum background is a must for valid collection of low level counting data. It is to the low level group that this paper is directed, in the hope that some of the comments made will be of assistance in helping them achieve the highest performance standards for their LS counter.

SAMPLE PREPARATION

The normal procedure followed in preparing samples for LS counting is to choose a sample-cocktail system which will require the least sample preparation and give the highest counting efficiency. In most cases, the sample is either dissolved, combusted or suspended as a solid in the cocktail media. Since most labelled samples to be counted have activity levels in the 10^3–10^5 disintegrations min^{-1} range, many weaknesses in the sample-cocktail system can be tolerated and still yield acceptable counting results.

Unfortunately, in low level LS counting the low activity levels of the majority of the samples require that the most demanding preparation techniques and efficient cocktails be utilised in order to obtain optimum conditions for data collection. The sample-cocktail systems that have found the most success have been those which have prepared the sample in a form miscible with or as an aromatic benzene-type solvent. Table 1 shows some examples of past endeavours of various researchers to prepare samples for low level radiocarbon dating. Sample methods (1) and (2) in Table 1 represent good examples of samples prepared in the gaseous state which are dissolved in counting solvents. Methods (4) and (5) are representative of samples prepared as organic compounds which are miscible with good counting solvents, whereas methods (3), (6) and (7) depict samples prepared as good counting solvents in themselves.

In a practical application, the selection of the sample form to be counted by liquid scintillation may be dictated by a number of considerations. For example, the initial form of the sample may play a decisive role in choosing the appropriate chemical conversion steps. Also to be considered will be the initial size of the sample and the chemical yields of the various steps. Still further consideration must be given to sample purity, possible isotopic fraction and sample contamination. Last, but not least, sample preparation time and overall costs will be of significance.

In general, one should initially strive to synthesise the LS sample to be in the form of the best LS counting solvents. These solvents fall into a broad category of aromatic-type molecules possessing low excitation energies and possessing non-

Table 1. Sample preparation for radiocarbon dating.

	Sample form	Solvent	References
(1)	Carbon dioxide (g)	Ethanol–toluene	Nystrom *et al.*[1]
(2)	Acetylene (g)	Ethanol–toluene	Audric and Long[2]
(3)	Toluene (l)	Toluene	Funt *et al.*[3]
(4)	Hexane–octane (l)	Ethanol–toluene	Arnold[4]
(5)	Paraldehyde (l)	Toluene	Leger *et al.*[5]
(6)	Benzene (l)	Toluene	Tamers[6]
(7)	Benzene (l)	Benzene	Noakes *et al.*[7]

bonding π-electrons. Benzene, toluene and 1,2,4-trimethylbenzene, in the order of increasing desirability, are the type of molecules that make good LS solvents. If syntheses of aromatic solvents are not feasible due to the above-stated considerations, then saturated hydrocarbons which are readily miscible with aromatic solvents should be considered as the next best selection.

Choice of solute or solutes (fluors) to be used in the cocktail will primarily be concerned with solute–solvent interaction and spectral response of the LS photomultiplier tubes (PMT's). Horrocks[8] has categorised solute–solvent interaction into solute solubility, solute self-quenching and solute self-absorption. Of these three, solute solubility is of foremost concern, especially when limited solubility is reached prior to culmination of the maximum energy transfer in the cocktail. This condition frequently arises when mixed solvents are employed in the cocktail and when the solute has limited solubility in the primary solvent.

Solute emission spectra should be matched to the spectral response of the LS PMT. With the newer bialkali PMT's with a maximum spectral response of 385 nm, a single solute of PPO or butyl-PBD may well make an ideal match, whereas with older-type PMT's with a spectral sensitivity of 410 nm, a wave shifter such as methyl-POPOP may be required.

On the custom design of the cocktail, it should be kept in mind that a more efficient cocktail will not only yield higher counting efficiencies but in most cases higher pulse height, thus enabling true scintillating pulses to be shifted above the thermionic and electronic background noise and therefore to be more accurately counted.

LS COUNTER VIALS

The sample vials utilised today in essentially all commercial LS counters have been standardised by the American National Standards Institute (ANSI) Code to dimensions of 60 mm height and 28 mm diameter which allows counting of sample volumes up to 20 ml. This agreement between LS counter manufacturers to standardise vial dimensions has proven to be indispensible for trouble-free vial–instrument interfacing and in the reduction of 'hang-ups' in the vial changing mechanism. In the case where smaller dimension vials are utilised, individual vial holders of ANSI dimensions must be utilised for each vial to minimise faulty operation.

Vials are constructed of a number of transparent to translucent materials ranging from glass, pyrex and quartz to plastics such as polyethylene, nylon and Teflon. Cap enclosures can be of a snap-cap, screw top or permanent nature and are fabricated from such materials as polyethylene, metal or white urea. Sealing devices within the cap enclosures such as aluminised cork or fibre liners or expandable flanges are utilised to prevent unduly high cocktail evaporation rates.

From what has been stated above, it would appear that any user of a commercial LS counter is somewhat restricted to the use of a standardised counting vial. In the case where low level counting is desired, this restriction may become intolerable, especially when sample volumes very much less than 20 ml are to be counted. It has long been recognised that a portion of the total LS counter background can be attributed to natural radioactivity of the vial material and to the interaction of external radiation with the sample–cocktail to produce detectable scintillating events. In order to minimise this background contribution, it is desirable to select vial material that has negligible amounts of natural radioactivity and to utilise the smallest cocktail

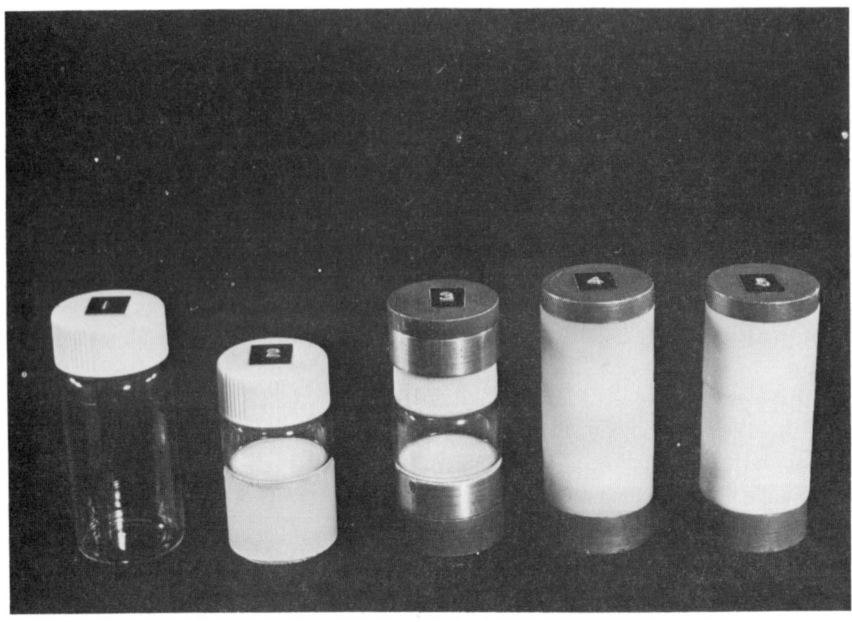

Fig. 1. Regular and small volume LS counting vials.

volume possible. Calf and Polach,[9] and Hartley and Church[10] have recently reported on the construction and advantages to be gained using small sample vials for low level counting.

Figure 1 shows a series of vials that have been utilised at the University of Georgia (UGa) Geochronology Laboratory when carrying out radiocarbon dating measurements by the benzene method and LS counting. Vial no. 1 is a standard commercial 20 ml vial of ANSI dimensions fabricated from borosilicate low ^{40}K glass with a white urea screw cap. Vial no. 2 is a modified no. 1 vial reduced by one-half the original vial weight to a volume of 5 ml. A polyethylene base is permanently fastened to the vial base in order to position it in the centre face of the two PMT's. Vial no. 3 is a standard vial modified similar to vial no. 2 but with a screw-attachable lead base and top to position and shield the vial from external radiation. Vial no. 4 is made of linear polyethylene and is designed to hold a cocktail volume of 5 ml. A screw-attachable lead base and top are utilised for positioning and shielding and an 'O' ring is designed into the top cover to enhance sealability. Vial no. 5 is of the same design as vial no. 4, but is made of Teflon.

Table 2 shows the background, counting efficiency and calculated E^2/B values for the above vials when counting a ^{14}C-labelled and non-labelled benzene sample in a benzene–PPO cocktail. The LS counter was set at balance point operation (to be discussed later) for radiocarbon with the lowest discriminator set above the E_{max} energy for tritium. Background values were calculated from 10–100 min counting periods with a stated 1 sigma (σ) standard deviation.

In Table 2, vials 1–4 show the effects of varying cocktail volume on background count rate and counting efficiency. By reducing the benzene volume from 20 ml to

Table 2. LS vial counting characteristics with varied volume and vial materials.

Vial no.	Sample–vial description	Background count (cpm)	Counting efficiency (%)	E^2/B
1	Std. 20 ml glass vial 20 ml benzene + PPO	9.691 ± 0.098	68	477
2	Std. 20 ml glass vial 15 ml benzene + PPO	7.452 ± 0.086	68	620
3	Std. 20 ml glass vial 10 ml benzene + PPO	5.944 ± 0.077	67	756
4	Std. 20 ml glass vial 5 ml benzene + PPO	4.663 ± 0.068	64	878
5	Mod. 5 ml glass vial + polyethylene base 5 ml benzene + PPO	4.532 ± 0.067	67	990
6	Mod. 5 ml glass vial + Pb base and top 5 ml benzene + PPO	3.820 ± 0.061	67	1175
7	Polyethylene 5 ml vial + Pb base and top 5 ml benzene + PPO	2.943 ± 0.054	66	1480
8	Teflon 5 ml vial + Pb base and top 5 ml benzene + PPO	2.277 ± 0.048	65	1855
9	Teflon 5 ml vial without Pb base and top 5 ml benzene + PPO	2.965 ± 0.054	65	1425

5 ml it was possible to almost halve the background count rate while only slightly reducing the counting efficiency. Vial 5 with a reduced vial glass, a polyethylene base and 5 ml of benzene gave a slightly lower background than vial 4 because of the reduced vial glass. The counting efficiency of vial 5 was slightly increased because of improved positioning of the sample relative to the PMT's. Vial 6 with reduced vial glass, lead base and top for shielding and positioning and 5 ml of benzene sample showed a decrease in background count rate of 26% from vial 5 which can only be attributed to the added lead shielding as the glass vials of 5 and 6 were identical. Vial 7 was constructed of linear polyethylene with lead base and top identical to vial 6. When counting a 5 ml benzene sample, the reduction in background count rate of 23% from vial 6 is directly related to the lower natural radioactivity of the polyethylene. Benzene did appear to interact with the polyethylene during a 48-h period causing the screw cap to swell and stick. Vial 8 was constructed of Teflon with lead base and top identical to vials 6 and 7. Using a 5 ml benzene sample, the lowest background count rates were observed with this vial.

Figure 2 shows a detailed drawing of the construction of the Teflon vial 6. The vial dimensions are 63 mm height and 27 mm diameter. The height is slightly higher (9 mm) than the standard vial, but this does not appear to interfere with the sample changing mechanism or elevator system for either a Beckman LS-100, a Packard 3300 or a Picker 220. The vial is constructed of four separate components. The sample

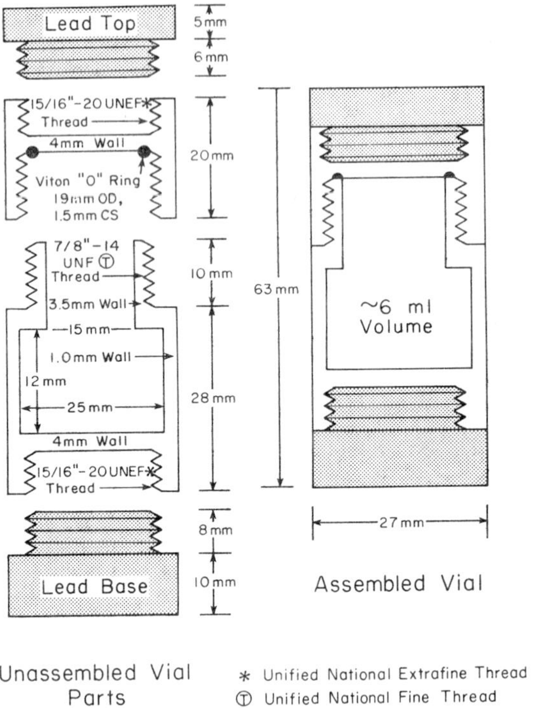

Fig. 2. Teflon LS counting vial.

holding compartment (vial and cap) is fabricated from a single piece of Teflon rod. A sample volume capacity of 6 ml was selected due to the limited sample volumes required for radiocarbon dating measurements at the UGa Geochronology Laboratory. Volumes of any size could be selected, but increasing cocktail capacity would necessitate diminishing the lead shielding. An 'O' ring is placed in the top shoulder of the vial enclosure. Sealability tests on the vial using a 5 ml benzene sample showed no solvent loss after a 96 h holding period. Wall thickness of the vial portion facing the PMT's is 1 mm which gives a translucent view of the cocktail. This wall thickness appears to reduce the counting efficiency of the Teflon vial slightly as compared with a comparable glass vial (Table 2). Benzene was not observed to inter-react or swell the Teflon cap or wall of the vial. A lead base of 18 mm height weighing 100 g is used to position the vial in the centremost portion of the PMT's and to act as added massive shielding. Positioning can account for a 3–5% increase in counting efficiency, especially when the sample volume is 5 ml or less. A top lead shield of 11 mm height weighing 64 g is used solely for vial shielding. A small amount of top and bottom vial shielding can be extremely effective in lowering background (Table 2), especially in elevator sample loading counters which have no shielding directly above or below the vial. Both top and bottom lead shields are detachable to permit more accurate weighing of the sample within the vial. From the above discussion it can be seen that in cases where low level radioactive samples are to be counted by liquid scintillation it is of distinct advantage to minimise the sample-cocktail volume in order to reduce instrument modifications. To effectively carry out this task, one can employ counting vials fabricated from Teflon equipped with 'O' rings for sealability and a lead top and bottom base for vial-phototube positioning and radiation shielding.

RADIATION SHIELDING

There are two types of shielding components employed in LS counters for lowering background caused by external radiation. The first and most commonly used in commercial LS counters is massive shielding. The second and less frequently employed is an electronic shield frequently referred to as an anti-coincidence guard.

(a) Massive shielding

In LS counters, as in most other radiation counters, the material of choice in constructing the massive shield is lead. This material is selected because of its good shielding qualities (higher atomic number, Z), its low content of natural radioactive constituents and its relatively low cost and ease of fabrication. Shield designs are primarily concerned with preventing as much incoming radiation as possible from interacting with the vial's cocktail media. In order to completely shield the vial it is necessary to shield not only the vial holder but also the two viewing PMT's which are in close proximity to the vial and, on occasions, some associated electronic component.

Figure 3 shows a typical top vial entry shield design that is employed in commercial

Fig. 3. Top sample vial entry LS counter lead shield.

LS counters. The thickness of the lead shield most often used today is approximately 2 in which results in a shield weight of from 200-300 lb, depending on individual shield design. At this thickness, approximately 90% of the normal terrestrial radiation originating from ^{40}K, Th and U decay daughters and the soft component from cosmic radiation is shielded. In order to increase this shielding level significantly, the shield thickness would have to be increased by 4 to 6 in, thereby more than tripling the shield weight and cost. Accordingly, the massive shields used today on commercial LS counters are not designed for optimum shielding levels, but as a prudent compromise between excessive production costs and acceptable instrument background performance.

From the above comments it can be seen that the dollar outlay for additional massive shielding and the necessary modifications to supporting cabinet structure can be excessive for only a modest expected background reduction. Certainly a less expensive approach and one that can be implemented without excessive modification in existing shield design.is the use of inner graded shield liners of cadmium and copper metal. Only a very thin layer of a few millimetres thickness is needed to stop the lead X-rays generated from the hard component terrestrial and cosmic radiation striking the lead shield. Reduction in background results from the reduction in the secondary radiation that is available to interact with the cocktail media within the vial. Consideration should also be given to the use of added massive shielding above and below the elevator sample entry portals of the shield as these two areas are completely unshielded in an automatic sample changing instrument. This can be accomplished by positioning lead shielding above and below the portal openings and/or the use of the lead vial shields that were discussed earlier.

(b) Electronic shielding

Electronic shielding is primarily employed when one wishes to reduce the background contribution caused by incoming hard terrestrial and cosmic radiation where massive shielding is essentially ineffective.

The electronic components making up the shield consist of one or more radiation detectors, a high voltage supply and electronic circuitry comprising pulse height discrimination, a pulse delay line and pulse gate. The radiation guard detectors most often employed are either of the crystalline NAI (T1) or plastic PVT type. Guard design is built around encompassing as much of the vial holder as possible and is preferably positioned within the lead shield in order to minimise the amount of costly detector material comprising the shield. The high voltage supplying the PMT's of the detectors can either be a separate unit or in some cases can be tapped off the existing LS high voltage system. Pulse height discrimination is used to reject guard detector pulses outside a specified energy range. Usually only a lower discriminator is utilised to eliminate low energy guard detector electronic noise. A delay line is needed to slow down the arrival time of the fast LS pulse at the electronic gate. A gate mechanism is used to restrict passage of any LS pulses for a specified period of time after having been previously activated by a guard pulse.

The electronic shield functions in the following manner. External radiation having enough energy to penetrate the guard shield detectors and LS vial cocktail media will generate pulses in both the guard and LS electronics. If the guard detector pulse is above the lower energy discriminator, it will be allowed to pass to and close the gate mechanism up to several microseconds. The liquid scintillation pulse generated from the same external radiation event will be delayed by a delay line so as to arrive at the gate slightly later than the guard pulse. If this time period is within the close time of the gate, the LS pulse will not be permitted to pass the gate and be counted. The gate functioning in this manner is referred to as an anti-coincidence gate as it is only active and acts upon pulses arriving within a specified coincidence time period.

The application of an electronic shield system to a commercial LS counter is difficult because of the inflexibility of the design of the massive shield to accommodate an internal guard detector. Accordingly, the electronic shielding systems that have been reported have utilised radiation guards positioned outside the lead shield. One approach has been to place one or several large plastic radiation detectors above the massive shield for the prime purpose of shielding cosmic radiation. In this mode of operation, the shield detectors are usually removed during placement of the sample in the LS counter and repositioned when counting is in progress. Another method which is a little less costly and non-restrictive for sample introduction is to employ an exterior detector alongside the massive shield. Figure 4 shows a 3 in × 3 in dia. NAI (T1) detector in this configuration. In this case little major reduction in background is accomplished, but background perturbations

Fig. 4. Side positioned electronic radiation guard.

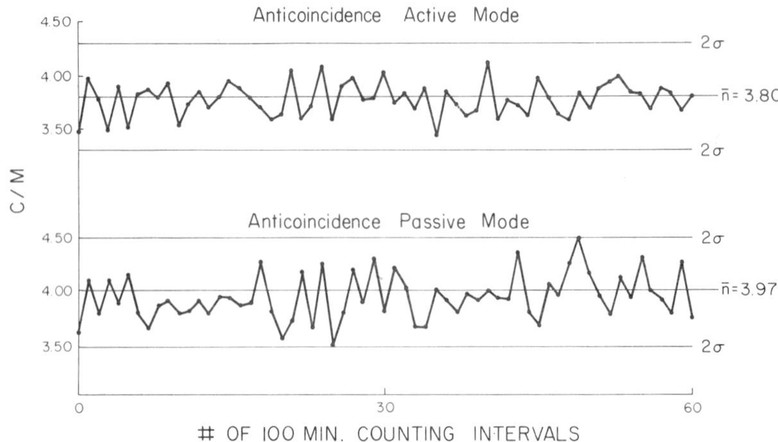

Fig. 5. LS counter background with side positioned electronic guard in the active and passive mode.

are reduced by shielding from cosmic radiation which arrives in bursts or showers. Figure 5 shows the results of two 4-day background measurements with a ^{14}C window setting on a 5 ml benzene-PPO sample with a side positioned anti-coincidence system in the active and passive mode.

To obtain the greatest advantage from electronic shielding for background reduction, the shield detector should be placed within the massive shield. Figure 6 shows a top view of a massive shield and guard detector in the process of fabrication. The dimensions of the shield are 5 ft long × 2 ft wide × 2 ft high with 6 in of Pb covering and a Cd–Cu inner lining. The guard detector shown is an NAI (T1) annulus of 8 in O.D. and 5 in long with a 3½ in dia. centre hole to accommodate a cylindrical vial holding up to 100 ml of sample. Counter background levels for ^{14}C-counting were less than 1 count min^{-1} using a 25 ml benzene-PPO sample, with electronic shielding accounting for 80% background reduction. E^2/B performance figures obtained were near 5000 for ^{14}C and 2000 for ^3H.

Fig. 6. Top view of massive shield and electronic guard detector in the process of fabrication.

PHOTOMULTIPLIER TUBES

In the last 25 years, LS counter development has relied quite heavily on the continuous improvement in superior PMT performance. As a result, today we have a number of excellent PMT's available which are designed specifically for the unique needs of LS counting. Two of the PMT's which are discussed below are typical of the ones found in today's LS counters.

One such PMT is the EMI 9634QB. This PMT has a diameter of 50 mm and is 125 mm long with a fused silica face of low ^{40}K content. It has a 13-stage venetian blind dynode construction with a Cs–K–Sb photocathode. It is operational over a temperature range from 60 °C to -5 °C which enables it to be used in refrigerated or ambient LS counters. Maximum spectral response is at 380 nm with a quantum efficiency of 28%. Operational voltage is 2500 V max. Another PMT of interest is the RCA 4501 of which there are several in the series. One of specific interest is the 4501–V4 PMT which has a convex-fused silica window and physical dimensions similar to the EMI 9635QB PMT. It is constructed with a K–Cs–Sb photocathode and a 12-stage in-line dynode string of which the first dynode secondary emitting surface is GaP. This PMT is also capable of either ambient or refrigerated operation with an operational voltage in excess of 2500 V. Maximum spectral response is 385 nm with a quantum efficiency of near 31%. Typical dark current noise (22 °C) is reported near 167 counts s^{-1}. Magnetic shielding of Mu metal or Mica C is utilised on both of these PMT's to prevent magnetic field electron focus distortion.

The utilisation of optimum performance PMT's is paramount to the successful achievement of low level LS counting. Selection of PMT type and performance can usually be made at the time of purchase if the instrument is to be new or if phototubes are to be replacements. The specifications should include the request for a matched PMT set having low dark current noise level and high quantum efficiency. One can also achieve the same goal if specific PMT specifications are lacking by requesting a counter performance parameter stated as E^2/B for a specific isotope and sample volume of interest, where E is counting efficiency in percent and B is background in counts min^{-1}. In the case of the purchase of a new commercial LS counter, the special low level specifications may be met with little or no extra charge depending on the zeal of the manufacturer to obtain the sale. The replacement of high performance PMT's in an existing LS counter will cost in the range of $ 1000 per matched set less old PMT exchange.

PMT background contribution can be divided between dark current noise or singles rate and cross talk. Dark current noise originates from the spurious release of electrons from the PMT photocathode or dynode string surfaces. PMT cross talk occurs when two PMT's are facing each other and are optically coupled as in the normal counting geometry, with an event in one PMT initiating a pulse in the other. Both of these contributions can be reduced by optimising the operational voltage of the PMT's to the radioisotopes one is counting.

Since ^{3}H is the lowest β-energy routinely counted in LS counting, and hence the most difficult to measure, all commercial LS counters are designed for achieving the highest counting efficiency for this radioisotope. In doing so the PMT's are usually made to operate at high voltages for maximum performance and by being so stressed enhance the singles rate and cross talk background contribution. If one is counting ^{3}H, then this operational voltage may be necessary, but if ^{14}C is of sole interest an advantage can be gained in background reduction without appreciable loss in counting efficiency by lowering the operational voltage in the range of 200 V.

A second procedure to specifically reduce cross talk can be implemented by taping the outer ¼ in peripheral photocathode face edge of the PMT with black tape as shown in Fig. 7. Most PMT's used in LS counters have their side walls taped to provide electrostatic insulation and to prevent light piping to the photocathode. By extending the tape over the edge of the PMT face, more efficient reduction in light piping is accomplished with only a small reduction in counting efficiency as the edge portion of the photocathode face is the least efficient face area of the PMT.

As an example of the performance figures that can be obtained from PMT's specially selected for low level counting, the data below was collected with a matched set of taped RCA 8850 PMT's operated at 1800 V. Both PMT's were measured for singles rate and found to be 4200 and 2200 counts min^{-1} respectively. Using a 50 ns coincidence gate, the coincidence singles rate contributing to background was measured at less than 1.2×10^{-2} count min^{-1}. With the two PMT's separated 1¼ in, a cross talk value of 0.850 ± 0.044 count min^{-1} was measured.

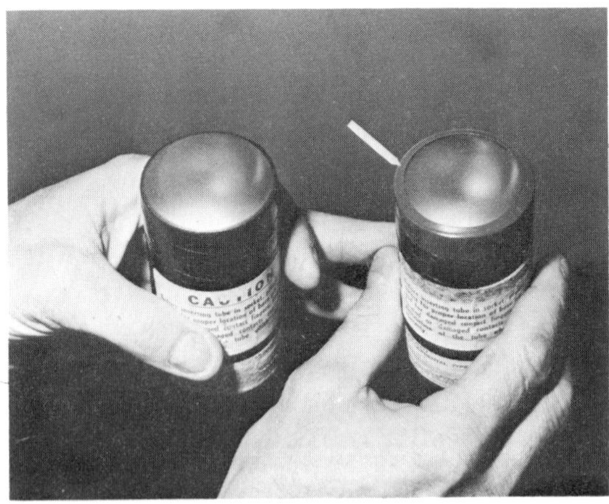

Fig. 7. Taped PMT face edge.

LS PULSE HEIGHT ANALYSIS

Since the earliest conception of LS counting, pulse height analysis (PHA) has been the major means for measuring both the decay energy and activity of a radioisotope within the liquid cocktail. Today's commercial LS counters usually employ up to three separate adjustable single channel pulse height analysers in order to make possible the simultaneous measurement of three different radioisotopes within a single vial. An additional two constant set single channel pulse height analysers are utilised where external standardisation is required for quench correction. The electronic means for PHA are either linear or logarithmic in response. The linear systems employ electronic circuitry such that the pulses generated are linearly proportional to the decay energy of the radioisotope, whereas the logarithmic systems are semi-logarithmic in response. Both types of PHA systems use fast coincidence gating of from 30–50 ns to lower background through reduction of the chance coincidence singles rate.

It is difficult to suggest any real major electronics alterations of the present-day commercial LS counters as they are all very finely engineered. However, if one does wish to enhance the instruments' capability for low level counting, the coincidence gate can be narrowed to 10–20 ns. Surprisingly enough, this action will not appreciably lower the counting efficiency of the instrument, but will measurably reduce the background. As a result, the E^2/B value for the instrument will be noticeably improved. A second suggested alteration would be to remove the radioactive source from the external standardisation system of the instrument if so equipped for quench correction. In most low level counting procedures, quench correction should not be a major concern since initial sample preparation should be selected so as to minimise this effect and most samples are compared to absolute standards. Removal of the source material can, in some LS instruments, reduce the background by as much as 2 counts min^{-1}. A third alteration would be to place the LS counter in an air-conditioned dehumidified room where a constant temperature somewhat below ambient temperature can be maintained. This is advisable for both ambient and refrigerated systems as the electronic circuitry will function in a more dependable and stable manner. The last suggestion would be to try and isolate the AC power source from any radio frequency (RF) interference and to minimise any sudden or intermittent voltage surges which can not only interrupt a long period count time but stress the total electronic system.

To achieve maximum LS counter stability during the long counting times required for low level counting or with minor variable sample quenching, the LS counter can be set up to operate in a balance point counting configuration. In this mode of operation, the discriminators, which form a window through which the scintillating events can be viewed, are set such that a major portion of the highest number of scintillation events are detectable. A small portion of events on the low and high energy sides is excluded under non-quenched conditions. Counting samples with minimum quench or at times when varying voltage fluctuations are present in the lines will cause the β-spectrum to shift to a low energy with quenching or lower PMT voltage, or higher with higher line voltage. In each case, regardless of whether there is a spectrum shift to the low or high side, a balance point configuration enables the count rate to be rather constant since the activity which is lost in one part of the spectrum due to spectral shift is added at the opposite end of the spectrum.

Balance point counting is only applicable to quench problems if they are of very minor importance or if a problem demands very stable counting characteristics of the LS counter, since this mode of counting necessitates a sizeable reduction in counting efficiency. In the balance point counting of ^{14}C one can usually realise a counting efficiency of only 60–70%

FAST PULSE-TIME LS COUNTING

As was stated earlier in the previous section, all commercial LS counters use pulse height discrimination as a means for utilising the coincidence counting technique. It is also possible to realise this same goal through time discrimination where fast pulse-time intervals are measured to achieve the coincidence mode of counting. This method has certain advantages over the conventional pulse height analysis systems, by providing lower component costs and achievement of a higher signal-to-noise ratio through more precise signal identification. The use of this instrumentation becomes even more practical in low level counting where single isotope counting does not require isotope identification through pulse height analysis.

Figure 8 shows a block diagram of such an LS counter with an anti-coincidence shield for low level counting. This instrument has been reported in detail by Noakes *et al.*[11,12] and therefore will be described only briefly.

A selected pair of matched RCA 8850 PMT's were used for the LS counting as well as four RCA 6199 PMT's for the NAI (T1) crystal shield. All additional components other than the PMT's were selected from NimBin electronics. The two preamplifiers (PA's), one for each PMT, are used for impedance matching to each delay line amplifier (DLA). The PA output is directed to two DLA's to provide for output pulse shaping and amplification. A dual sum and invert amplifier (DSI) collects DLA output pulses for recorder display. The DLA pulses are also collected by two timing single channel analysers (T-SCA) which initiate a fast logic pulse of less than 5 ns rise time. A delay setting of the two T-SCA's provides two of the inputs for the timing pulse height converter (T-PHC) such that if two pulses leave the DLA's simultaneously the resulting T-SCA output pulse will be separated by 35 ns. The T-SCA pulse with the least delay serves as a start pulse for the T-PHC and the other pulse serves as a stop gate. If the time interval separating the start and stop pulse

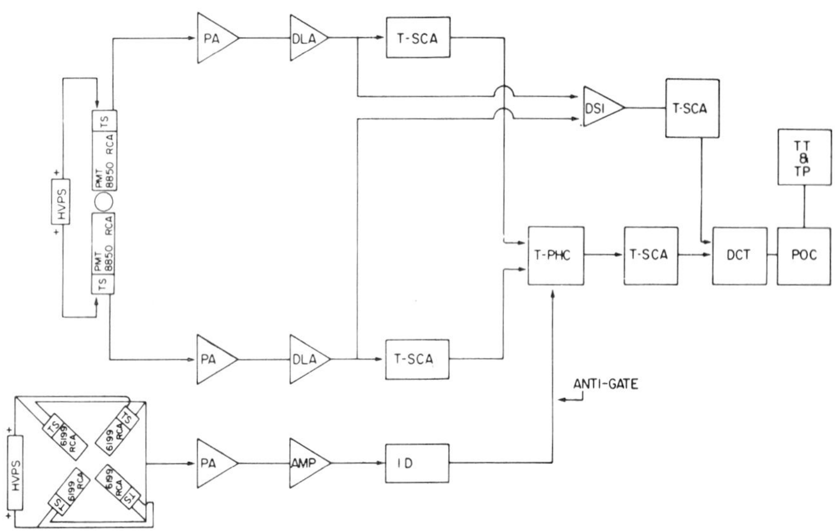

Fig. 8. Fast pulse-time measurement LS counter.

is less than the selected gate range, an output pulse is generated. If, however, a stop pulse arrives after the set gate range, no output pulse will be generated. Also, if an output pulse from the anti-coincidence shield arrives during the gate range, no pulse will be generated. The T-PHC puts out a bipolar pulse proportional to the arrival time of the two T-SCA's. A third T-SCA performs time–pulse height discrimination which is fed to a dual counter-timer (DCT) operated by a print-out controller (POC) and tabulated by teletype and punch tape (TT&TP).

The performance capability of the instrument is shown in Table 3, using a 25 ml benzene sample with a PPO scintillator.

Table 3. Performance capabilities of a fast pulse-time measurement LS counter.

Window (%)[a]	Background (cpm)[b]	Efficiency (%)[c]	E^2/B[c]	E^2/B per ml[c]
100.0	4.20 ± 0.5	96.0 (64)	2194 (975)	100 (44)
95.0	3.55 ± 0.41	91.2 (61)	2332 (1041)	106 (47)
88.8	2.11 ± 0.40	85.2 (57)	3440 (1540)	156 (70)
75.9	1.22 ± 0.26	72.9 (48)	4356 (1889)	198 (86)
64.7	0.79 ± 0.24	62.1 (41)	4882 (2170)	222 (98)
55.3	0.71 ± 0.24	53.1 (29)	3971 (1764)	180 (80)

[a] Relative % efficiency.
[b] All errors expressed in cpm at 2% confidence level.
[c] Numbers in parentheses are for ^3H; those without parentheses are for ^{14}C.

PULSE SHAPE LS COUNTING

The application of pulse shape analyses (PSA) to LS counting has only recently been reported in the literature. McDowell *et al.*[13,14] have applied this method to low level α LS counting where ^{239}Pu was counted at 100% efficiency with a background of 0.01 to 0.005 count min^{-1}. Separation of the α-radiation from the background of β, γ and cosmic radiation was accomplished through preferential measurement of the long-term α-decay component of the liquid scintillator. Sipp and Miehe[15] have also used PSA for investigating the fluorescence self-absorption and time resolution in liquid scintillators. This study has verified that differences in scintillation pulse shape can be measured where the radiation interaction probability is constant over the scintillator as in γ-rays and where the interaction is localised as in α- and β-rays. From the above-mentioned work, it is evident that more interest in the future will be directed to PSA for LS counting.

SUMMARY

Low level LS counting requires that the most rigorous standards be maintained in sample preparation, instrument stability and low background operation. In order to achieve these goals, one must pay particular attention to sample chemistry and cocktail selection in order to minimise quench and obtain the highest counting efficiency. The use of special low volume vials, fabricated for positive seal and with material of low cocktail permeability and low natural radioactivity, is one of the most economical means for lowering background and reducing sample loss during long counting periods. Acquisition of select high quantum yield low noise PMT's is an excellent way to achieve high counting efficiency and lower electronic-generated background. Radiation shielding both by massive and electronic guards is the most important means for reducing background caused by external radiation sources. Pulse height discrimination and fast pulse time circuitry can both be applied with fast coincidence counting for background reduction. PSA can be applied to β, γ and β–γ counting and as a means for background reduction.

REFERENCES

1. R. F. Nystrom, W. H. Yanks and W. F. Brown, *J. Amer. Chem. Soc.* **70**, 441 (1948).
2. B. N. Audric and J. V. P. Long, *Nature (London)* **173**, 992 (1954).
3. B. L. Funt, S. Sobering, R. W. Pringle and W. Tarchinetz, *Nature (London)* **175**, 1042 (1955).
4. J. R. Arnold, *Science* **119**, 155 (1954).
5. C. Leger, G. Delebrias, L. Pichat and C. Baret, *Commissart A. L. Energy Atomic Rapport* No. 1075 (1958).
6. M. A. Tamers, *Science* **132**, 668 (1960).
7. J. E. Noakes, S. M. Kim and J. J. Stipp, *Int. Conf. Radiocarbon Dating and Tritium Proc.* (USAEC Cont. 650652), Pullman, Washington, 1965, pp. 68–92.
8. D. L. Horrocks, *Applications of Liquid Scintillation Counting,* Academic Press, London and New York, 1974, pp. 35–64.
9. G. E. Calf and H. A. Polach, in *Liquid Scintillation Counting – Recent Advances* (eds. P. E. Stanley and B. A. Scoggins), Academic Press, London, 1974, pp. 223–34.
10. P. E. Hartley and V. E. Church, in *Liquid Scintillation Counting – Recent Advances* (eds. P. E. Stanley and B. A. Scoggins), Academic Press, London, 1974, pp. 67–76.
11. J. E. Noakes, M. P. Neary and J. D. Spaulding, *Nucl. Instrum. Methods* **109**, 177 (1973).
12. J. E. Noakes, M. P. Neary and J. D. Spaulding, in *Liquid Scintillation Counting – Recent Advances* (eds. P. E. Stanley and B. A. Scoggins), Academic Press, London, 1974, pp. 125–37.
13. W. J. McDowell, J. H. Thorngate and D. J. Christian, *Health Phys.* **27**, 123 (1974).
14. W. J. McDowell, *IEEE Transactions in Nuclear Science* **NS-22**, 649 (1975).
15. B. Sipp and J. A. Miehe, *Nucl. Instrum. Methods* **114**, 255 (1974).

DISCUSSION

R. Evans: If Professor Noakes would care to expound further on the present position and further possibilities of pulse shape analysis, I, for one, and, I suspect a good number of others, would be very interested to hear him.

J. E. Noakes: I'm afraid I don't have much more information on this recent technique but I can confirm that it allows differentiation of α-, β- and γ-pulses.

D. Case: I feel that lesser mortals doing lesser things are astonished when they realise how stringent the hygiene factors must be in your laboratories and I wonder if you would comment on this. Second, it would be interesting to hear more of the actual count rates you are dealing with in experimental samples.

J. E. Noakes: Most samples we count in our laboratory relate to radiocarbon dating. Surprisingly, our primary concern with these submitted samples is not contamination but that the samples are truly representative of the material to be dated. Sample contamination of foreign carbon can occur both during field collection and in laboratory processing. An awareness of this hazard is the best means for eliminating it.

Samples of modern carbon origin which possess a specific activity of approximately 14 dpm per gram carbon represent the highest count rate encountered in radiocarbon dating. Too often, when a sample is of extensive age, we are forced to place high credence in count rates of a few tenths to hundredths of a count per minute. In order to carry out these measurements with a high degree of accuracy all counting is done at a 95% confidence level which necessitates extensive counting times and extreme stability of the LS counter.

R. P. Parker: Would Professor Noakes – and the Chairman – comment on the relative merits of LS counters and gas counters for low level work?

J. E. Noakes: In the past it has been assumed that gas counters possessed superior counting capabilities when counting low activity samples. This assumption has arisen from the fact that comparisons were always made between commercial LS counters and specially built gas counters. If one makes a comparison between two specially built LS and gas counters with comparable state of the art electronics and massive and anti-coincidence shielding, one finds that the counting characteristics such as counting efficiency, background and stability are very comparable. In the case where large sample material is available for counting, the LS counter has a decidedly higher sensitivity due to its ability to accommodate and count more sample thereby yielding a high count rate/unit time of counting.

R. Burleigh: Many, perhaps even most, laboratories setting out for the first time on low level work or those replacing worn out existing equipment would now opt for LS counting rather than gas counting. There are several practical reasons for this; for example, the greater control afforded by LS counters in which standards and background samples can be directly alternated with the unknowns allowing much more continuous monitoring of counter stability. Also the need for massive shielding is avoided. Gas counter shields for low level work have not only themselves to be free from traces of radioactivity, implying the need for aged (or non-radiogenic) lead or pre-Atomic age steel, but are generally so massive as to pose considerable structural problems in the buildings in which they are housed. The chemistry of sample preparation requires a comparable expenditure of effort in both cases, but LS samples once prepared can be much more easily stored for possible remeasurement. One could go on to list several other advantages of the LS counter. For radiocarbon dating the cut-off is determined by sample age. Samples near the far age limit of the method ($\sim$40,000 y) or very small samples ($\ll$1 g C) are still best measured by gas counting. However, most laboratories normally work in the range where commercially available LS counters can be used 'off the shelf' without modification and the fact that this is so, whereas gas counters suitable for low level work have to be specially constructed, is also an important consideration. Finally, experience shows that for a number of reasons such as rapid sample changing and as a result more continuous counting, the output of measurements from the LS counter is higher than that of the gas counter.

F. H. Kendall: With respect to diurnal variation due to radon in granite areas, our experience shows that use of flushing gas (N_2 or air) improves stability and possibly reduces background very slightly, e.g. by 0.2 count min^{-1}

D. Bowyer: How important is cross talk between PMT's in determining backgrounds? What methods might be used to further reduce backgrounds from this source?

J. E. Noakes: In low level counting where other factors such as external radiation, etc. have been already reduced by shielding or electronic means, cross talk can account for a large part of the remaining background of the counter. There are several means for reducing this background contribution. One is to tape the outer peripheral edge of the PMT's face thereby restricting the light piping effect of the glass envelope of the PMT's. A second means is to use a counting vial designed so as to restrict the viewing of the two PMT's to just the sample to be counted.

B. H. Laney: Since commercial LS spectrometers are designed to measure a diversity of samples over a wide range of energies, conditions may be less than optimum for a specific application. I concur with your recommended changes when the instrument is to be used exclusively for ^{14}C-dating. Operation of the phototubes at a low voltage (less than 1800 V) will eliminate potential electroluminescence. Taping of the perimeter of the photocathode about 1 cm will reduce phototube cross talk. However, cross talk can be eliminated electronically without degrading tritium efficiency, by the method I described in Chapter 5.

Events caused by phototube cross talk may also be eliminated by increasing the coincidence threshold. As shown earlier, increasing the lower level eliminates cross talk events when lesser PHA, but not summed PHA, is employed.

Removal of the external standard source and addition of shielding above and below the elevator are effective measures to reduce background induced in the sample.

Chapter 16

Evaporation Losses of Organic Samples from Liquid Scintillation Counting Vials

R. Burleigh and A. Hewson

Research Laboratory, The British Museum, London, England

INTRODUCTION

There is a wide variety of liquid scintillation counting vials at present in use and their characteristics have been reviewed both in this country and abroad.[1,2] The choice of vial for a particular application is dependent on factors such as background count rate, efficiency and cost. One area which has not been subject to much investigation is the rate of loss of material from the vials during the period for which the sample is counted. In many applications this is unimportant as counting times are limited to a few minutes or perhaps hours. However, in low level applications, counting may extend over much longer periods and, moreover, standards must be kept, semi-permanently, in vials identical with sample vials. This is certainly true of the British Museum Radiocarbon Dating Laboratory where typical samples may be counted intermittently over a period of several weeks and standards may be kept for a year or more. Detailed descriptions of our methods have been given elsewhere.[3,4]

With this in mind we embarked on a series of experiments to evaluate three vial designs (Fig. 1) which have each been used for low level ^{14}C measurements. Design A, in everyday use at the British Museum Laboratory, is the standard 20 ml, low-potassium borosilicate glass vial as supplied by several manufacturers. Design B, developed by the Low Level Measurements Laboratory, Harwell, uses the standard vial but a different closure described by Otlet and Slade[5] of Harwell in these words: 'A good seal is now achieved with an indium foil washer (0.008 cm) held firmly to the specially ground vial top by a rubber washer (0.165 cm) used in place of the usual cork/aluminium pressure pad. A Teflon washer (0.026 cm) in between prevents damage to the indium when tightening the screw cap.' Finally, design C is a Teflon vial sealed with a Viton 'O' ring held in an aluminium cap and was developed in Australia by Calf and Polach[6] specifically for low activity ^{14}C samples.

EXPERIMENTAL WORK

As a pilot experiment, six standard vials (design A) were taken and filled with 5 ml of benzene and 10 ml of toluene. These quantities were chosen to approximate

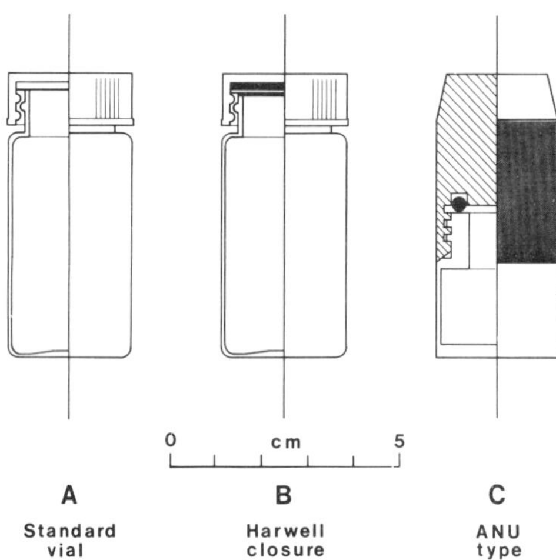

Fig. 1. Counting vials used for low level ^{14}C measurements. (Copyright: British Museum.)

the counting cocktail in use at the British Museum Laboratory and were adhered to in all subsequent evaporation measurements. Five vials were cooled to 0 °C by placing in the sample changer of the laboratory's Packard 3315 counter and the sixth was left in the open laboratory at about 20 °C. They were then weighed at intervals and the results were as summarised in Tables 1(a) and 1(b). These indicated that solution loss was occurring at a rate of about 0.5 mg per day at 0 °C which if left uncorrected in a sample counted for one month would in a typical case give rise to an extra

Table 1(a). Cap weight increase (mg).

	Days				
	4	12	35	104	769
Mean (0 °C)	12.1	22.2	45.0	52.9	72.9
S.D.	1.2	2.0	9.1	4.1	6.9
Mean (20 °C)	2.2	3.2	5.0	16.7	69.0

Table 1(b). Solution loss (mg).

	Days				
	4	12	35	104	769
Mean (0 °C)	2.9	13.7	24.7	45.4	159.7
S.D.	0.4	6.9	10.6	22.0	33.5
Mean (20 °C)	9.4	20.1	57.2	147.6	1051.1

Table 2(a). Cap weight increase (mg).

	Days												
	2	3.8	8.7	14.8	20.7	36	39	42	52.8	70	72	77	79
Mean (0°C)	2.5	4.2	9.3	19.2	17.9	24.3	27.4	26.2	29.0	34.3	34.6	34.7	34.6
S.D.	2.0	2.3	2.9	3.5	3.2	3.5	6.8	4.0	3.5	3.3	3.3	3.2	3.2
Mean (20°C)	2.2	5.0	2.8	0.2	-3.1	-0.5	4.9	3.5	2.6	14.7	12.9	10.7	9.3
S.D.	0.7	4.7	0.6	0.4	0.3	1.5	9.7	1.3	0.4	4.1	0.5	0.4	0.3

Table 2 (b). Solution loss (mg).

	Days												
	2	3.8	8.7	14.8	20.7	36	39	42	52.8	70	72	77	79
Mean (0°C)	7.9	9.1	11.4	23.0	28.4	32.6	35.9	37.7	51.2	60.0	61.0	62.1	64.6
S.D.	8.2	8.4	8.3	15.3	19.2	20.3	17.8	16.5	17.7	19.3	18.8	19.0	19.4
Mean (20°C)	10.6	19.2	33.4	41.1	47.8	67.9	78.7	86.8	103.0	126.0	135.0	144.0	147.0
S.D.	16.3	22.8	23.4	25.6	27.5	31.6	25.2	23.3	24.4	33.9	30.9	31.6	31.8

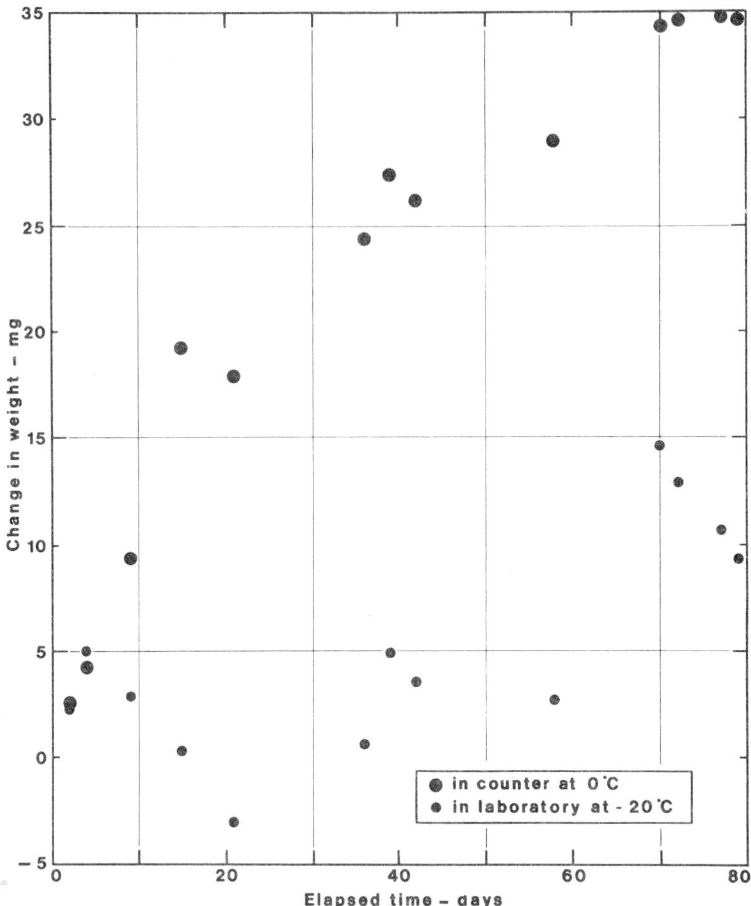

Fig. 2. Change in weight of standard vial caps. (Copyright: British Museum.)

radiocarbon date measurement error of about 10 years. However, two other features were also apparent: (i) the standard deviation of the amount of solution loss increased with time, indicating that some vials consistently lost more material than others; (ii) the vial caps increased in weight during the course of the experiment and it is conceivable that a substantial proportion of the material lost from the solution was being absorbed by the cap, although the rate of increase of the cap weight was not maintained for the whole of the experiment.

To investigate this more thoroughly the experiment was repeated with ten samples, five maintained at 0 °C and five at 20 °C, and more frequent weighings. The results are presented in Tables 2(a) and 2(b) and Figs. 2 and 3 and it is clear that for the samples at 20 °C, solution loss is considerably greater than the cap weight increase. Preliminary results of experiments at present in progress indicate that the increase in cap weight for samples at 0 °C is caused by the high humidity in the Packard counter. In the early part of this experiment, several vials lost weight at a rate substantially in excess of the trend before settling down. This appeared to be due to a phenomenon which Otlet[7] has christened 'ratcheting', where the vial cap eases open sometime after it is originally tightened.

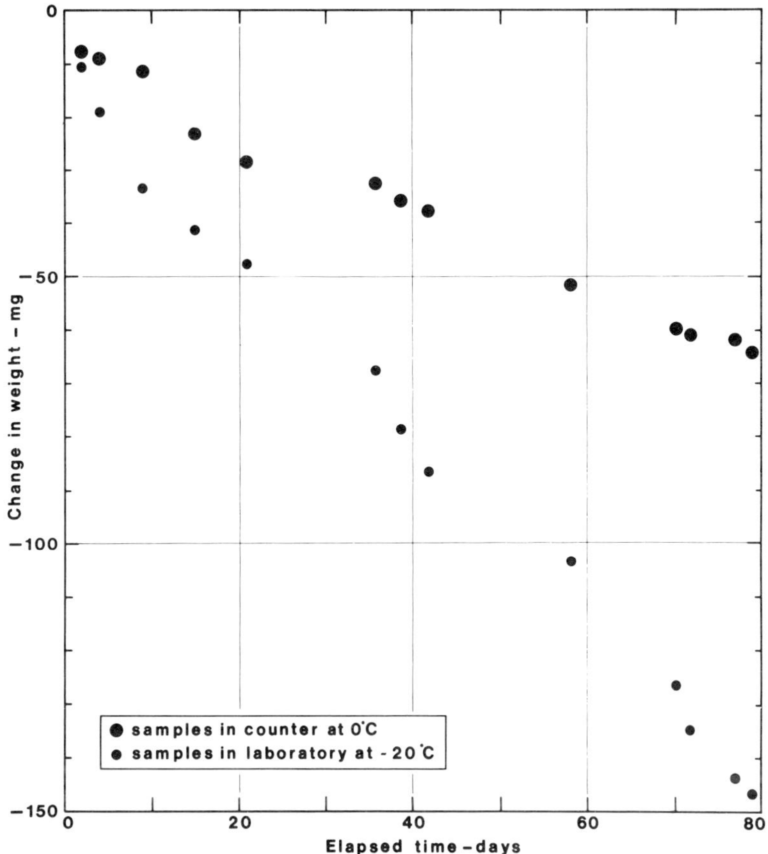

Fig. 3. Sample weight loss from standard vials. (Copyright: British Museum.)

Next the vial caps from the samples maintained at 0 °C were removed and allowed to stand in the laboratory for several weeks, their weights being remeasured at intervals. The results (Table 3) showed that the caps lost about 50 mg in 3 weeks and then hovered around their new weight.

Finally, five each of designs A and B, and the single example of design C available to us, were each filled, sealed and left in the counter for 28 days. On reweighing (Table 4) the Harwell-developed design B proved to have performed best with a mean loss of 0.7 mg which was substantially better than either of its rivals. The result for design C was disappointing, particularly in view of the claims of Calf and Polach that

Table 3. Cap weight loss in laboratory (mg).

	Hours								
	22.6	65	142	185	331	472	591	723	840
Mean	13.6	22.4	28.7	34.1	42.7	48.8	47.6	35.9	49.4
S.D.	3.9	3.9	4.8	4.5	4.5	5.0	5.0	7.2	4.9

Table 4. Solution loss (mg).

	Design		
	A	B	C
Mean	11.3	0.7	13.5
S.D.	2.9	0.6	—

the loss rate should be 0.5 mg per week at 18 °C. It was noticeable that the 'O' ring increased in weight by 7 mg during the experiment and the solid aluminium cap by 3 mg. This is consistent with the results for an adapted cap produced in the British Museum Laboratory which was sealed by a Viton 'O' ring held in an aluminium plug. The leakage rate for this design proved to be unacceptable and incidentally the 'O' ring and the aluminium plug were observed to gain weight in a similar manner to design C.

CONCLUSIONS

The sealing qualities of the standard vials (design A) are not entirely satisfactory over long periods due to variable loss rates and the risk of occasional more severe leakage. The overall weight of a sample sealed in a vial is an unreliable guide to the leakage rate as the cap weight changes with environment. Design B, although a somewhat more cumbersome system, provided a highly effective seal but it is possible that the indium itself contributes to the background count rate.[a] Possibly the best arrangement for low level ^{14}C counting would be a Teflon vial with its advantages of low background and high counting efficiency sealed in a manner similar to the Harwell design. It is our intention to produce and test such a vial in the near future.

ACKNOWLEDGEMENTS

We would like to thank Mr. R. L. Otlet of AERE Harwell for useful discussions and for kindly allowing us to use his vial sealing system in our experiments.

REFERENCES

1. K. Painter, in *Liquid Scintillation Counting – Recent Developments* (eds. P. E. Stanley and B. A. Scoggins), Academic Press, London, 1974, p. 431.
2. K. Painter and M. J. Gezing, in *Liquid Scintillation Counting,* Vol. 3 (eds. M. A. Crook and P. Johnson), Heyden, London, 1974, p. 34.
3. R. Burleigh, in *Liquid Scintillation Counting,* Vol. 2 (eds. M. A. Crook, P. Johnson and B. Scales), Heyden, London, 1972, p. 139.
4. R. Burleigh, in *Liquid Scintillation Counting*, Vol. 3 (eds. M. A. Crook and P. Johnson), Heyden, London, 1974, p. 295.
5. R. L. Otlet and B. S. Slade, *Radiocarbon* **16**(2), 178 (1974).

[a] ^{115}In, 96% natural abundance, 6×10^4 years half-life, emits β-particles with an energy of 0.6 MeV.

6. G. E. Calf and H. A. Polach, in *Liquid Scintillation Counting – Recent Developments* (eds. P. E. Stanley and B. A. Scoggins), Academic Press, London, 1974, p. 223.
7. R. L. Otlet, personal communication.

DISCUSSION

C. McEvoy: At the A.R.C. Institute in which I work, we have completely sealed glass vials in which we wish to keep standards; thus we have no diffusion loss as the entire vial is sealed glass. Would this be a better way of sealing your samples and keeping your standards?

A. Hewson: These would not be suitable for our particular application as the vials used are selected for their low background and are re-used.

D. I. Chapman: Your said that a loss of 50 mg from vials in 100 days was equivalent to an error of 10 years in 5,000 years in your dating procedures. Since you consider it important to reduce this loss, are you therefore claiming an accuracy of 10 years when dating a sample about 5,000 years old?

A. Hewson: No, this is a 10 year systematic error in addition to the random measurement error of typically ± 40 years.

Chapter 17

Desktop Calculator-Assisted Liquid Scintillation Spectrometry of ^{3}H and ^{14}C with Special Consideration of Low Level Activities and Spectrometer Instability

P. Jordan and K. May

Radiochemical Laboratory, Swiss Federal Institute of Technology, Zurich, Switzerland

SUMMARY

For the liquid scintillation measurement of ^{3}H and ^{14}C in single- and dual-labelled samples with the same spectrometer settings, which are desirable for routine work, the lower discriminator level of the ^{14}C channel has to be set down to a value for which the tritium crosstalk α amounts to about 15%. Under these conditions, the short time variations of α are important for the overall attainable precision. A simple method for the automatic computation of drift corrections based on the continuous measurement of a standard sample is presented. Furthermore, for low activities an appreciable improvement of the precision may be achieved by systematic re-use of the glass vials together with the cumulated measurements of their individual backgrounds. The precision of low activity measurements is also enhanced by correction of background counting rates for quenching.

In order to simplify operating conditions in routine work, it is advantageous to use the same measuring channels for single- and dual-labelled samples. In doing so a further advantage is gained: All single-labelled samples are also treated as dual-labelled and thus cross contamination is automatically detected. The considerations which follow were developed in order to meet this experimental situation; they apply to LSC coincidence spectrometers with external standardisation. Three "Mark I" (Nuclear Chicago, Searle) spectrometers with Ba-133 external standard were at our disposal. They showed very similar, but not identical, characteristics, and were operated with channel settings chosen to make them interchangeable in routine work, with proper consideration of their individual response at the data reduction stage. Data output was by teletype (TTY) on punched tape. Calculations were done off-line using a programmable calculator (Hewlett Packard 9820A) equipped with tape reader, plotter and TTY with tape punch. The improvements suggested are, however, not restricted to this instrumental set-up.

A general study of the overall efficiency, taking into consideration the prevailing working conditions (relative quota of samples of different types, as $^{3}H/^{14}C$, single/double, high level/low level, and so on) as well as basic principles of counting statistics, shows that, in order to optimise the output of the laboratory, about 15% of the ^{3}H-activity has to be admitted in the ^{14}C-channel of the spectrometers. On the other hand, as the required long-term consistency of the measurements is set at a fairly high level, say 2% for the r.m.s.-deviation of single measurement, it ensures that the instrumental parameters f_T (^{3}H efficiency in ^{3}H-channel), f_C(^{14}C efficiency in ^{14}C-channel), α (^{3}H in ^{14}C-channel/^{3}H in ^{3}H-channel) and β (^{14}C in ^{3}H-channel/^{14}C in ^{14}C-channel), which are functions of γ (external standard in standard channel/external standard in ^{14}C-channel), must always be known precisely. Thus, the fitting of analytical functions to the experimental data with a sufficiently high degree of approximation necessitates the use of polynomials of up to the seventh degree. Under such conditions, special care has to be taken in view of the short-term instabilities of the spectrometers, which are variations of gain and discrimination, and affect mainly the stability of α. A constant re-evaluation of this parameter would represent a prohibitive amount of work. It appeared, however, that the variations of α may be satisfactorily corrected by an appropriate shift of the whole α-curve along the γ-axis (Fig. 1), which constitutes an important simplification of the situation. In practice, the correction method consists of determining the value $(\alpha, \gamma)_i$ of a ^{3}H standard together with each group of samples to be measured. The calculator automatically finds by iteration the shift $\Delta\gamma_i$ of the function α which corresponds to the

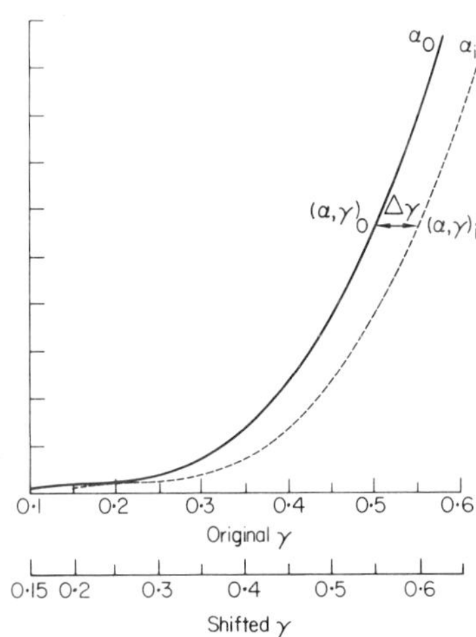

Fig. 1. Correction for drift of α.

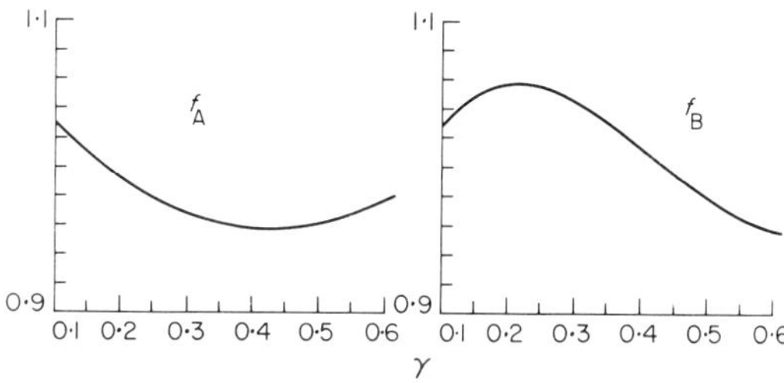

Fig. 2. Quench effect on background.

actual state of the spectrometer. This method of correction requires a minimal amount of work and measuring time. The component of error eliminated in this way amounts typically to about 1.5%.

In the search for further possibilities to improve the measurement of low activities by simple means, two points were found, by which some more precision may be gained. A first improvement consists in taking into account the dependence of the background on the degree of quenching of the samples. Figure 2 shows typical curves for the background in the ^{3}H- and ^{14}C-channels. A second improvement may be achieved by re-using systematically the same glass vials and calculating cumulated background mean values for each vial, which tend asymptotically to constant figures of higher statistical weight. The increment of precision gained thereby amounts typically to about 1%. Detailed computation programs are available upon request.

SECTION V

ERRORS IN LIQUID SCINTILLATION COUNTING

Chapter 18

Accuracy and Merit in Liquid Scintillation Counting

Lloyd A. Currie

Analytical Chemistry Division, National Bureau of Standards, Washington, U.S.A.

INTRODUCTION – STATES OF THE ART

The practical accuracy of any analytical method is revealed when a number of laboratories cooperate in the analysis of a common (standard) sample of complex composition. Such was the case, for example, in a recent intercomparison organised by the National Bureau of Standards, for the analysis of calcium in human serum. Two sets of results are shown in Fig. 1. Those to the left of the diagram were obtained by laboratories using a standard reference material (SRM) and adhering to the carefully defined protocol of the reference method (RM) for calcium.[1,2] When the same sample was sent 'blind' to hospital and commercial laboratories, the results to the right of Fig. 1 were obtained: three-quarters of the (12) results were in error by more than the desired medical limit of uncertainty, and one-third of these could have resulted in dangerous mis-diagnoses. Upon investigation, it was learned that the relatively large errors were connected with factors other than the particular method of analysis employed.

This example is representative of what one finds also in the field of radiochemical analysis and liquid scintillation counting. The precision and accuracy of measurements on difficult samples – such as complex biological materials or those containing trace levels of radioactivity – tend to be far more dependent upon the skill and knowledge of the scientist than upon the specific analytical method. Regrettably, erroneous results for just such samples are likely to have the greatest practical consequence.

In order to assess the current performance in the field, the literature was examined for actual limits of intra- and interlaboratory errors as deduced from replication, analyses of standards and intercomparison exercises. As shown in Fig. 2, it was necessary to partition the measurements into three broad classes according to skill of the scientist ('laboratory') (*and* the quality of his equipment), the complexity of the sample and the level of radioactivity. The relative error is seen to vary by about a decade from class to class, and the inaccuracy, as measured by the interlaboratory errors, is many-fold greater than the imprecision. The importance of Fig. 2 is that it indicates what is currently possible with liquid scintillation counting – i.e. class 1 (exemplary) – given sufficient skill and care; at the same time, it indicates what is

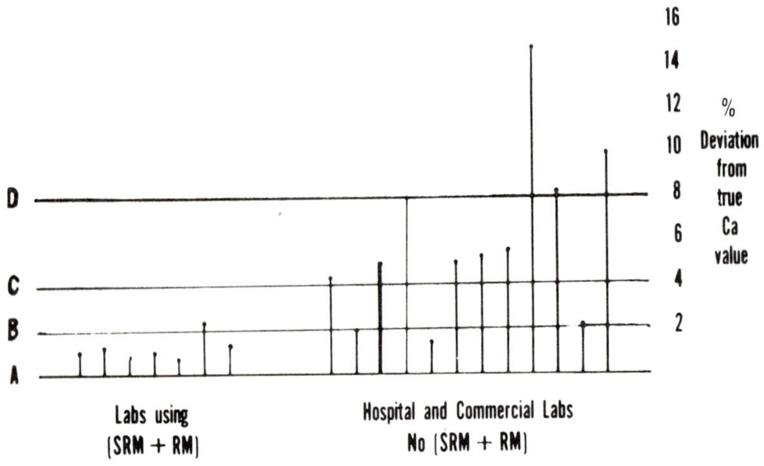

Fig. 1. Calcium in human serum.[1,2] Results obtained in connection with the development of a referee method by the National Bureau of Standards. Seven results to the left of the diagram were obtained through the use of the referee method (RM) and a standard reference material (SRM); those to the right (same sample) involved neither.

actually achieved under less optimal circumstances. The potential benefits of paying close attention to careful sample preparation and possible sources of error – to be discussed below – are thus made manifest; and the danger of basing uncertainty intervals and performance (merit) wholly upon Poisson counting statistics becomes evident. Illustrations of the three performance classes will be given in the following paragraphs.

Vaninbroukx and Stanef[3] reviewed results of high accuracy liquid scintillation counting (*class 1*), typically as performed by national standardising laboratories. These authors quoted overall calibration errors (inaccuracies) ranging from 0.1% for α-emitters to 1–3% for low-energy β-emitters. Imprecision was generally two to ten times smaller. Typical of *class 2* are the many laboratories engaged in careful measurements of ^{14}C or ^{3}H in natural materials at low and intermediate levels. For example, an IAEA-sponsored intercomparison (35 laboratories) of tritium-spiked water samples (10–250 TU) yielded relative standard deviations ranging from 2.6% to 25%, depending upon the concentration and method employed: gas or liquid scintillation counting, with or without electrolytic enrichment.[4] A significant conclusion from the study was that a residual error of about 3.5% arose from the enrichment process itself. The importance of intercomparisons and standard sample analysis for radiocarbon ages was cited by Kim,[5] who reported a relative error of 5% from the analysis of 80 interlaboratory cross-check samples. Inter- and

Laboratory; Procedure	**Sample**	**Relative Error** *intralab*	*interlab*
Exemplary	Pure radionuclides ideal preparations $(S \gg B)$	$\stackrel{\sim}{<}$0.3%	$\stackrel{\sim}{<}$2%
Experienced	Restricted nuclides and matrices, careful preparation	$\stackrel{\sim}{<}$5%	$\stackrel{\sim}{<}$20%
Less experienced	Varied and mixed nuclides, complex matrices $(S \stackrel{\sim}{<} B)$	$\stackrel{\sim}{<}$50%	$\stackrel{\sim}{<} 10^2$

Fig. 2. States of the art. Liquid scintillation performance classes.

intralaboratory variations in measurements of natural radiocarbon were evaluated also by Polach,[6] who demonstrated the significantly greater dispersion for comparative measurements *among* laboratories – apparently due to contamination and isotopic fractionation during sample preparation.

The results of routine inter- or intralaboratory comparisons should not be interpreted as limiting. When higher accuracy is demanded, it may be met by taking special precautions: Glass[7] noted the achievement of better than 0.5% accuracy ('95% confidence limits') for industrial ^{14}C and ^{3}H measurements, and Downes[8] succeeded in achieving 0.4% agreement among relative measurements in which ^{14}C was employed to determine the mean diameter of wool fibres.

Radiochemical measurements at low levels in complex matrices (*class 3*) present difficulties, particularly when several radionuclides are present and when less experienced workers participate. Because of the importance of such measurements, however, a number of intercalibration exercises have been sponsored by the International Laboratory of Marine Radioactivity of the International Atomic Energy Agency. Though the results of these exercises may appear discouraging, the chief objective is being accomplished: education and improvement of the performance of the participants. One example, taken from one of the earliest exercises,[9] concerns the measurement of ^{106}Ru (and other fission products) in seawater. The 22 positive results which were reported for sample SW-1-2 ranged from 0.3 to 62 pCi/kg! Restricting one's attention to the 16 laboratories reporting relative uncertainties smaller than 10% narrows the range to about a factor of six. (Serious conceptual or experimental systematic errors on the part of four of the participants is believed to account for most of this residual range.)

The foregoing examples demonstrate the importance of adequate internal and external quality control (QC) in order to achieve and maintain the desired level of accuracy. External QC, in the form of 'blind' measurements of standard materials and interlaboratory comparisons, provides a direct measure of practical accuracy in the field.[10,11] Internal QC, on the other hand, is a prerequisite for the production of stable, meaningful measurements by any laboratory. Both are essential for protection against blunders (mistakes in experiment *or* theory) which may occur at any step of the measurement process.

The great utility of control chart techniques for internal QC in liquid scintillation counting is well illustrated in the article by Wyld.[12] Redundancy can play a very important role in signalling systematic error, as shown by Glass[7] and Mueller[13] through the use of the double ratio technique. Computation of the sample counting efficiency via two independent measures – sample channels ratio (SCR), and automatic external standard counts (AES) or external standard channels ratio (ESCR) – provides protection against occasional blunders (misprints) as well as systematic error (assumed homogeneity or electron density). Redundancy (replication) is important also with respect to random error. Many a pitfall awaits those who would assume that random error is determined strictly by Poisson counting statistics.[14]

In the following two sections we shall review some of the primary sources of error in liquid scintillation counting and examine the question of merit, particularly as related to optimisation and the selection of alternative counting systems. In addition to specific references which will be given in the next section, there are a number of texts and review articles which treat the problem of liquid scintillation errors. Apart from the basic text by Birks,[16] most of the volumes are proceedings of international conferences which have been held more or less annually since 1969.[16–21] The volume edited by Bransome[15] has been especially helpful because of its subdivision in accordance with the various aspects of the measurement process. Major sections on 'The Labeled Sample' and 'Quenching', for example, contain a wealth of information about individual sources of error. Excellent summaries and guides to the literature are found in the reviews by Parmentier and ten Haaf[22] and Vaninbroukx and Stanef.[3]

THE MEASUREMENT PROCESS; SOURCES OF ERROR

A systematic assessment of the physics and chemistry underlying the overall liquid scintillation measurement process would be required to reveal all potential error sources. Such an assessment is complementary to analytical quality control (replication, measurement of standards, intercomparison exercises, etc.) in that causative factors would be identified and to some extent regulated. Principal steps of the measurement process include *sampling, sample preparation* – including chemical or physical purification and final preparation for measurement – *instrumental measurement,* and *data reduction and reporting.* Throughout, one must face the problem of *stability*: stability of the measuring device, stability of the sample and stability of the background. Obviously, several of the measurement steps are not peculiar to liquid scintillation counting, but they must nevertheless be considered if one is concerned with overall accuracy. Poisson counting statistics, similarly, is common to all methods of radioactivity measurement (pulse counting); and for all, it sets the lower limit to overall imprecision.

Rather than attempt an assessment of errors in each step of the measurement process, we shall, in the following text, focus upon the special problem in liquid scintillation counting: the fact that the sample itself is an integral part of the measurement device. The chemical and physical state of the sample, as well as its stability, can thus exert a profound effect on the instrument calibration (detection efficiency) and the background rate. The central issue of errors in the measurement step will therefore be discussed, after some brief comments on sampling and preparation.

Sampling and sample preparation

Sampling errors set a basic limit to the accuracy of any measurement process. Only when the entire sample is taken for analysis may such errors be presumed absent. For the most part, especially in biological experiments, the entire sample may not (even in principle) be taken, and sampling variability or even gross sampling errors (blunders) may far exceed all other sources of error. To cite just one example: measurements by Tyler *et al.*[23] of ^{14}C activity in animal tissue exhibited 2.5 times more random error in sampling than in counting. These authors demonstrated also the dependence of sampling error on sample size, as well as the effect of operator experience on overall accuracy.

Sample preparation generally leads to more reliable measurement if purification steps are involved. Apart from contamination and efficiency and stability questions, which will be discussed below, chemical (or physical) separation may, however, lead to losses and altered isotope ratios – particularly for ^{3}H and ^{14}C. If uncorrected, such factors can produce bias; and even if corrected, overall imprecision will be somewhat increased. Recovery (yield) correction via 'carrier' or stable- or radioisotope dilution is well known to analysts and to radiochemists. Stable strontium or ^{85}Sr, for example, has long been used in the radiochemical determination of ^{90}Sr. Isotope fractionation problems are, perhaps, less well known, except to those who routinely measure low levels of ^{3}H and ^{14}C. In the radiocarbon dating community fractionation effects are regularly corrected by mass spectrometric measurements of δ^{13}C.[24] Without such corrections, radiocarbon measurements may exhibit a limiting (minimum) variability of at least 0.5%. Random variations in the isotopic enrichment factor for electrolytically enriched tritium have been shown to limit the overall accuracy of low-level liquid scintillation measurements of tritium to about 3%, largely due to differences among electrolysis cells.[4] Because of this source of error, Sauzay and Schell[25] found improved accuracy *without* enrichment for liquid scintillation counting of water samples having tritium activities exceeding 500 TU.

Measurement

The major opportunity for error in liquid scintillation counting is in connection with the measurement step. Under ideal circumstances the error is entirely random and may be computed from Poisson counting statistics. In real circumstances, however, the counting mixture may be unstable or it may be otherwise non-identical with that used in calibration. Consequently, assumed efficiencies and backgrounds may be incorrect.

(1) Instability. This occurs when the counting mixture changes its composition as a result of (i) phase separation or precipitation, or (ii) losses to the vial walls (adsorption), through the walls (diffusion) or to the surrounding atmosphere (evaporation). Besides composition-related variability, one must sometimes pay attention to other sources, such as electronic instability, luminescence decay and external background variations. The stability of heterogeneous counting preparations (suspensions, colloids) has been studied by Fox,[26] van der Laarse[27] and others. Visual appearance and apparent (physical) stability have been shown to be unreliable as indicators of counting stability, particularly for tritium. As a result, Fox introduced the Instrument Merit Stability Quotient (MISQ) as a measure of the quality of aqueous colloids of varying composition, where MISQ takes into account the observed *counting* stability, detection efficiency and volume percentage of water.

Adsorption of a radioactive solute is troublesome in that it changes the effective solid angle from about 4π steradians to 2π, thus decreasing the efficiency by a factor of two. Amounts and rates of adsorption depend upon the nature of the vial surface, the scintillation cocktail, and the chemical structure and specific activity of the labelled species.[28] In a concise review of the topic, Litt and Carter[29] demonstrated the prevalence and complexity of adsorption phenomena. They noted particularly the benefits from 'the addition of a pinch of carrier' and the use of plastic vials.

Though advantageous with respect to the adsorption problem, certain plastic vials have been responsible for long-term (∿days) drifts in sample, external standard and background counting rates because of their permeability.[25,30,31] Sample losses occur because of diffusion of the labelled species through the wall, and the background increases with time because the wall becomes a plastic scintillator, due to solvent plus solute diffusion. Rates of loss through polyethylene vials were reported by Horrocks[31] as 130 mg/day for (tritiated) toluene and 5 mg/day for water. (The recommendations for controlling diffusion-induced errors, presented at the end of Horrocks' paper are well worth the reader's attention.)

Losses may occur also through evaporation. The problem is most serious when low activity levels dictate extended measurements, as in radiocarbon dating. For example, at the Australian National University, sample counts are generally accumulated for 1000 min each over a period of one week. The high rate of evaporation with commercially available vials – ∿ 25 mg/day – led Polach[32] to design a special small volume vial having a machined PTFE stopper. The rate of loss of solvent was thereby decreased to ∿ 10 mg/week. Incidentally, when samples having different levels of activity remain together in the same sample changing chamber, one sample's losses become another sample's contamination.

(2) Efficiency changes or quenching. This must be anticipated whenever the sample composition or configuration differs in any way from that of the calibration standard. 'Quenching', which generally signifies decreased intrinsic detection efficiency, arises from divers sources. Perhaps the most serious blunders (needless systematic errors) in liquid scintillation counting come about when improper quenching corrections are applied. In order to appreciate the nature of such errors, we must consider some of the causes and monitoring (correction) methods for quenching: (i) *causes* – 'chemical quenching' (impurity competition for the excitation energy of the solvent molecules), 'colour quenching' (molecular absorption of the fluorescence radiation) and 'photon quenching' (decreased photon yield due to self-absorption of the primary nuclear radiation); and (ii) *monitoring* – internal standard, sample channels ratio (SCR), automatic external standard count rate (AES) and external standard channels ratio (ESCR). Cerenkov counting has one major advantage over ordinary liquid scintillation counting, namely immunity to chemical quenching. The following paragraphs will illustrate a few of the pitfalls in quench monitoring; for further information the reader is advised to examine the comprehensive accounts of Parker and Elrick,[33] Neary and Budd [34] and Peng.[35]

Errors may arise from a mismatch between quenching mechanism and monitoring method. If the degree of quenching is small and if the sample is in complete solution and stable, however, there should be little difficulty; all methods ought to yield similar results. Low count rates may limit the usefulness of the SCR method, however, because of the correspondingly large (Poisson) counting errors. The AES method has suffered problems in the past because of irreproducible source positioning and sensitivity to sample characteristics (volume, electron density) and vial characteristics.

Channels ratio quench corrections for homogeneous solutions derive from the fact that decreased efficiency commonly accompanies changes in the amplitude or shape of the scintillation pulse height distribution. Unless the radionuclide, quenching agent and monitoring method are all fixed, however, faulty conclusions may result from the use of a single quenching curve. The major problem arises when one fails to distinguish between the effects of chemical and colour quenching, particularly if the external monitor is employed. The difference between colour versus chemical quenching curves (ESCR) can be quite significant both for ^{14}C and ^{3}H, as shown in the work of Noujaim *et al.*[36] Although these authors found little difference in SCR curves (colour versus chemical), it has been conclusively demonstrated[34] that for ^{14}C and higher energy β-emitters, significant systematic error will result if one does not pay attention to the type of quenching in highly quenched solutions, Unfortunately, in practice one may not know whether quenching is primarily chemical or due largely to coloured material, particularly when biological samples are involved. A promising approach to the problem is to require concordance among two or more different quenching monitors. Lange,[37] for example, computed ^{14}C efficiencies of 35.2% and 24.5% using the ESCR and 'chemical' or 'colour' quench correction curves, respectively, where chlorophyll was the quenching agent. Computations based upon AES curves yielded 30.7% and 25.2%, respectively. Concluding, from the consistency between results, that the quenching was largely due to colour was indeed the correct decision: the actual efficiency was 24.9%.

Errors due to heterogeneous counting samples (suspensions, precipitates, etc.) may also be signalled through the use of redundant quench monitoring. This is one of the major features of the 'double ratio' method as proposed by Bush[38] and as illustrated by Glass.[7] When sample counts are sufficiently large, a plot of SCR versus ESCR serves to monitor homogeneity. Two important points must be borne in mind with respect to heterogeneous samples, however: (i) they may not *appear* to be heterogeneous;[27] and (ii) erroneous efficiencies can result from the use of *either* (SCR or ESCR) quenching correction curve.[13]

Addition of an internal standard will generally yield a correct measure of counting efficiency, provided the standard has the same environment as the unknown sample, i.e. the specific activity and chemical composition of the added standard must be such that it produces no significant changes in the nature (volume, phases, etc.) of the sample. If more than one phase is present, it is essential that the added material distribute itself in the same manner as the sample radioactivity. The problem becomes acute for tritium whose radiation is so readily absorbed.[39] In an investigation of three-component (toluene, water, triton) emulsion counting of tritium, for example, van der Laarse[27] observed counting efficiencies which differed by as much as a factor of three for tritiated toluene versus tritiated water. Similarly, misleading results due to unequal phase distribution of ^{14}C-labelled components in blood have been reported by Laurencot and Hempstead.[40]

Most of the foregoing sources of systematic error may be avoided through combustion if one is concerned with the measurement of ^{3}H or ^{14}C.[23,41] Additional benefits include the reduction or elimination of chemiluminescence, and increased counting efficiency. Perhaps the most notable example of these benefits is the radiochemical conversion of the sample – ^{3}H or ^{14}C – to a scintillation solvent, benzene. Quantitative benzene synthesis has dramatically altered the role of liquid scintillation counting in radiocarbon dating.[42,43]

(3) The blank. So long as sample activity levels are comparable to one another and large compared to the background, errors from contamination and background variations are of little importance. Low-activity samples, however, demand that one pay attention to numerous additional sources of blank effects and background variations which otherwise would be inconsequential. Vial materials must be selected to yield the lowest possible background.[44,45] Low-potassium glass or quartz or plastic (polyethylene, nylon) vials seem the most attractive in this respect. To achieve the best background accuracy, Polach[32] found it necessary to determine discrete background rates for individual, numbered vials. Sample combustion apparatus can lead to inter-sample contamination due to 'memory' effects;[41] and diffusion and/or evaporative losses have been responsible for sample contamination in radiocarbon dating.[46] 'Reverse' contamination (decreased sample activity) may also occur in the pretreatment of radiocarbon samples with organic solvents through exchange reactions.[47] When the lowest levels of activity are of concern, unusual sources of contamination must be considered: water samples which were exposed to the atmosphere during processing, within about 15 km of deuterium-moderated reactors, exhibited tritium contamination of 3 to 6 pCi/litre.[48]

Photoluminescence and chemiluminescence are of special importance in low-level liquid scintillation counting. The former may be minimised by dark-adapting samples for an adequate period of time, and the latter by sample preparation techniques which decrease the likelihood of chemiluminescent reactions. A thorough discussion of the causes and controls for these phenomena is given in Section V of Ref. 16. One caution: chemiluminescence may persist for a period of weeks.[49] Although luminescence phenomena consist of single photon events, interference occurs even in coincidence counting systems, through pulse pile up and accidental coincidences.

Other potential sources of background error include background variations with time,[25] assumed background (baseline) shape under low amplitude α-peaks,[50] quenching effects[39] and interfering nuclear reactions.[62] The last two deserve some amplification. With respect to quenching, Scales[39] has shown that the background for ^{3}H and ^{14}C measurements also decreases with increasing quenching. The variations with counting efficiency, however, are non-linear, and depend upon the nature of the quenching – i.e. chemical versus colour. An ancillary background effect observed by Scales, in his direct measurements of dissolved tissues, was the variable contribution due to *natural* radioactivity (^{40}K in the tissues). Errors due to an interfering nuclear reaction were encountered by Horrocks[62] in his measurements of the production of fission-tritium from nuclear power reactors. In order to accurately determine the fission yield of ^{3}H from ^{235}U, Horrocks found it necessary to reduce the lithium impurity in normal uranium to less than 5 ppm, because of the interfering reaction, Li(n, α)^{3}H.

(4) Example – ^{37}Ar. Departing slightly from the literal topic of this symposium, I should like to illustrate quenching and blank problems which we have encountered in the measurement of ^{37}Ar by *gas* proportional counting. In certain respects there are, in fact, close analogies between internal gas and liquid scintillation counting in terms of causes and monitoring of efficiency changes. In both cases, the sample is an integral part of the detector, so its properties can affect the efficiency and the background. Efficiency monitoring may in each case be based upon an internal standard, an external standard or the shape of the sample spectrum.

In our laboratory, it was necessary to measure extremely small concentrations ($\sim$ 0.003 dpm/litre Ar) of ^{37}Ar – a 35-day radionuclide which decays by electron capture – in connection with a study of atmospheric reactions of cosmic rays and tropospheric mixing.[51] Long measurement times ($\sim$ weeks) necessitated (quench) corrections for the shifting pulse-height spectra. Shifts in the 2.82 keV full energy

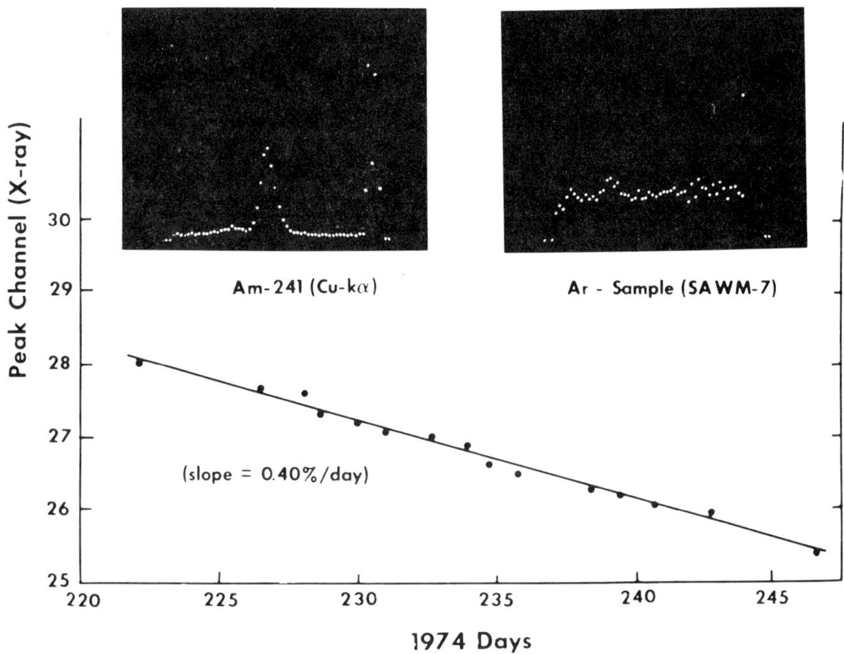

Fig. 3. Quench monitoring of ^{37}Ar. Curve shows drift in gain over 25 days. Counting gas gain and resolution were monitored via an external standard source (^{241}Am, left insert). Argon sample was far too weak to employ the sample for internal gain monitoring (SCR); right insert shows ^{37}Ar peak accumulated in 25 days of counting.

peak, though slow (~ 0.4%/day) were brought about by composition changes in the counting gas resulting from gradual degassing of internal counter components. Figure 3 shows the quench behaviour using an external standard (^{241}Am). Over the 25-day period covered by the plot, we monitored the peak position of the secondary (Cu-Kα) X-ray induced in the brass wall of the gas counter. From the position of the copper X-ray peak, we deduced that of the argon peak and corrected for the quenching accordingly. The two spectra included in the figure represent a 5-min observation of the copper X-ray and a 25-*day* observation of the ^{37}Ar full energy peak for one of our atmospheric samples. As with SCR-monitoring and low-level liquid scintillation counting, it is obvious that the ^{37}Ar sample peak was far too weak to provide the 14 points on the quenching curve.*

It is instructive to examine the blank effects in the ^{37}Ar experiment, for nearly all possible types were encountered. As is evident from the sample spectrum in Fig. 3, baseline shape assumptions were important; background variations with time and counter were observed; and four distinct kinds of radioactive contamination were encountered. (i) *Radiochemical* contamination (^{3}H) came from argon isotopic enrich-

*The external source ^{241}Am was used to monitor the position of the ^{37}Ar peak. As with liquid scintillation counting of α-peaks, efficiency was maintained constant by correspondingly shifting the peak integration (energy) 'window'.

ment facilities which were proximate to ^3He production facilities; (ii) *isochemical* contaminants (^{222}Rn, ^{85}Kr) were present in the original atmospheric samples and/or the 'reagents' (charcoal, CaC_2); (iii) *isotopic* contamination (^{39}Ar) also came from the original atmospheric sample, due largely to cosmic ray interactions and influenced by the isotopic enrichment process; finally, (iv) *isonuclidic* contamination (^{37}Ar) came from atmospheric or underground nuclear tests as well as from neutron reactions with our CaC_2 reagent via the reaction $^{40}Ca(n, \alpha)^{37}Ar$. Discussion of steps taken to minimise errors from the foregoing background variations and contamination sources is beyond the scope of this chapter, but the magnitude of the problem may be perceived by the fact that the concentration of artificial atmospheric ^{37}Ar, following a nuclear test, was frequently 100 times the natural concentration.[52]

ON THE QUESTION OF MERIT*

Conventional measures of performance

Merit and *sensitivity* are two of the most transparent, yet ambiguous, terms currently in use in chemical and radiochemical analysis. Qualitatively, the meanings are obvious: *sensitivity* relates to the ability to detect or quantify, and *merit* is a measure of relative sensitivity.

Both terms, in turn, are tied implicitly to the structure of the measurement process and to its component errors. Explicit definitions, however, depend upon the user. In spectrochemical analysis, for example, *sensitivity* is defined as the slope of the calibration function, whereas for nuclear measurements (liquid scintillation counting, activation analysis, etc.) the same term is frequently used to specify a minimum detectable quantity.[45,53] *Merit*, likewise, takes on numerous meanings when defined by different workers; the significance of some of the more popular 'figures of merit' and their relationship to the underlying errors and assumptions will be examined below.

Despite the ambiguities, the central aim remains clear: we wish to provide an objective and quantitative means for assessing the capacity of a given procedure to detect or to measure (with a given relative standard deviation) the radionuclide (or element or compound) of interest. We wish also a quantitative measure of the relative capabilities of alternative procedures or that of a given procedure as a function of controllable variables, such as sample volume or discriminator level. These ends will be met in an easy-to-apply, yet objective, graphical procedure which may be used for the practical *planning* of radiochemical measurements and for the selection of optimal procedures or optimal values for experimental variables. Sources of error, as discussed in the preceding section, together with Poisson counting statistics, will provide the basis for the evaluation of merit in liquid scintillation counting.

To begin the evaluation of counting methods, it will be instructive to examine some of the expressions commonly used to denote merit. One finds, in perusing the liquid scintillation and radiochemical literature, the following terms (among others): Figure (or Factor) of Merit (F),[13,41,54-56] Merit Number,[27] Merit Value and Merit Stability Quotient.[26] The term Figure of Merit, *per se*, is equated to a bewildering array of expressions ranging from S^2/B to $\sqrt{\epsilon V}$ (S = sample counts or rate, B = background counts

*Definitions for symbols used in this section will be found in Table 1.

or rate, ϵ = counting efficiency and V = sample volume). To these is added the Background Equivalent Activity (BEA = B/ϵ or $B/[\epsilon V]$) which is equivalent to the 'absolute' or 'concentration sensitivity' as used by Gelsema *et al.*[53] With the exception of the Merit Stability Quotient (MISQ), the expressions used to deduce the relative merit of alternative radiochemical measurement systems are based on the premise that random errors arise from Poisson counting statistics only, and that systematic errors are negligible. As has already been seen, such (non-Poisson) errors cannot always be ignored. Their importance in terms of colloid stability has been recognised and incorporated into MISQ by Fox,[26] and their significance in radiocarbon dating was treated by Currie[57] by resolving F into Poisson and non-Poisson components. The use of *BEA* as a measure of merit is predicated on the assumption that signals much smaller than the background may be unreliable, due to background variability arising from external sources. Though *BEA* is a helpful guide, it is in fact essential to assess the magnitude and *variability* of the *blank* if reliable conclusions are to be drawn.

The variety of expressions for merit, based upon counting statistics alone, obtains from three sources: (i) whether total activity or specific activity (per unit mass or volume) is of interest; (ii) whether the relative (inverse) variance or standard deviation is considered (e.g. S^2/B versus $S/\sqrt{B}$); and (iii) whether $S<<B$ or $S>>B$ is assumed. The first two assumptions should cause little difficulty in any particular case, *provided* the conventions in use are made clear. The last assumption, however, may lead to some peculiar difficulties, especially from the arbitrary application of the most commonly used Figure of Merit, ϵ^2/B. ϵ^2/B (or $[\epsilon V]^2/B$), of course, derives from one of the asymptotes of the inverse (relative) Poisson variance:

$$F = \frac{1}{\phi^2} = \frac{S^2}{S + 2B} \left| \begin{array}{ll} \propto \epsilon^2/B & (S<<B) \\ \propto \epsilon & (S>>B) \end{array} \right. \tag{1}$$

Apparent paradoxes which arise from ϵ^2/B (or $\epsilon/\sqrt{B}$) are seen, for example, in reviews by Calf[58] and Cameron.[59] In the former case, direct counting of tritium as water (in dioxane) was shown to have a higher figure of merit (0.6 versus 0.4) than the benzene synthesis method, yet the benzene method was capable of quantitatively measuring a slightly smaller amount of tritium (300 TU versus 310 TU) than the water–dioxane method (in 500 min). The same situation occurs in Table III of Cameron's review for low-background gas (GM) tritium counters: #13 (Sweden) versus #15 (UKAEA). Here, one finds also the phenomenon of 'reversal' – where one method (counter) is superior at one level of activity (#13, at 500 TU), and the other is superior at a different level of activity (#15, at 5000 TU). (Determination of the activity level at which reversal takes place is one of the important questions which will be discussed later in this section.)

A closely related problem arises when, in the comparison of two methods, $S<B$ for one, while $S>B$ for the other. Then, clearly, neither of the limiting expressions yields a valid measure of relative merit. One solution which has frequently been taken is to calculate absolute or relative merit directly from Eqn. (1), by *assuming* a given level of activity, hence a given value for S. Such an approach is not entirely satisfactory, however, for it is not clear what particular activity level ought to be pre-selected for the computation. In the following sub-section we shall investigate an approach which is independent of the foregoing constraints and which can be applied singly or simultaneously to a collection of alternative procedures.

Table 1. Symbols and relationships.

t = counting times
R_B = background rate, B = background counts = $R_B t$
R_S = net (sample) rate, S = sample counts = $R_S t$

ϵ = measurement efficiency (calibration factor)[a]
A = activity = R_S/ϵ
BEA = background equivalent activity = R_B/ϵ

ρ = reduced activity = S/B = A/BEA

ϕ = relative standard deviation = σ_A/A
F = merit = $1/\phi^2$

Subscripts: P, B, ϵ = variance or merit component due to Poisson, background or calibration errors, respectively.

[a] ϵ represents the overall calibration factor, including both counting efficiency and chemical recovery; when specific activity is of concern, ϵ *must* incorporate also sample mass (m) or volume (V).

A consistent approach to merit

(1) Reduced activity. The primary aim is to develop a quantitative measure of performance which requires no prior assumption of the relative magnitude of S and B or of activity level. One such measure is the detection limit (A_D), given the counting time (t); another is the quantitative (or determination) limit (A_Q). Expressions for A_D and A_Q[60] are as follows:

$$A_D = \frac{1}{\epsilon t}(2.71 + 4.65\sqrt{B}) \tag{2}$$

$$A_Q = \frac{50}{\epsilon t}(1 + [1 + B/12.5]^{1/2}) \tag{3}$$

(A_D assumes 5% risks – errors of the first and second kinds – and paired observations of sample and background; A_Q assumes 10% relative standard deviation for quantitation.) Once a procedure has been specified – i.e. R_B, ϵ fixed – $A_{D,Q}$ are functions of counting time only. One such function, A_Q (t) for ^{32}P measured by liquid scintillation counting, is shown in Fig. 4.

Equations (2) and (3) and Fig. 4 are unsatisfactory in that separate computations or curves are required for each procedure of interest. The deficiency may be overcome, however, by the substitution of dimensionless ('reduced') coordinates, in which the activity level (ordinate) is normalised with the BEA, and the time (abscissa) with the mean interval between background counts ($1/R_B$). A universal curve results which permits us to immediately deduce for any procedure the 'reduced' quantitative activity ($\rho_Q = A_Q/BEA$) as a function of the 'reduced' time ($\tau = t/t_B = tR_B = B$). The folly of assuming a fixed expression (e.g. ϵ^2/B) to compute a single, numerical value for relative or absolute merit is implicit in the changing slope of the curve (Fig. 5). Merit must be viewed as a two-dimensional (or two-compo-

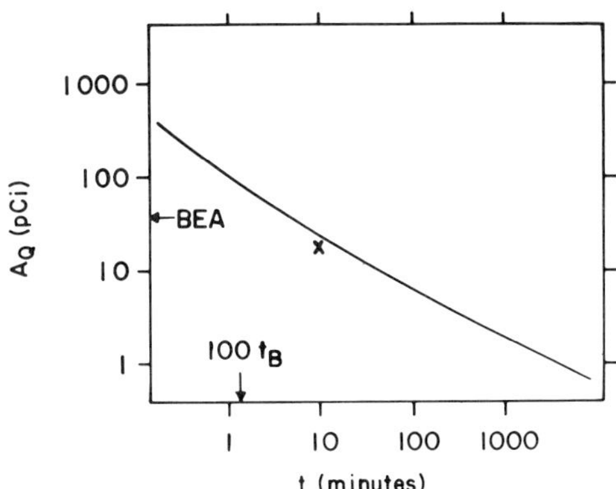

Fig. 4. Quantitation limit (A_Q, pCi) for liquid scintillation counting (L) of ^{32}P versus counting time (t). t_B equals the mean interval between background pulses ($1/R_B$); *BEA* equals the Background Equivalent Activity (40 pCi). Curve shows, for example, that for t = 10 min, A_Q = 22.7 pCi. Point (x) – 20 pCi – has poorer precision (∿ 11%) as it lies below the curve. Assumed background and efficiency data are given in Table 2 (Method L).

nent) quantity in counting experiments; only at the asymptotes ($B \to 0$, slope → -1, or $B \to \infty$, slope → -1/2) does a single value suffice. An added benefit of the dimensionless representation (Fig. 5) is explicit indication of the point at which $S = B$.

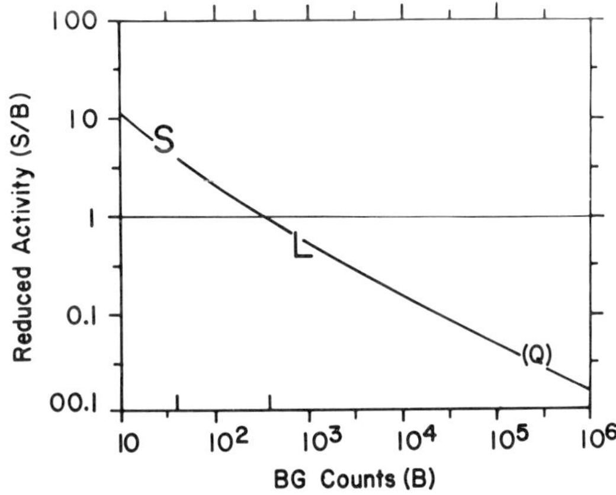

Fig. 5. Reduced activity plot, ρ_Q versus B. The curve is universal, but the points characterise the performance of liquid scintillation (L) and $2\pi\beta$ solid (S) counting methods for ^{32}P when the activity equals 20 pCi and the counting time 10 min. Note that $S > B$ for S, but $S < B$ for L. (Input data from Table 2.)

We see that $\rho_Q > 1$ ($S_Q > B$) if $B < 300$ counts, and *vice versa*. (For the detection limit, ρ_D = 1 at B = 27 counts.) These crossing points (300 counts, 27 counts) are helpful in judging the adequacy of either of the asymptotic expressions for merit, given the counting time.

Figure 5 may be used also to evaluate the relative performance of alternative methods. Points 'S' and 'L' refer to solid (external, $2\pi\beta$) and liquid scintillation counting of ^{32}P, using counting efficiencies and background rates as given in Table 2. The points have been normalised to 20 pCi and 10 min. (Thus, the coordinates at L are y = 20 pCi/BEA = 20 pCi/40 pCi = 0.50, and x = (80 counts min^{-1}) (10 min) = 800 counts.) Comparing the positions of the points (L,S) with the curve, it is evident that S has the greater merit, *for this particular choice of activity and counting time.* S, lying slightly above the curve, has better than 'quantitative' precision (ϕ_S = 9.4%), whereas L, which lies below the curve, has poorer precision (ϕ_L = 11.2%). Levels for which each of the two methods would be quantitative (ϕ = 10%) derive directly from the curve (t = 10 min):

$$A_Q = \rho_Q \cdot BEA = \begin{cases} (0.566)\ (40.0) = 22.7 \text{ pCi (for L)} \\ (4.74)\ (3.86) = 18.3 \text{ pCi (for S)} \end{cases}$$

The effects of changes in activity levels or counting times correspond simply to translations of the points (L,S) along the y- or x-axes, respectively. Note that for the 10 min counting time selected $\rho(S) > 1$ whilst $\rho(L) < 1$ – i.e. the net signal exceeds the background for the one procedure *at the same time* that the background exceeds the net signal for the other. This example should serve to illustrate (i) the utility of the dimensionless representation, as well as (ii) the potential pitfall in generally applying *either* FOM asymptote when comparing alternative procedures.

(2) Higher dimensions; non-Poisson errors. In order to provide a general treatment to the question of measurement performance, it has been necessary to increase the dimensionality from one (scalar, asymptotic-F) to two (ρ, B). Already implicit in the discussion – through the introduction of A_D and A_Q – is a third dimension, related to the experimental precision (ϕ). A contour plot is given in Fig. 6a, for the evaluation of performance at various levels of precision. (Curves C, D and Q represent the critical (decision) level, detection limit and quantitation limit, respectively, as defined by Currie.[60]

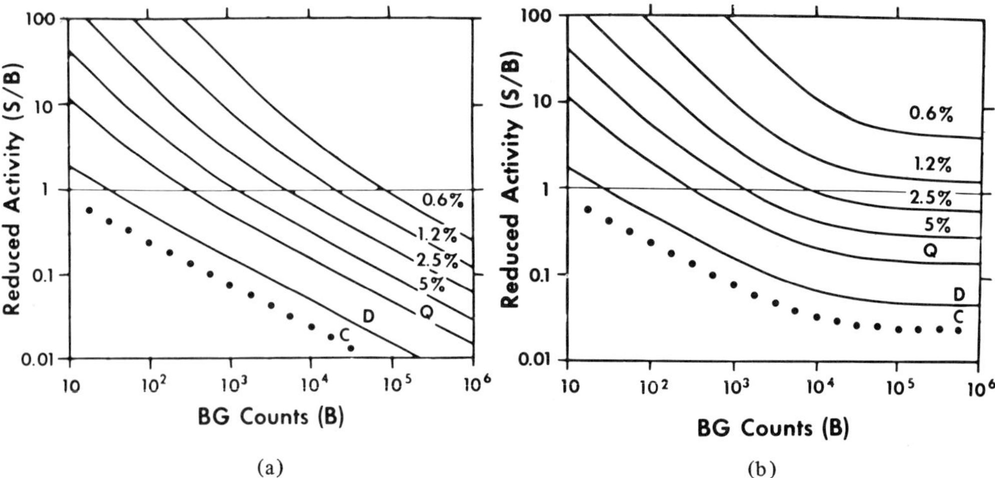

Fig. 6. Reduced activity curves. Contour plots are presented for reduced activity ($S/B = \rho$) versus background counts (B) and counting precision (ϕ). Part (a) includes Poisson errors only; part (b) incorporates additional random error (0.50% for counting efficiency, 1.0% for background variability).

Two additional parameters, ϕ_B and ϕ_ϵ, have not yet been taken into account. These represent non-Poisson random errors in the background and calibration factor, which cannot be ignored in any real experimental situation. As four- or five-dimensional plots (even contour plots!) are quite beyond the capability of this author, a particular 'cut' of the five-dimensional figure has been selected, as shown in Fig. 6b. The values selected for ϕ_B and ϕ_ϵ (1% and 0.5%) are, in fact, identical to those used in Fig. 1 of Currie.[57] These values were believed to be representative of about the best obtained in routine radiocarbon dating. As such, they may serve as a useful guide for a limiting merit diagram for real counting experiments, just as Fig. 6a serves as a limiting diagram for ideal counting experiments (Poisson errors only). If the *overall* performance of any specific counting experiment is to be evaluated, the corresponding values for ϕ_B and ϕ_ϵ must be determined. These may then be incorporated into a figure similar to Fig. 6b using Eqn. (4), which follows from the definition of reduced activity (ρ) and error propagation:

$$\phi^2 = \left[\frac{1}{B}\left(\frac{\rho + 2}{\rho^2}\right)\right] + 2\left(\frac{\phi_B}{\rho}\right)^2 + \phi^2_\epsilon$$

where the first term (in parentheses) refers to the Poisson component (contours in Fig. 6a).

The most important point to be noted in Eqn. (4) is the limiting effect of ϕ_B and ϕ_ϵ. i.e. as counting time increases ($B = R_B t$ increases) the Poisson component approaches zero. The overall relative variance (ϕ^2) is then limited by the last two terms. The relative standard deviation of the calibration factor (ϕ_ϵ), which depends upon random errors in efficiency (quenching and quench-corrections), isotopic enrichment, chemical recovery, etc., sets a fixed limit for the best achievable precision. The contribution from the relative standard deviation for the blank (ϕ_B), however, varies with the reduced activity level ($\rho = S/B = A/BEA$), being negligible for $S >> B$ but becoming all-important for signals small compared to the blank ($\rho << 1$). For example, if one were attempting to measure quantitatively a sample whose activity was 10% of the *BEA* ($\rho = 0.1$), non-Poisson contributions would be significant unless $\phi_\epsilon < 10\%$ and $\phi_B < 0.7\%$ – the latter limit (ϕ_B) being given by the constraint that the second term in Eqn. (4) be $< (10\%)^2$. That such bounds on ϕ_B and ϕ_ϵ are not easy to achieve was evident in the discussion of liquid scintillation errors in the preceding section.

(3) A general definition and asymptotic relations. In keeping with the original intent of the Figure of Merit in counting experiments we may conclude that procedures having equal precision possess equal merit. The contours in Fig. 6 are therefore equimerit contours. The index, F, when treated like a statistical weight, may be set equal to the reciprocal variance ($1/\phi^2$). F, for a given procedure, or relative F's for alternative procedures, therefore require specification of B, ρ, ϕ_B and ϕ_ϵ (Eqn. 4). Changing activity levels ($\rho = S/B$) or changing counting times ($B = R_B t$) will generally alter *relative* as well as absolute F's.

Two important conclusions derive from the geometry of the points and contours in Fig. 6: (i) the relative positions of points representing different procedures are invariant with time or activity level – they are fixed by background rate and background equivalent activity (or counting efficiency) alone; (ii) the critical angle (slope) for points of equal overall merit (cf. Fig. 6b) varies continuously from -45° (slope = -1) to -0°. The first conclusion yields an extremely convenient device for evaluating relative merit and experiment adequacy, which will be utilised subsequently. The second conclusion yields limiting expressions for equal merit, as shown in Fig. 7.

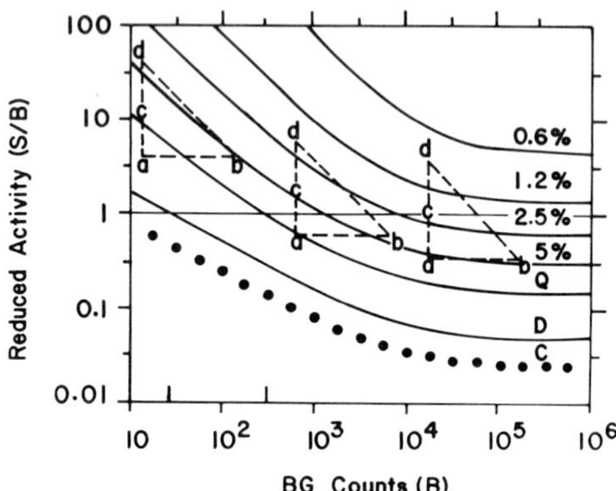

Fig. 7. Merit. Construct a-b-c-d, superposed in three different locations on Fig. 6b, shows changing relations (comparative merit) among procedures. (b,d) have equal counting efficiency, ϵ ; (b,c), equal ϵ^2/B; and (a,b), equal background equivalent activity (or B/ϵ).

The construct, a-b-c-d, in Fig. 7 is taken to represent four alternative counting procedures, the relative positions of which are fixed by the respective R_B's and *BEA* 's. Three locations, respresenting translations in y (activity level) and x (time), have been given for the construct. Changing relations among the relative merit (F) or precision thus become clear:

Location-1 (upper left):	$F_b \approx F_d > F_c > F_a$
Location-2 (intermediate):	$F_d > F_b = F_c > F_a$
Location-3 (lower right):	$F_d > F_c > F_b \approx F_a$

The transition from small B and large ρ is thus accompanied by a gradual change in the limiting expression for equal merit from: (i) ϵ = const. (b,d), slope = -1; to (ii) ϵ^2/B = const. (b,c), slope = -1/2; to (iii) $\epsilon/B = 1/BEA$ = const. (b,a) slope = -0. (If ϕ_B and ϕ_ϵ were always neglible, the contours would have limiting slopes of -1 and -1/2, as in Fig. 6a.) The inequalities are worth noting: for two procedures of equal efficiency (b,d), the one having the smaller background (d) is generally preferred, because $F_d \geqslant F_b$; for two procedures of equal ϵ^2/B (b,c), the one having the larger background is preferred, because $F_b \geqslant F_c$ *provided* non-Poisson errors are negligible.

The practical evaluation of counting experiments

(1) ^{32}P: merit of alternative methods of measurement. The foregoing approach to merit and the resulting contour plots can be readily adapted for very rapid planning and comparative evaluation of counting procedures. This may be accomplished by using the basic contour plot (Fig. 6), together with an array of points representing the invariant *relative* positions of the procedures in question, and x-, y- or z- (contour) translations of the entire array. (The array may be constructed simply by plotting y = $1/BEA$ versus x = R_B for each of the procedures in question.)

The investigation of merit in ^{32}P counting has been carried further in Fig. 8, where two *independent* arrays of points are given for (i) activity (pCi – methods S, C, L), and (ii) specific activity (pCi/ml – methods © and Ⓛ). The first set has been normalised to 20 pCi, the second to 20 pCi/ml; both sets have t = 10 min. Assumed input data are given in Table 2. (Note that the background and

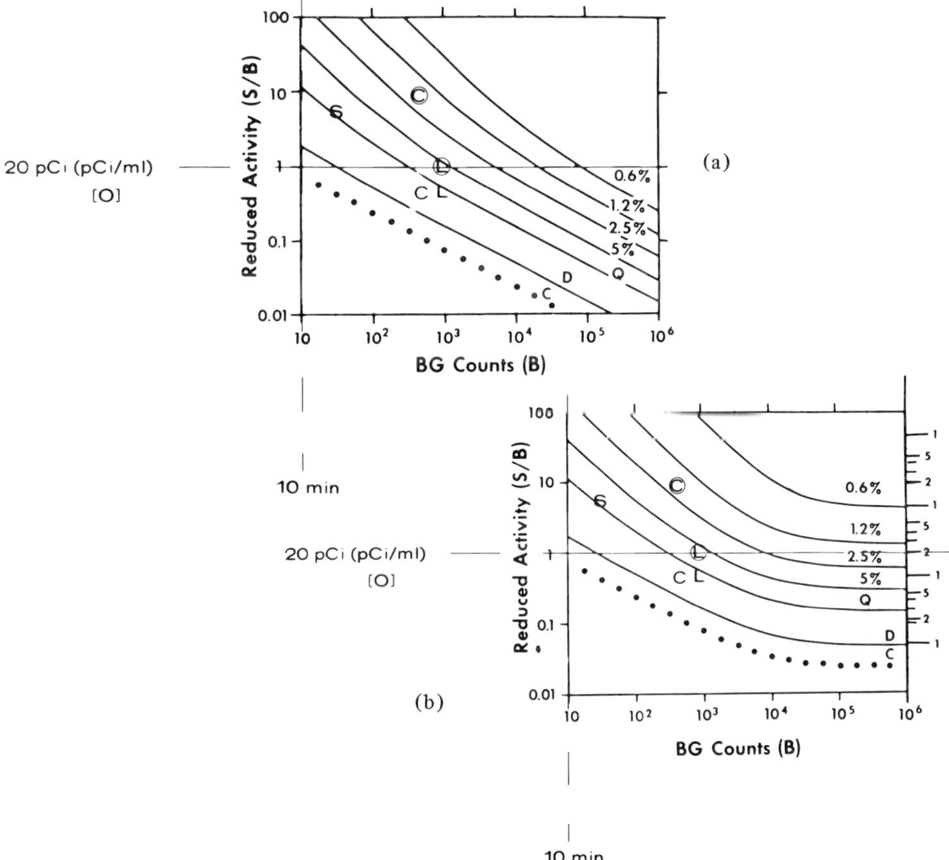

Fig. 8. Comparative performance of solid, liquid scintillation and Cerenkov counting of ^{32}P. Circled points refer to specific activity (pCi/ml), others to total activity (pCi). Normalisation set at 20 pCi (or pCi/ml) (y-axis) and 10 min (x-axis); input data from Table 2. Part (a) uses contours from Fig. 6a (Poisson errors only); part (b) uses contours from Fig. 6b. (Scale at left is reduced activity; scale at right, pCi. Abscissa in this and following figures indicates background counts as well as counting time (min).)

Table 2. ^{32}P Merit: input data.

	Counting method[a]				
	L	C	S	Ⓛ	©
Counting efficiency (%)	90	40	35	90	40
Background rate (counts min^{-1})	80	40	3.0	80	40
Sample volume (ml)	-	-	-	2	20
BEA (pCi)	40.0	45.0	3.86	-	-
BEA (pCi/ml)	-	-	-	20.0	2.25

[a]L, C, S = liquid scintillation, Cerenkov and solid ($2\pi\beta$) counting for total activity (pCi); Ⓛ, © = liquid scintillation, Cerenkov counting for specific activity (pCi/ml).

efficiencies used are assumed, plausible values, used here strictly for the sake of illustration.) For total activity measurements (S, C, L), we have already noted the somewhat better performance of S compared to L (Fig. 5 discussion). From Figure 8a it is apparent that Cerenkov counting (C) yields slightly poorer performance than either of the other methods.* On the other hand, upon examination of the points denoting specific activity, ©, Ⓛ, we find that Cerenkov counting is considerably better than liquid scintillation counting; ϕ (©) ∿ 2%, ϕ (Ⓛ) ∿ 5%. We see also from the figure that ρ (Ⓛ) $\approx$ 1, whereas ρ (©) $\approx$ 10. Therefore the normalisation activity (20 pCi/ml) equals the background equivalent activity for Ⓛ, so neither asymptotic expression for F would be appropriate. For ©, $S >> B$, so the asymptotic form, $F \propto \epsilon V$, would be acceptable.

Figure 8b differs from Fig. 8a (i) by the addition of an activity scale at the right (set at 20 pCi or pCi/ml for $\rho = 1$), and (ii) through the use of contours incorporating the effects of non-counting errors, like Fig. 6b. Performance as deduced from Fig. 8b is scarcely different from that deduced from Fig. 8a because of the relatively small numbers of counts – i.e. Poisson errors are still limiting. The effects of large increases in counting time, however, are very different. Figure 8b shows, for example, that the limiting precision ($t \to \infty$) for both Cerenkov (C) and liquid scintillation counting (L) of 20 pCi of ^{32}P is ∿ 4% due to the assumed non-counting errors.

The reversal of relative merit with activity level is illustrated in Fig. 9 which corresponds to a one decade vertical translation of the array of points to 200 pCi (still 10 min). Now, the precision of L is somewhat better than that of S, though the precision of both is improved in comparison with Fig. 8b (∿2-3% rather than ∿ 10%). Obviously, therefore, there must exist an intermediate activity level for which L and S have equal merit. The magnitude of that intermediate level may be found, by vertical translation, to be about 40 pCi, at which point both procedures have a precision of 6%.

Actually, a continuous series of points of equal merit (for L and S) exists. It is characterised by a functional relationship involving counting time (t), activity level (A) and precision (ϕ), such that all three parameters are the same for the two procedures (L,S). Equal

*Note that curve D (detection limit) corresponds to a relative standard deviation of about 30%.

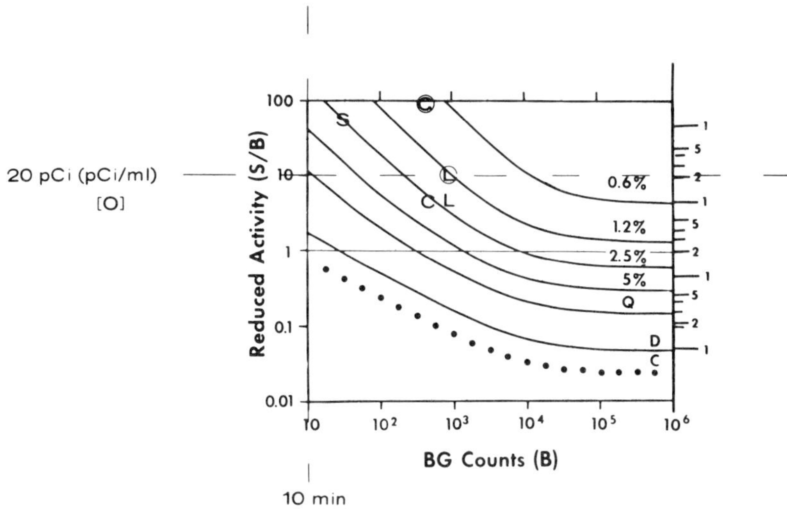

Fig. 9. Comparative performance, ^{32}P. Array of Fig. 8b translated one decade vertically to new normalisation of 200 pCi (or pCi/ml). Reversal of merit is exhibited by points L and S: Fig. 8b, $\phi_S < \phi_L$; Fig. 9, $\phi_S > \phi_L$ (ϕ = relative standard deviation).

merit values of t, A, ϕ may be derived from translations along the x-axis (t) or along the y-axis (A) or along a contour. The result of such an analysis is given in Fig. 10. $2\pi\beta$ (S) counting has greater merit than liquid scintillation counting (L) for all samples whose activity is less than about 35 pCi. Liquid scintillation counting is better for large activities, but the point of (merit) reversal increases gradually with counting time, having approximately doubled by the time that t – 100 min.

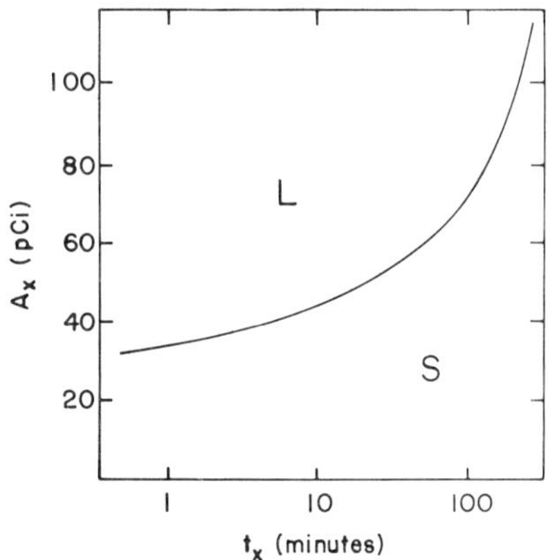

Fig. 10. Equimerit curve for ^{32}P (L, S). Curve for which the counting precision is identical using liquid scintillation (L) or solid, $2\pi\beta$ (S) methods. 'Crossing' activity level (A_x) versus time (t_x) may be used for selecting preferred method depending upon activity level or counting time. Liquid scintillation counting (L) is better for space above the curve; S is better for the space below.

The quantitative aspects of merit reversal, as exhibited in Fig. 10, may be of considerable importance in the planning of experiments or the selection of alternative procedures. None of the conventional expressions for figure of merit are adequate for this task, for they do not even *provide* for the reversal of relative merit.

To conclude the discussion of ^{32}P, let us examine a less subtle, but equally practical, question: 'What counting time is required to measure a given level of activity with a desired precision?' This question is addressed in Fig. 11 for the quantitative determination (ϕ = 10%) of 4 pCi/ml of ^{32}P by liquid scintillation counting. The result, t = 100 min, obtains from a downward translation of the array of points to 4 pCi/ml followed by a translation to the right until Ⓛ intersects the Q contour.

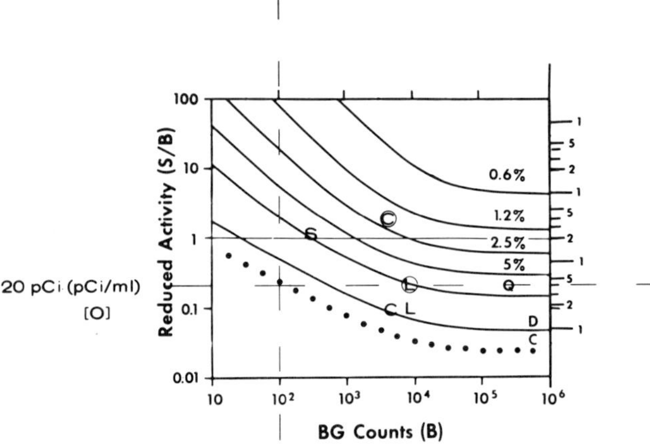

Fig. 11. Array translation for quantitative determination. Estimation of t_Q for 4 pCi/ml of ^{32}P by liquid scintillation counting.

(2) Tritium: effects of experimental variables on merit. An array of points or a continuous trace (locus) may be superposed on the reduced activity diagram in order to assess the effects of controllable variables such as sample volume, or discriminator or gain settings. Such an approach reveals variable adjustments having significance, and it makes possible the selection of settings yielding maximum merit.

Observed characteristics for two state-of-the-art low-level tritium counters (gas proportional and liquid scintillation) are listed in Table 3. For the latter, results obtained with four different volume–discriminator combinations have been tabulated. A five-point array, based on the data in Table 3, is shown in Fig. 12a – normalised to 200 pCi/kg H_2O (62 TU) and t = 10^3 min.* It is immediately evident that G, a and A are all approximately equivalent, exhibiting a precision of ∿5%; b and B are clearly superior, with B having slightly greater merit. We conclude, therefore, for 62 TU and 10^3 min, that the effects of discriminator changes (a→A, b→B) for the liquid scintillation system are relatively unimportant, whereas increased sample volume (a→b, A→B) improves merit despite the attendant increase in background. Two points should be stressed in this particular comparison: (i)

*The activity normalisation chosen (62 TU) is somewhat less than the tritium concentrations of northern hemisphere surface waters (100–200 TU), but considerably more than the concentrations (1–5 TU) found in the ocean. (See the introductory section of Noakes *et al.*[45] for discussion and references.)

Table 3. Tritium merit: input data.

	Counting method[a]				
	G	a	A	b	B
Counting efficiency (%)	90	34	53	34	53
Background rate (counts min^{-1})	0.20	0.89	1.72	1.68	2.98
Sample volume (ml)	–	10	10	50	50
Water equivalent (g)	1.61	6.87	6.87	33.4	33.4
BEA (pCi/kg)[b]	62.2	172	213	66.6	75.8

[a] G = gas proportional counter (NBS counter, approximate characteristics), a, A, b, B = liquid scintillation counter,[45] with varying volumes and efficiences (from discriminator changes) as shown in table.
[b] 1 TU (Tritium Unit) = 3.24 pCi/kg H_2O.

though G, a and A possess equal merit, the gas counter requires only about 25% as much sample as the (10 ml) liquid scintillation counter; (ii) use of $(\epsilon V)^2/B$ as a measure of merit would be incorrect, for the signal exceeds the background ($\rho > 1$) for all five of the methods.

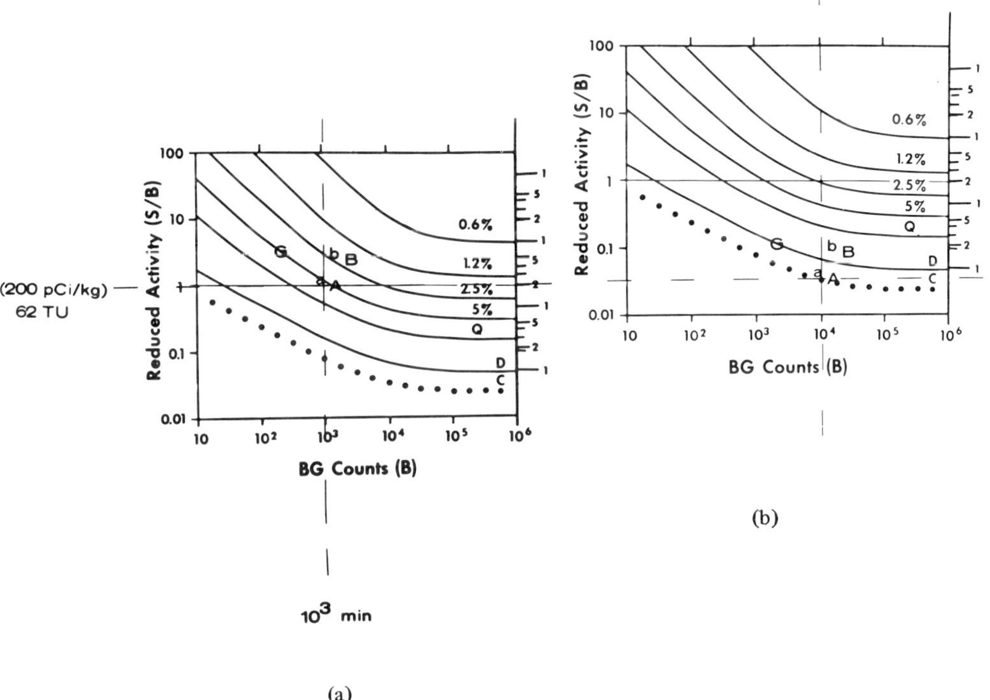

Fig. 12. Comparative merit for tritium methods. Comparison of gas counting (G) and liquid scintillation counting (a, A, b, B) of tritium. Liquid scintillation counting is illustrated for four different efficiency-sample volume choices. (See Table 3 for symbol identification.) Part (a) is normalised to 200 pCi/kg H_2O (62 TU) and 1000 min; part (b) exhibits the lowest feasible detection limit (for method G, 1 week of counting). Note the change in relative merit and optimum variable (discriminator, volume) combination.

A second comparison of interest treats the case where the gas counter is pushed to its limit. In Fig. 12b we evaluate the smallest detectable tritium concentration (counter G) when the maximum feasible counting time (10^4 min ∿ 1 week) is employed. This concentration is seen to be about 7 pCi/kg (∿2 TU), where the signal is just slightly greater than one-tenth of the background. G is no longer equivalent to a and A, however. In fact, it is now intermediate between a, A and b, B in performance; a, A lie below the detection limit, and b, B may be measured with a precision of about 20%. A final observation: the relative merit of b, B is reversed. The larger volume liquid scintillation samples still provide the best merit, but now the small efficiency member (b) has a slight edge. $(\epsilon V)^2/B$ would be appropriate for ranking the five methods based upon Poisson errors alone, since $\rho \ll 1$. However, B (background *counts*) is so large, especially for the liquid scintillation measurements, that the effects of background reproducibility must also be considered.[61]

CONCLUSION

The utility and general meaning of Figure of Merit (F) are clear. F is obviously extremely useful for procedure selection and variable optimisation, and it must surely be defined as a measure of signal/noise or relative precision. Difficulties arise, however, once numerical values are assigned to F – even for the assessment of *relative* merit – because of equivocal or improperly used asymptotic expressions.

Taking F to be $1/\phi^2$ means that variance components translate directly into F components (eqn. 4), and that F must be viewed as a function of A, ϵ, R_B, t, ϕ_B and ϕ_ϵ. Transformation to reduced coordinates (ρ, B), however, permits F_P (Poisson component) to be represented by a single diagram (Fig. 6a). This, in turn, permits the merit of any number of alternative procedures to be simultaneously evaluated. Figures 6a and b and Eqns. (3) and (4) yield the following caveats:

(i) A_Q (given t) or t_Q (given A) may be used a a single index of performance. No prior assumption of activity level or S/B is required.

(ii) Validity of alternative, asymptotic expressions for F relate to contour slopes:

- $-45°$ implies $F \propto \epsilon$
- $-30°$ implies $F \propto \epsilon^2/B$
- $-0°$ implies $F \propto (\epsilon/B)^2 \propto (1/BEA)^2$

(iii) The approach to zero slope establishes maximum useful counting times and minimum useful ρ's (S/B).

(iv) No single asymptotic measure of merit (ϵ, ϵ^2/B, $1/BEA$) may be generally relied upon for comparing alternative procedures, for different procedures may simultaneously be in different asymptotic regions (cf. Fig. 5).

(v) Unless two procedures have identical BEA's a range of A's *will* exist for which $S_1 > B_1$ at the same time that $S_2 < B_2$.

(vi) If the slope of the line connecting points representing two procedures lies between $-45°$ and $0°$, there *will* exist a range of A's and t's over which merit reversal will take place (cf. Figs. 7-10).

ACKNOWLEDGEMENT

Helpful discussions with B. M. Coursey, W. B. Mann and J. E. Noakes concerning the literature, problems and capabilities of liquid scintillation counting are gratefully acknowledged.

REFERENCES

1. H. T. Yolken, in *Proceedings of the 6th Materials Research Symposium,* NBS Spec. Publ. 408, U.S. Government Printing Office, Washington, 1975, pp. 237-45.
2. J. P. Cali, J. Mandel, L. Moore and D. S. Young, *A Referee Method for the Determination of Calcium in Serum,* NBS Spec. Publ. 260-36, U.S. Government Printing Office, Washington, 1972.
3. R. Vaninbroukx and I. Stanef, *Nucl. Instrum. Methods* **112**, 111-16 (1973).
4. T. Florkowski, B. R. Payne and G. Sauzay, *Intern. J. Appl. Radn. Isotopes* **21**, 453-8 (1970).
5. S. M. Kim, in *Organic Scintillators and Liquid Scintillation Counting* (eds. D. L. Horrocks and C. T. Peng), Academic Press, New York, 1971, p. 965.
6. H. A. Polach, in *Proceedings of the Eighth International Radiocarbon Dating Conference* (eds. T. A. Rafter and T. Grant-Taylor), Royal Society of New Zealand, 1973, p. H92.
7. D. S. Glass, in *Organic Scintillators and Liquid Scintillation Counting* (eds. D. L. Horrocks and C. T. Peng), Academic Press, New York, 1971, p. 803.
8. A. M. Downes, in *Organic Scintillators and Liquid Scintillation Counting* (eds. D. L. Horrocks and C. T. Peng), Academic Press, New York, 1971, p. 1031.
9. R. Fukai, S. Ballestra and C. N. Murray, in *Radioactivity in the Sea*, Publ. No. 35, International Atomic Energy Agency, Vienna, 1973.
10. P. D. LaFleur (ed.), *Accuracy in Trace Analysis: Sampling, Sample Handling and Analysis,* 7th Materials Research Symposium, NBS, Gaithersburg, Maryland, 1974.
11. O. Suschny and D. M. Richman, *IAEA Bull.* **16** (4), 11 (1974).
12. G. E. A. Wyld, in *Liquid Scintillation Counting* (ed. E. D. Bransome, Jr.), Grune & Stratton, New York, 1970, p. 69.
13. E. B. Mueller, in *Liquid Scintillation Counting* (ed. E. D. Bransome, Jr.), Grune & Stratton, New York, 1970, p. 181.
14. L. A. Currie, *Nucl. Instrum. Methods* **100**, 387-95 (1972).
15. J. B. Birks, *The Theory and Practice of Scintillation Counting,* Pergamon Press, Oxford, 1964.
16. E. D. Bransome, Jr. (ed.), *Liquid Scintillation Counting,* Grune & Stratton, New York, 1970.
17. D. L. Horrocks and C. T. Peng (eds.), *Organic Scintillators and Liquid Scintillation Counting,* Academic Press, New York, 1971.
18. A. Dyer (ed.), *Liquid Scintillation Counting,* Vol. 1, Heyden, London, 1971.
19. B. Scales (ed.), *Liquid Scintillation Counting,* Vol. 2, Heyden, London, 1972.
20. M. A. Crook and P. Johnson (eds.) *Liquid Scintillation Counting,* Vol. 3, Heyden, London, 1974.
21. P. E. Stanley and B. A. Scoggins (eds.), *Liquid Scintillation Counting: Recent Developments,* Academic Press, New York, 1974.
22. J. H. Parmentier and F. E. L. ten Haaf, *Intern. J. Appl. Radn. Isotopes,* **20**, 305-34 (1969).
23. T. R. Tyler, L. R. Chapin, R. F. Alvaro and J. L. Cox, *J. Radioanalytical Chem.* **12**, 475-82 (1972).
24. R. F. Flint and E. S. Deevey, *Radiocarbon,* **3**, v (1961).
25. G. Sauzay and W. R. Schell, *Intern. J. Appl. Radn. Isotopes* **23**, 25-33 (1972).
26. B. W. Fox, *Intern. J. Appl. Radn. Isotopes* **25**, 209-15 (1974).
27. J. D. van der Laarse, *Intern. J. Appl. Radn. Isotopes* **18**, 485-91 (1967).
28. D. C. Wigfield and V. Srinivasan, *Intern. J. Appl. Radn. Isotopes* **25**, 473 (1974).
29. G. J. Litt and H. Carter, in *Liquid Scintillation Counting* (ed. E. D. Bransome, Jr.), Grune & Stratton, New York, 1970, p. 156.
30. K. D. Neame, *Intern. J. Appl. Radn. Isotopes* **26**, 393-7 (1975).
31. D. L. Horrocks, *Intern. J. Appl. Radn. Isotopes* **26**, 243-56 (1975).
32. H. A. Polach, *Atomic Energy in Australia* **12**, No. 3.(1969).
33. R. P. Parker and R. H. Elrick, in *Liquid Scintillation Counting* (ed. E. D. Bransome, Jr.), Grune & Stratton, New York, 1970, p. 110.
34. M. P. Neary and A. L. Budd, in *Liquid Scintillation Counting* (ed. E. D. Bransome, Jr.), Grune & Stratton, New York, 1970, p. 273.
35. C. T. Peng, in *Liquid Scintillation Counting* (ed. E. D. Bransome, Jr.), Grune & Stratton, New York, 1970, p. 283.

36. A. Noujaim, C. Ediss and L. Wiebe, in *Organic Scintillators and Liquid Scintillation Counting*, (eds. D. L. Horrocks and C. T. Peng), Academic Press, New York, 1971, p. 705.
37. J. F. Lang, in *Organic Scintillators and Liquid Scintillation Counting* (eds. D. L. Horrocks and C. T. Peng), Academic Press, New York, 1971, p. 823.
38. E. T. Bush, *Intern. J. Appl. Radn. Isotopes* **19**, 447 (1968).
39. B. Scales, *Anal. Biochem.* **5**, 489-96 (1963).
40. H. J. Laurencot and J. L. Hempstead, in *Organic Scintillators and Liquid Scintillation Counting* (eds. D. L. Horrocks and C. T. Peng), Academic Press, New York, 1971, p. 635.
41. J. D. Davidson, V. T. Oliverio and J. I. Peterson, in *Liquid Scintillation Counting* (ed. E. D. Bransome, Jr.), Grune & Stratton, New York, 1970, p. 222.
42. H. A. Polach, J. Gower and I. Fraser, in *Proceedings of the Eighth International Radiocarbon Dating Conference* (eds. T. A. Rafter and T. Grant-Taylor), Royal Society of New Zealand, 1973, p. B36.
43. J. E. Noakes, S. M. Kim and J. J. Stipp, in *Radiocarbon and Tritium Dating,* USAEC Report CONF-650652, U.S. Atomic Energy Commission, Washington, 1965, p. 68.
44. Y. Kobayashi and D. V. Maudsley, in *Liquid Scintillation Counting* (ed. E. D. Bransome, Jr.), Grune & Stratton, New York, 1970, p. 76.
45. J. E. Noakes, M. P. Neary and J. D. Spaulding, *Nucl. Instrum. Methods* **109**, 177-87 (1973).
46. D. D. Harkness and H. W. Wilson, in *Proceedings of the Eighth International Radiocarbon Dating Conference* (eds. T. A. Rafter and T. Grant-Taylor), Royal Society of New Zealand, 1973, p. B101.
47. H. S. Jansen, in *Proceedings of the Eighth International Radiocarbon Dating Conference* (eds. T. A. Rafter and T. Grant-Taylor), Royal Society of New Zealand, 1973, p. B63.
48. H. G. Östlund, R. M. Brown and A. E. Bainbridge, *Tellus* **16**, 131 (1964).
49. J. Winkelman and G. Slater, *Anal. Biochem.* **20**, 365 (1967).
50. D. L. Horrocks, *Nucl. Instrum. Methods* **117**, 589-95 (1974).
51. W. Rutherford, J. Evans and L. A. Currie, *Anal. Chem.* **48**, 607 (1976).
52. H. Loosli, H. Oeschger, R. Studer, M. Wahlen and W. Wiest, in *Proc. Noble Gases Symposium 1973*, ERDA-TIC: CONF-730915, p. 24.
53. W. J. Gelsema, C. L. de Ligny, J. B. Luten and F. G. A. Vossenberg, *Intern. J. Appl. Radn. Isotopes* **26**, 443-50 (1975).
54. D. E. Watt and D. Ramsden, *High Sensitivity Counting Techniques,* Pergamon Press, Oxford, 1964.
55. B. Francois and M. Ghizzo, *J. Nucl. Med. Biol.* **1**, 147-52 (1974).
56. D. A. Kalbhen and A. Rezvani, in *Organic Scintillators and Liquid Scintillation Counting* (eds. D. L. Horrocks and C. T. Peng), Academic Press, New York, 1971, p. 149.
57. L. A. Currie, in *Proceedings of the Eighth International Radiocarbon Dating Conference* (eds. T. A. Rafter and T. Grant-Taylor), Royal Society of New Zealand, 1973, p. H1.
58. G. E. Calf, in *Organic Scintillators and Liquid Scintillation Counting* (eds. D. L. Horrocks and C. T. Peng), Academic Press, New York, 1971, p. 719.
59. J. F. Cameron, in *Radioactive Dating and Methods of Low-Level Counting*, International Atomic Energy Agency, Vienna, 1967, p. 543.
60. L. A. Currie, *Anal. Chem.* **40**, 586 (1968).
61. B. H. Laney, in *Tritium Symposium* (ed. A. Moghissi), Messenger Graphics, Las Vegas, Nevada, 1973, pp. 156-70.
62. D. L. Horrocks, in *Developments in Applied Spectroscopy*, Vol. 9 (eds. E. L. Grove and A. J. Perkins), Plenum Press, London, New York, 1971, p. 403.

Chapter 19

Assessment of the Significance of Low Count Rates

D. E. Case, D. J. Barnfield and P. R. Reeves

Imperial Chemical Industries Ltd., Pharmaceuticals Division, Macclesfield, Cheshire, England

When considering possible errors in scintillation counting, one is naturally concerned with those errors which are associated with the measurement of levels of radioactivity in samples of various kinds. We wish to focus attention on a topic which might therefore be considered out of place in this symposium; we are concerned to define how to reach the conclusion that a sample contains no significant level of radioactivity above the background level.

Frequently, metabolism and tissue residue studies of medicinal and veterinary drugs at some stage involve the measurement of very low count rates of soft β-emitters in tissues and biological fluids. In studies designed to evaluate tissue residues, low count rates are often the norm rather than the exception.

In any overall assessment of the significance of low count rates, three questions must always be posed:

(i) Is there a significant level of radioactivity to be found in the sample as a result of the prior administration of the labelled compound?

(ii) Is the radioactivity present due to the drug or its metabolites, or to incorporation of the radioactive atom(s) into normal endogenous metabolic pathways?

(iii) Is the level of radioactivity present of any pharmacological or toxicological significance?

The second question, where relevant, can lead to very extensive and complex metabolic studies.[1,2] The third has to be answered by logical argument based on acceptable levels in edible tissues for example, or on the known toxicological and pharmacological properties of the drug and its metabolites. We wish for the moment to consider only the first question.

The necessity to evaluate very low count rates arises either by stumbling unexpectedly into the predicament or by setting out deliberately to assay samples anticipated to contain low or zero levels.

An example of the first situation arose with 'Vivalan'[a] (viloxazine), an anti-depressant whose metabolism has been studied in animals and in man.[3,4] This compound was labelled in three different positions with ^{14}C. One of these radiolabelled forms gave rise to blood levels in dogs with a terminal phase having an extraordinarily long half-life.

[a] 'Vivalan' is a Trade Mark, the property of I.C.I. Ltd.

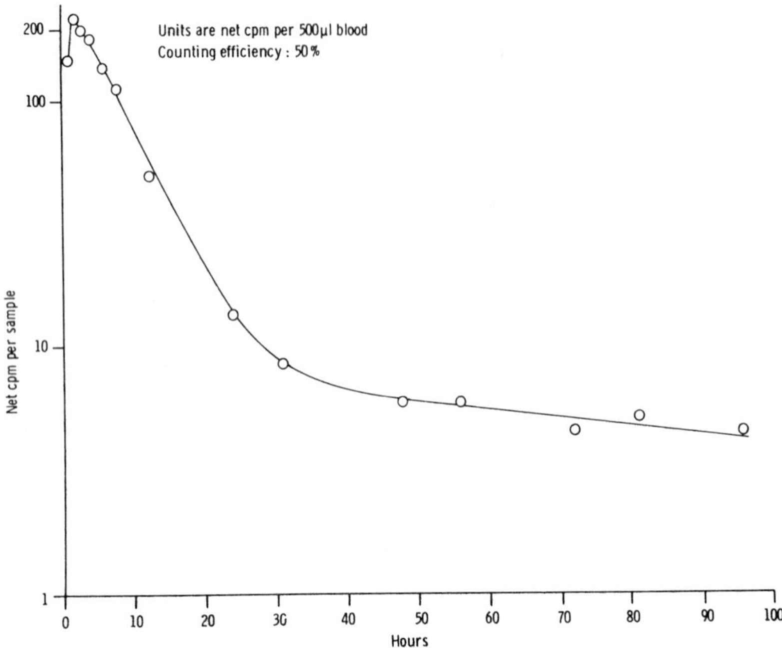

Fig. 1. ^{14}C blood level profile in an animal dosed orally with 40 µCi.

This was shown not to occur when the ^{14}C-label was placed in a different position in the molecule and the effects are due to metabolic incorporation of the ^{14}C due to degradation of the molecule by a very minor metabolic pathway.[2] In this work we were unexpectedly having to measure levels of radioactivity in blood corresponding to 4–40 counts min^{-1} above background (Fig. 1), and particular attention was required in assessing the data.

Let us now consider the situation in which one deliberately sets out to measure very low or zero levels of radioactivity. Data will be presented which is derived from work with a veterinary product in cows. The objectives of the study were to monitor the elimination of the drug and its metabolites from the animal and to define the clearance from edible tissues and milk. The molecule involved could not be satisfactorily labelled with ^{3}H and was eventually prepared with ^{14}C at a high isotopic abundance at one position.

Preliminary studies indicated that any residues present would be of a very low order and that the count rates to be measured would be near to, or indistinguishable from, background. Our conclusion from all the data subsequently obtained was that this product did not in fact give rise to any detectable residual level of radioactive components in the edible tissues. How does one reach this conclusion?

We have been disconcerted to learn the different arbitrary methods used in some groups to define a level of significance above background. Some workers will only accept a count rate as being significant if it is twice the background count rate, or 100 counts min^{-1} above background, or at some other arbitrarily chosen level. This to us seems attractive only in the measure of convenience conferred and several of the problems we have faced in the past would have evaporated had we applied such arbitrary limits. Publications often contain the expressions ‘not detectable’, or ‘not significant’, without further explanation.

There are mathematical expressions describing the error function associated with a net sample count rate which are derived from consideration of the Poisson distribution of radioactive events. These are of value in the design of an experiment, in that such expressions predict the minimum error associated with a given set of counting parameters, but do not take into account other sources of error, introduced, for example, by the sample preparation procedures. Having performed the experiment and counted the samples, our concern becomes the *total* variance of the results, which is inevitably larger than that derived from Poisson statistics, and can only be estimated from the results observed.

To develop the problem as we see it and to present our own methods it is useful to use data from the residue study mentioned previously. Milk samples were counted in 'Instagel', at least six aliquots being counted in new vials for at least 100 min on an Intertechnique Model SL30 with a counting efficiency of 80%. Background data, generated by using predose samples from the same animal, gave comparatively high but reproducible values (Table 1).

Table 1. Mean count rates for predose milk.

Cow no.	Mean background (counts min^{-1}) (n = 9)	S. D.
1	56.84	0.66
2	54.87	0.69
3	56.74	0.53
4	56.01	0.84
5	56.17	1.13
6	55.49	0.62
7	55.35	0.88
8	56.20	1.87
9	55.95	0.90
10	55.94	0.58

Samples contained 10 ml 'Instagel' + 10 ml milk.
Counting time: 100 min per sample.

The background data for one particular cow is shown in more detail in Table 2. The levels of radioactivity found in the milk from this animal declined rapidly after

Table 2. Cow No. 10: Background count rates (counts min^{-1})

Sample	Count rate	Sample	Count rate
1	55.57	6	56.14
2	55.75	7	56.44
3	54.84	8	55.43
4	56.20	9	56.52
5	56.56		
Mean =	55.94	S.D. =	0.58

Samples contained 10 ml 'Instagel' + 10 ml predose milk.
Counting time: 100 min per sample.

dosing to levels which were clearly at or near the background value (Table 3). The mean values and standard deviations derived from this data have been calculated for clarity (Table 4). The problem remaining is to determine which samples contain levels of radioactivity which are significantly in excess of the background value. At what point in time can it be concluded that no detectable level of drug or metabolite is present?

The method employed in our laboratories to assess low count rates is based upon the assumption that the variance of the replicate blanks is the same as the variance of the assayed samples. This must be a very reasonable assumption at extremely low count rates and under conditions where rigorous attention is paid to sample preparation. This assumption cannot be made, of course, where very much higher count rates are involved.

Making the assumptions that the samples are independent and normally distributed, and that the sample standard deviation (S) estimates the population standard deviation

Table 3. ^{14}C found in milk samples (six replicate aliquots taken from each sample).

Sample	Collection period (h)	Net counts min^{-1} per aliquot	
1	0–4	256.4	266.0
		257.6	264.2
		261.0	267.3
2	4–8	182.3	195.0
		185.6	187.4
		192.9	188.4
3	8–24	30.3	29.1
		30.3	28.1
		29.8	30.6
4	24–36	5.2	6.9
		7.1	7.5
		7.6	7.7
5	36–48	0.2	1.2
		0.6	-1.7
		-0.3	0.8

Samples contained 10 ml 'Instagel' + 10 ml milk, and were counted for 100 min/ 10^4 counts.

Table 4. ^{14}C found in milk samples (summary).

Sample	Collection period (h)	Mean net counts min^{-1}	S.D.
1	0–4	262.1	4.5
2	4–8	188.6	4.7
3	8–24	29.5	1.0
4	24–36	7.0	0.9
5	36–48	0.1	1.0
6	48–60	−0.6	0.3
7	60–72	0.6	0.4

(σ) with n degrees of freedom, then the variance of the difference between sample gross count and the background is equal to the sum of the variances. Thus,

Variance of the difference = variance of assayed samples + variance of blanks, i.e.

$$\frac{S_A^2}{N_A} + \frac{S_B^2}{N_B}$$

where N_A refers to the number of assayed samples and N_B refers to the number of blanks.

Hence,

$$\text{Standard error of difference} = \sqrt{\frac{S_A^2}{N_A} + \frac{S_B^2}{N_B}} \qquad (1)$$

with associated degrees of freedom $(N_A-1) + (N_B-1)$.

Frequently there are insufficient replicates of the assayed samples to allow a reliable estimate of S_A and the use of Eqn. (1) above. However, blanks and assayed samples are prepared and processed in an identical manner and produce count rates of the same order when extremely low count rates are involved. Thus it is reasonable to assume that both have the same variance, whence, in Eqn. (1), S_A becomes equal to S_B.

Thus Eqn. (1) reduces to

$$\text{Standard error of difference} = S_B\sqrt{\frac{1}{N_A} + \frac{1}{N_B}} \qquad (2)$$

where the number of degrees of freedom is (N_B-1).

A more accurate measure of error may be obtained by combining estimates of error from blanks and assayed samples when this is possible, i.e. by using the expression

$$\text{S.D.} = \sqrt{\frac{\Sigma(N_i-1)\,S_i^2}{\Sigma(N_i-1)}}$$

where N_i is the number of replicates in the ith group of observations.

Normally, however, Eqn. (2) is used and this requires a satisfactory estimation of the background variance. It is worth noting, of course, that the entire problem with which one is concerned hinges upon the accurate determination of the background count rate and its variance.

Using the milk data previously presented, the following data is used in the example which follows: number of background samples = 9, number of replicates of the assayed sample = 6, degrees of freedom = 8, mean background count rate = 55.94 counts min^{-1}, standard deviation for the latter = 0.58.

The calculation using Eqn. (2) above is thus

$$\text{Standard error of difference} = S_B \sqrt{\frac{1}{N_A} + \frac{1}{N_B}}$$

$$= 0.58 \sqrt{\frac{1}{6} + \frac{1}{9}} = 0.31 \text{ count min}^{-1}$$

This has now to be multiplied by the Student's t-value for 8 degrees of freedom ($p = 0.01$, single sided) which is 2.90. Hence the confidence limit, i.e. the limit of detection at the confidence level specified, is 0.90 count min^{-1} above background.

In applying this treatment to the experimental data previously presented (Table 4), we now understand that at 99% confidence limits a mean count rate 0.90 count min^{-1} above the mean background value of 55.94 is significant. By inspection it can be concluded that net count rates in excess of 0.90 count min^{-1} are significantly above background. From Table 4 it is clear that samples 1 to 4 contain significant levels of radioactivity and that the remainder do not.

The method of assessment described may of course be evaluated by adding known amounts of radioactivity to control blanks and determining experimentally whether the levels at which radioactivity can apparently be measured reliably agree with those predicted.

As an example, data are given in Table 5 for blood samples spiked with known amounts of the ^{14}C-labelled compound. It will be seen that levels in excess of 2 disintegrations min^{-1} can apparently be detected reliably. Analysis of the background data (Table 6) indicates a limit of detection corresponding to 1.84 disintegrations min^{-1}. This is consistent with the experimental results (Table 5) which become unreliable below this level. These data were generated using an oxygen flask combustion method based upon that of Kalberer and Rutschmann,[5] the samples being counted for 100 min in a Packard 2211 Tri-Carb Scintillation Counter.

Table 5. Experimental determination of limits of detection, 0.5 ml whole blood spiked with ^{14}C standard.

Nominal dpm added	Net dpm found
0.5	-0.3
1.0	-0.2
2.0	2.4
5.0	6.0
10.0	9.8

The procedure described for blood was repeated for several tissues, with the results presented in Table 7. By calculation of the limit of detection from blank data, and comparison with the 'dpm found' figure, one may determine, for each tissue, the standard for which a significant result was just obtained. The experimentally determined limit of detection must thus lie between this standard and the next smaller one (for

Table 6. Limit of detection predicted from blank data.

Number of samples = 6 Combustion aliquot factor = 1.5
Counting time = 100 min Mean counting efficiency = 70%
Mean blank = 29.44 ± 0.72 counts min^{-1} (S.D. : n = 19)
Student t-value (p = 0.01), with 18 degrees of freedom = 2.55

Calculated limit of detection

$$= 2.55 \times 0.72 \sqrt{\frac{1}{19} + \frac{1}{6}} \text{ counts min}^{-1}$$

= 0.86 net counts min^{-1} per aliquot = 1.29 net counts min^{-1} per sample
= 1.84 dpm per sample

Table 7. Experimental estimation of limits of detection.

Standard added (dpm)	dpm found in spiked sample[a]				
	Blood	Muscle	Liver	Kidney	Fat
0.5	-0.3	NA	NA	NA	NA
1	-0.2	NA	NA	NA	NA
2	2.4	5.9	1.9	2.6	2.3
5	6.0	7.4	5.0	5.1	5.3
10	9.8	11.5	8.6	11.3	11.2
25	NA	27.0	25.2	26.0	23.3
Calculated limit of detection from blanks (dpm)	1.8	3.2	2.0	2.8	2.9
Actual dpm in '1st sig. sample'	2	2	5	5	5
Actual dpm in 'last non-sig. sample'	1	(1)	2	2	2

[a] All results are means of 6 replicates, each counted for 100 min.
NA: Not assayed.

which a significant result was not quite obtained). It can be seen that, with the exception of muscle, the calculated limit of detection lies in this range and that the practically and theoretically estimated limits are in good agreement.

The result for muscle exemplifies the importance of the questions mentioned earlier. For the 2 disintegrations min^{-1} standard, there is a significant result. We are more fortunate than usual in being able to state that it is obviously inaccurate, but the result remains significant nonetheless.

Knowing the correct answer, one is in a position to question the source of this inaccurate result. Is the error due to a faulty standard preparation, contamination or an inappropriate background determination?

Had it not been a standard, then subject to the latter two qualifications it would have been accepted at face value. The ultimate importance of this result would then have been evaluated in terms of its biological significance. The answers to such questions will be obtainable not from statistical considerations but from commonsense and experience.

When concerned with problems of very low count rates, one may be exhorted to remove the problem by using a higher specific activity, a higher dose or a different labelled isotope. This, unfortunately, is not always possible.

In some circumstances, one may be in the fortunate position of being able to predict, from toxicological data, or guidelines laid down by regulatory authorities, the minimum tissue levels which may be acceptable. This enables one to calculate a corresponding level of radioactivity below which the corresponding levels are of no concern. If this level is sufficiently removed from background values, any statistical analysis may be dispensed with.

We would like to consider that scintillation counting is a technique amenable to statistical analysis and to base our calculations upon a rational treatment of the data. It is fascinating to find that very little has been published in this area and we would welcome discussion of the method presented and of any other treatment currently used in assessing the significance of very low count rates.

ACKNOWLEDGMENT

The authors wish to acknowledge the invaluable advice and continuing assistance of S. H. Ellis of the Statistics Group, I.C.I. Pharmaceuticals Division.

REFERENCES

1. J. C. Potter, J. E. Loeffler, R. D. Collins, R. Young and A. C. Page, *J. Agric. Fd. Chem.* **21**, 163 (1973) and references cited therein.
2. D. E. Case, *Xenobiotica* **5**, 133 (1975).
3. D. E. Case, H. Illston, P. R. Reeves, B. Shuker and P. Simons, *Xenobiotica* **5**, 83 (1975).
4. D. E. Case and P. R. Reeves, *Xenobiotica* **5**, 113 (1975).
5. F. Kalberer and J. Rutschmann, *Helv. Chim. Acta* **44**, 1956 (1961).

DISCUSSION

L. A. Currie: With regard to measurement of detection limits and statistical detection powers and errors of the first and second kind, my congratulations on a rational and objective approach to the problem of detection in counting. It would also be valuable to consider the question of detection power, i.e. given a certain level of radioactivity, what is the probability of detecting it according to your detection criterion? In this way one may encompass both errors of the first kind (probability of false detection) and of the second kind (probability of false non-detection) in identifying the detection limit.

P. Johnson: You began by showing a slide in which you briefly mentioned the possible significance of low count rates in tissues and body fluids, although your talk rightly concentrated on the main topic under consideration, which is the techniques involved in measuring and interpreting low count rates. You also emphasised that not everyone can attain the high specific activities or administer the high doses that they might wish to in order to obtain reasonable count rates, such as in a residue study, and that assessment of low count rates is then essential. I agree. However, there is also a danger, which I feel should be emphasised, when material of high specific activity or high doses can be used. This danger is that, in setting a very low absolute level of detectable counts, data will be generated from which a legislative authority may give a significance to a few photons detected in a scintillation counter which is out of all proportion to the biological significance; in fact. the biological significance may be zero.

D. E. Case: In reply to your last comment, I must say that I agree entirely. In all studies of this kind the biological and toxicological significance of the results must be assessed rationally.

Often, the amounts detected may be totally insignificant in terms of safety evaluation. However, in some circumstances, we may be asked, having concluded that no detectable level is present, what level *could* be present which you are not quite able to detect by your methods? We feel, therefore, that some method of assessment is necessary; whatever the outcome, commonsense must be used in evaluating the real toxicological significance of the results.

B. R. Twite: The low count levels in the milk samples occurred 24 h or longer after dosing the radiolabelled drug. This means that your control (blank) samples were 24 h older than the counted samples. At the very low count levels measured, perhaps the difference in age between the samples could be responsible for the difference in counts observed. As it is possible to obtain a statistical difference in counts between samples which in fact does not exist, is it not more valid to quote residue levels as less than a certain value, the value being set to give a greater certainty that a true difference in counts does exist (e.g. twice background)?

D. E. Case: We concluded that the use of predose milk taken from each cow was the best possible blank; it is, of course, impossible to determine the absolute blank at a time point after administration of the labelled compound. It was also found that background data for different cows (Table 1) were remarkably consistent despite the fact that these samples were of slightly different ages when actually counted. I would object to your proposal to use a 'certain value' (e.g. twice background) as this falls into the arbitrary kind of practice which we find difficult to justify on any theoretical grounds.

J. E. Noakes: Milk samples are analysed for natural radiocarbon content at the University of Georgia by freeze drying and pallete spraying and combusting samples to CO_2 followed by conversion to C_6H_6. Perhaps greater sensitivity could be gained in your present mode of counting raw milk samples if enough material is present to concentrate the samples in this way. If a combustion apparatus is available using sample conversion to CO_2 and collecting in an amine (2–methoxyethylamine), this could also be counted with good efficiency and perhaps with lower background, thereby increasing the sensitivity of your analytical method.

Chapter 20

Errors in dpm Measurement in Liquid Scintillation Counting with External Standardisation

E. G. Cummins* and C. G. Horne

Packard Instrument Ltd., 13–17 Church Road, Caversham, Berkshire, England

INTRODUCTION

Liquid scintillation counting is not an absolute measuring technique. The results obtained from any system need to be qualified by a determination of the efficiency at which the samples were measured. Basically, samples may be classified as either heterogeneous or homogeneous. Measurements on heterogeneous samples including pieces of filter paper or microcellular filtration discs are affected by self-absorption which is not easy to make allowances for. This paper deals with homogeneous samples and the accuracy of the efficiency measurement in order to obtain correct values of disintegrations min^{-1}.

There are three methods of monitoring efficiency available in liquid scintillation counting: (i) internal standardisation – this is the only primary technique but is very slow because it requires double counting, and is relatively expensive; (ii) the sample channels ratio technique – this takes place as the sample is being counted but has several recognised disadvantages, notably the occurrence of statistical inaccuracies when measuring low activity samples; (iii) external standardisation – originally carried out by external standard count (ESC) methods and later by external standard ratio (ESR) measurement. The latter is probably the most maligned of the three techniques because there were a number of instrument problems in earlier years and some confusion among users, but is now certainly the most suitable method for most types of sample.

Since

efficiency = f_1 (quench)

and external standard = f_2 (quench)

it follows that efficiency = f_3 (external standard)

This relationship forms the basis of a quench correction curve. The idea of using a γ-emitting isotope to produce a Compton electron spectrum in a liquid scintillation counter was first applied to LSC in the early 1960's.[1–4] Any external standardisation system should include the following essential characteristics:

**Present address:* Burkard Scientific Ltd., Rickmansworth, Hertfordshire, England.

(i) wide dynamic range – normally to cover tritium efficiencies down to 0.0%;
(ii) volume independence – over the designed range, normally 8–18 ml;
(iii) count rate independence;
(iv) quencher independence – providing the sample remains stable in solution.

Desirable features are:

(v) cocktail independence;
(vi) vial independence.

In current commercially available liquid scintillation counters, three types of radionuclide are used as external standard sources: ^{133}Ba with a γ photo-electric peak at 360 keV and an X-ray peak at 80 keV; ^{137}Cs with a γ photo-electric peak at 662 keV and an X-ray peak at 33 keV; ^{226}Ra, a complex spectrum with a large number of daughter peaks with energy up to 2.8 keV. These spectra are shown in Fig. 1 as they would appear in a standard liquid scintillation counter using unquenched toluene–PPO–dimethyl-POPOP cocktail.

EXPERIMENTAL

Ten millilitres of cocktail (5 g l^{-1} PPO, 0.5 g l^{-1} dimethyl-POPOP) was made up with toluene and counted in standard low-potassium glass vials. A series quenched with carbon tetrachloride was made up for ^{3}H and ^{14}C (NBS certified ± 1% ^{3}H, ± 3% ^{14}C). Figures 2, 3 and 4 show the quench correction curves obtained with these samples on each of three different instruments. In each case the manufacturer's normal wide window counting condition was selected.

In order to assess the dynamic range of an external standard system, the indicated external standard ratio was plotted against the volume of quenching agent employed, as shown in Fig. 5. These results indicate that ^{133}Ba has too wide a dynamic range, that of ^{137}Cs is too narrow, while that of ^{226}Ra is most convenient to use. This ties in with Figs. 2, 3 and 4 which show ^{133}Ba with efficiency zero at ESR 0.2 and ^{137}Cs with efficiency 15% and 30% at ESR zero.

To assess the effects of volume on the liquid scintillation counter, a series of standards of from 5–20 ml volume was constructed using the stock cocktail and radioactive label. The effect on counting efficiency of changing volume is shown

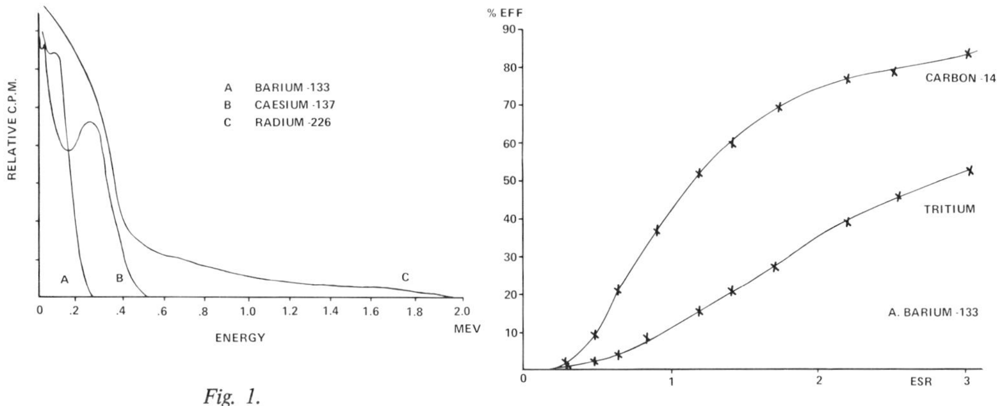

Fig. 1.

Fig. 2.

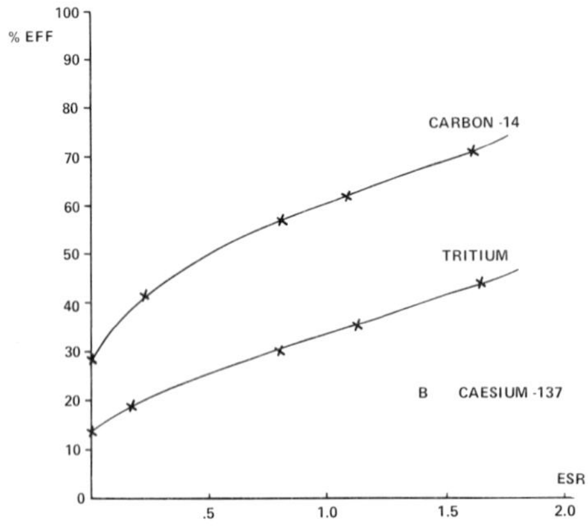

% EFF
CARBON -14
TRITIUM
B CAESIUM -137
ESR

Fig. 3.

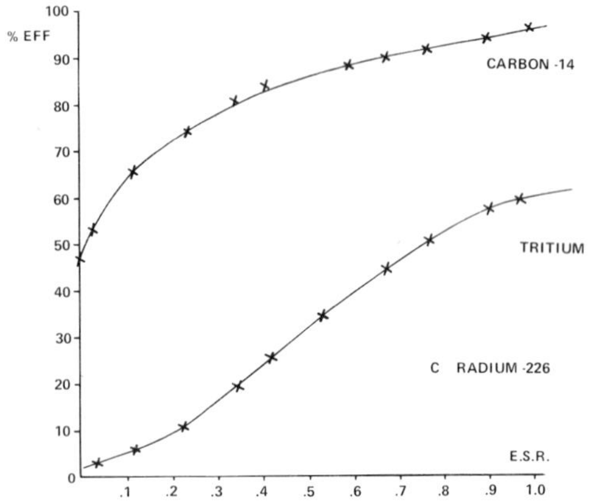

% EFF
CARBON -14
TRITIUM
C RADIUM -226
E.S.R.

Fig. 4.

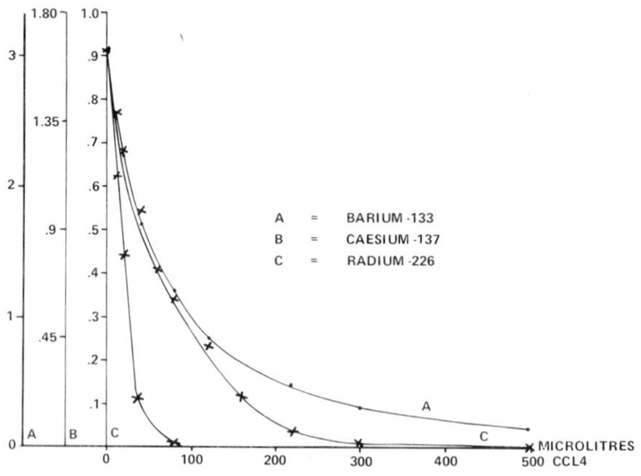

A = BARIUM -133
B = CAESIUM -137
C = RADIUM -226
MICROLITRES
500 CCL4

Fig. 5.

% EFF
CARBON -14
TRITIUM
ml

Fig. 6.

ESR
ESC
X1000
ESC
ESR
EFFICIENCY
MONITOR
V'S VOLUME
ESR = x
ESC = •
20 ml

Fig. 7.

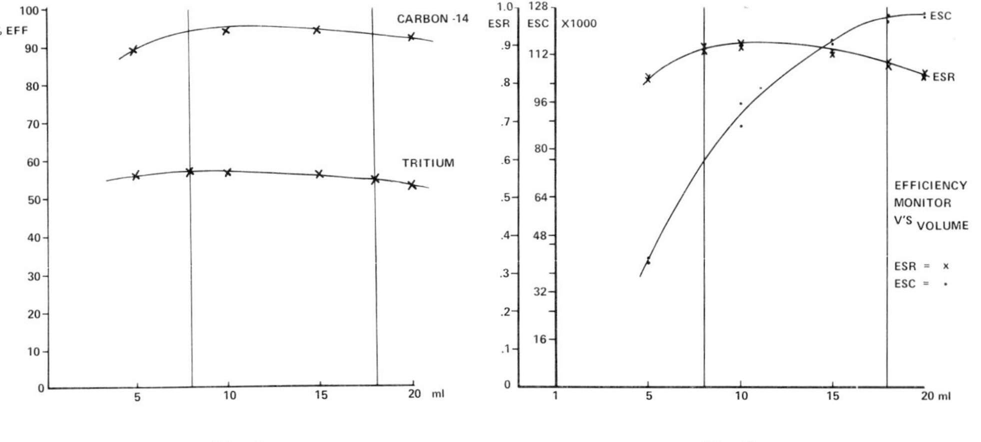

in Fig. 6. Between 8–18 ml, the efficiency for ^{14}C and ^{3}H changes by 3%. The effect of volume changes on the two external standard methods is compared in Fig. 7, where it can be seen that the external standard count curve differs considerably from the efficiency volume curve, whereas the external standard ratio curve is very similar to it. The comparative accuracy in disintegrations min^{-1} assessment using ESR and ESC with volume change is shown in Fig. 8. Data relating to the indicated counting efficiency was obtained from a standard quench curve using samples of 10 ml volume. It can be seen, therefore, that even small changes in volume or sample height caused by vial imperfection could cause appreciable errors in disintegrations min^{-1} accuracy.

The ability of an instrument to be independent of count rate tests not only the external standard system but all of the counting electronics. Figures 9 and 10 illustrate results obtained by increasing activity at constant volume. In the case of ^{3}H, ^{133}Ba and ^{226}Ra show very similar results, whereas ^{137}Cs seems to be less accurate as the count rate increases. In the case of ^{14}C, both ^{226}Ra and ^{133}Ba initially are very similar but then diverge, the results for ^{226}Ra being similar for both ^{3}H and ^{14}C. In the case of ^{137}Cs, the results are obviously completely different. (The results for ^{137}Cs would appear to be completely reversed from those found by Gogan and Gogan,[5] who found that with a caesium external standard source, in the case of ^{3}H the error was positive and in the case of ^{14}C the error was negative.)

		8ml	DPM ERROR	18 ml	DPM ERROR
CARBON-14	ESC IND / EFF	758K / 49%	+ 16.3%	126K / 62%	-12%
	ACT EFF	57%		54.6%	
	ESR IND / EFF	.88 / 56.5%	+ .8%	.85 / 55%	- .7%
	ACT EFF	57%		54.6	
TRITIUM	ESC IND / EFF	76K / 89%	+ 5.6%	126K / 97%	- 4.7%
	ACT EFF	94		92.5	
	ESR IND / EFF	.88 / 93.5	+ .5%	.85 / 93	- .6%
	ACT EFF	94		92.5	

Fig. 8.

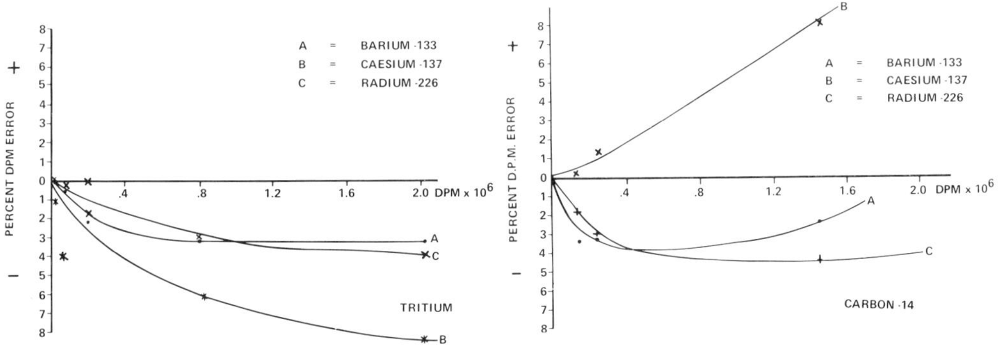

Fig. 9. *Fig. 10.*

The study of the effect of introducing different quenching agents produced the results shown in Fig. 11. Samples were prepared using 10 ml of a standard toluene cocktail and quenched with acetic acid, chloroform, red dye and yellow dye. In each case the amount of quenching agent added was such that all samples should count at a similar efficiency and be measured at approximately 70% of the maximum ESR for that nuclide. It can be seen that the overall accuracy of all three instruments is quite similar in the case of acetic acid, chloroform and yellow dye, but with the red dye the results differ widely. The ^{133}Ba source appears to show much better accuracy with ^{14}C samples than with ^{3}H. The ^{137}Cs source appears to give high results for ^{14}C samples, corresponding with the high results obtained with ^{14}C in the increasing count experiment. In the case of ^{226}Ra, the results are consistent for chemical quenchers, but in the case of the ^{3}H red sample the answer is very different from that obtained with the other two isotopes. Since the same sample was measured in all three instruments, any error in the orginal number of known disintegrations min^{-1} in the sample should show a similar error with all three isotopes, i.e. they should all be high or low.

The next experiment was designed to test the practicality of using the quench curve obtained for one solvent scintillator system for samples contained in another type of cocktail. Four standard commercial cocktails were used (10 ml):Instafluor® (® Registered Trade Name, Packard Instrument Co.), a simple xylene-based system designed to count organic samples at high efficiency; Instagel®, a complex emulsion cocktail, xylene-based, designed to cope with the presence of large quantities of water; Permafluor V®, a toluene-based cocktail with a high scintillator concentration designed for use with sample oxidisers; Permafluor II®, a dioxane-based cocktail similar to Bray's solution. The same amount (45 μlitres) of quenching agent was added to each cocktail. From the ^{3}H results shown in Fig. 12 it can be seen that the Permafluor II® assessments were all low suggesting either that a toluene quench curve cannot be used for dioxane samples or that the original activity in the sample was low. In the ^{14}C results (Fig. 12) only Instagel appears to give a consistent figure; therefore, it can be assumed that for ^{14}C a different quench curve is probably needed for each different scintillator system. The overall errors with the ^{133}Ba source are quite wide, particularly in the case of Instafluor. The

PLOTTING TABLE

CHEMICAL AND COLOUR

^{3}H(1)	A	B	C
ACETIC ACID	–	– 0.4	– 1.6
CHLOROFORM	– 2.0	+ 1.5	+ 0.8
RED	+ 9.8	+ 7.8	– 6.3
YELLOW	– 1.8	+ 2.4	– 1.4
^{14}C(2)			
ACETIC ACID	– 2.5	+ 0.8	– 1.2
CHLOROFORM	0	+ 8.8	0
RED	– 0.1	+ 8.6	– 1.8
YELLOW	– 1.4	+ 5.8	– 3.7

Fig. 11.

PLOTTING TABLE

^{3}H(1)	A	B	C
INSTA FLUOR	+ 4.6	+ 2.4	+ 0.2
INSTA GEL	– 2.7	– 4.5	– 0.2
PERMAFLUOR V	– 7.7	– 6.5	– 0.8
PERMAFLUOR II	– 8.8	– 6.5	– 5.9
^{14}C(2)			
INSTA FLUOR	+ 1.7	+ 1.0	+ 9.1
INSTA GEL	0	+ 0.8	– 0.7
PERMAFLUOR V	– 6.3	+ 4.9	+ 2.0
PERMAFLUOR II	+ 0.4	– 0.9	+ 7.6

Fig. 12.

results from ^{137}Cs are much more consistent than those of ^{133}Ba or ^{226}Ra. ^{226}Ra shows good consistency with xylene- and toluene-based cocktails, except in the case of Instafluor for ^{14}C, but does not seem to like dioxane-based cocktails.

REVIEW OF TWO EARLIER PAPERS

The ability to be able to use vials of different materials is important in some laboratories. Two publications have examined the use of polyethylene vials with toluene-based cocktails. Gogan and Gogan,[5] and Rauschenbach and Simon[6] examined instruments using all three nuclides as external standard sources (Fig. 13). In the case of ^{133}Ba it was found that the external standard ratio changed by 0.4% per hour. In the case of ^{137}Cs it was found that the external standard ratio changed by 0.8% per hour. Both groups found that no similar change appeared in an instrument using ^{226}Ra. The actual error in disintegrations min^{-1} assessment would depend on the individual quench correction curve and the level of quench in the sample, but in an overnight run it could be as much as 8% for an instrument using ^{133}Ba or 16% for an instrument using ^{137}Cs. When dioxane-based cocktails are used in polyethylene vials no similar drift occurs in an instrument. No instrument shows any change using glass vials with any type of solvent system. Horrocks[7] recently published a method of using ^{137}Cs for polyethylene vials using toluene-based cocktails which reduces the drift to much lower levels, but this is done by sacrificing dynamic range. With any type of external standard source it is necessary to construct a different quench curve for each type of vial in use – glass, polyethylene, or minivial.

NUCLIDE	G. and G.[a]	R. and S.[b]
^{133}Ba		0.5% h^{-1}
^{137}Cs	0.8% h^{-1}	0.8% h^{-1}
^{226}Ra	*M*	*M*

[a] Gogan and Gogan Ref. 5.
[b] Rauschenbach and Simon Ref. 6.

Fig. 13. External standard ratio shift with polyethylene vials.

CONCLUSION

It has been shown that instruments with different isotopes as external standards differ widely in that all external standard sources do not show the same performance, whereas with two machines using the same isotope as an external standard source it is possible that the characteristics will only differ slightly. Therefore, in their own interest users and prospective users of liquid scintillation counting should thoroughly investigate the suitability of an instrument for their requirements. The fact that many instruments have an external standard does not mean that all such instruments are the same. Its suitability has to be examined closely if disintegrations min^{-1} errors are to be kept to a minimum.

REFERENCES

1. W. J. Kaufman *et al., Proceedings of the University of New Mexico Conference on Organic Scintillators*, TID 7612, 1960, pp. 251–5.
2. D. G. Fleishman *et al., Instr. Exptl. Tech.* 472–4 (1962).
3. T. Higashimura *et al., Intern. J. Appl. Radn. Isotopes* **13**, 308–9 (1962).
4. H. E. Hobbs, *Nature (London)* **200** (4913), 1283–4 (1963).
5. F. Gogan and Gogan, *Anal. Biochem.* **60** (2), 363 (1974).
6. P. Rauschenbach and H. Simon, *Anal. Chem.* **256**, 119 (1971).
7. D. L. Horrocks, *Intern. J. Appl. Radn. Isotopes* **26**, 243–56 (1975).

DISCUSSION

B. H. Laney: Did you test various instruments or one instrument with three sources?

C. G. Horne: The tests were done on three commercially available instruments using manufacturers' recommended counting conditions and external standard windows.

B. H. Laney: The phenomenon of the external standard ratio (ESR) drift with polyethylene vials is caused by additional scintillations produced by the external standard in the vial walls when permeated by the solvent. Consequently, the vial becomes a plastic scintillator producing weak scintillations proportional to the γ-excitation. I reported these observations in *Tritium*, Messenger Graphics (1971) p. 166. In the figure, ^{226}Ra, ^{137}Cs and ^{133}Ba external standard sources are compared in the same instrument with the same instrument settings. The ESR drifts 15 to 25% within a day when the external standard windows are set to produce statistically significant ESR's over a wide (25:1) change in pulse heights. When the lower level discriminators of both ESR windows are raised to exclude the scintillations produced in the vial walls, the ESR drift with freshly prepared samples is less than 2% with each of the three external standard sources. (Fig. 14).

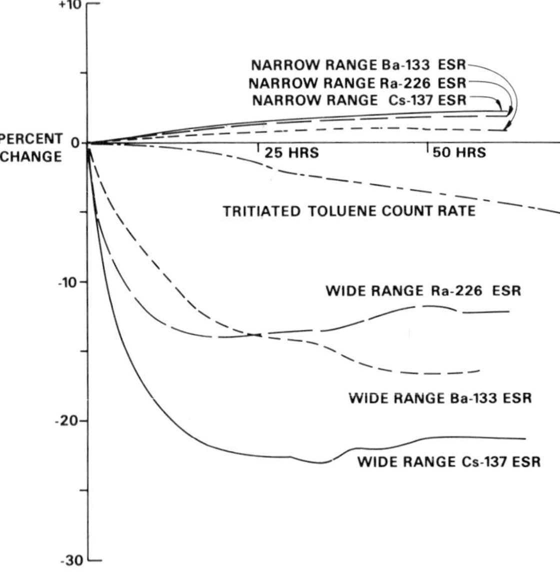

Fig. 14.

C. G. Horne: The settings chosen by a manufacturer for his instrument are a compromise to attempt to meet the essential requirements as previously discussed. Therefore, although it is possible that an instrument equipped with any type of external standard source can be adjusted to handle with accuracy one characteristic requirement, it is certainly done to the detriment of the performance of the system to other essential requirements. The slides which Mr Laney has shown have no practical consequence to the man in the laboratory using liquid scintillation counting as another analytical technique. He is not in a position to readjust his instrument daily for every type of sample which he may encounter that day.

B. Morris: Having shown that a different vial and/or scintillation cocktail requires a different quench correction curve (external standard channels ratio method), have you investigated the effect of scintillant ageing and batch to batch variation on the application of any one correction curve to any one vial/cocktail system?

C. G. Horne: No.

T. J. Rising: We have prepared quenched standards for a number of scintillators and biological samples such as urine and plasma and used Radium as the AES source. Over a period of about three months we have checked the curves and found good agreement between scintillators such as Permafluor and Nuclear Enterprise 260.

P. Gresham: No curve is reliable in the long term.

R. P. Parker: Important samples justify a separate curve generated at the time of the experiment.

D. J. Barnfield: Can you explain to me, as a non-physicist, why there are such large discrepancies associated with the use of ^{137}Cs with samples of high activity?

C. G. Horne: I think that in such cases we should look at the instrument as a whole, and consider the electronics involved. Instruments using ^{137}Cs as external standard tend to utilise logarithmic amplification and those using ^{133}Ba and ^{226}Ra tend to utilise linear amplification, the answer to your question lies in that area.

A. Dyer: It would be interesting to speculate on what might have happened if you had not put in dimethyl-POPOP.

C. G. Horne: We attempted to use materials which are the ones most commonly used in practice.

Chapter 21

Temperature Compensation in Liquid Scintillation Counters

F. E. L. ten Haaf and J. G. H. Kuipers

N. V. Philips Gloeilampenfabrieken, Eindhoven, The Netherlands

INTRODUCTION

Temperature, and changes in temperature, may affect the performance of a liquid scintillation counter in various ways. For this reason, temperature-controlled cooling has long been a standard feature of liquid scintillation counting systems. Originally, its main objective was to reduce background. The older photomultiplier types used to exhibit high thermal noise count rates at room temperature, which contributed appreciably to background, even in coincidence systems. This effect could be considerably reduced by cooling the detector system to a temperature around 0°C. However, since the introduction of modern tubes with bialkali photocathodes and low electron transit time spread, cooling is no longer necessary to obtain a low background. With these tubes, and using fast coincidence circuitry, the count rate due to accidentally coinciding thermal noise pulses is typically less than 0.1 count min^{-1} at room temperature which can be considered negligible for most practical purposes.

From the point of view of sample composition, counting at ambient temperature often offers the advantage of better solubility for many chemical compounds. Cooling, however, remains useful to improve background stability in low level counting applications and, occasionally, to 'stabilise' otherwise unstable inhomogeneous samples or to reduce persistent chemiluminescence.

Apart from those already mentioned, three temperature-sensitive parameters remain to be considered (it is assumed that the influence of temperature on the electronic circuits used in the counting system – such as the pulse amplifiers, the discriminators and the high voltage supply for the photomultiplier tubes – is reduced to negligible proportions by means of conventional electronic techniques). These are the scintillation efficiency of the sample, the photocathode efficiency and the gain of the electron multiplier section of the photomultiplier tubes. The relevant literature has been reviewed by Birks.[1] Although it appears difficult to assign a specific temperature coefficient to any of these parameters, the general trend is a small negative value in the range of temperatures that is of interest in liquid scintillation counting. As a consequence, the counting efficiency varies with temperature, increasing if the temperature of the system is lowered. Although this effect normally is not large enough to be valuable in obtaining substantially better counting statistics or shorter counting times, it is often large enough to cause inaccurate measuring results if the temperature of the counting system is not kept constant within a few degrees Centigrade. This tends to limit the accuracy of ambient temperature systems, unless operated in a controlled environment. In temperature-controlled counting systems, the effect may make recalibration necessary if the set temperature of the system is altered. The objective of this study has been to investigate the feasibility of compensating for the effects mentioned by introducing a temperature-dependent element in the overall gain of the system.

PRELIMINARY CONSIDERATIONS

Almost from the outset it seemed clear that it would be impossible to compensate for the temperature influence on the scintillator efficiency as each scintillator would generally require a different amount of compensation. The same consideration appears to be valid with respect to the photocathode efficiency. As shown, among others, by Murray and Manning,[3] the temperature coefficient of the photocathode efficiency strongly depends on the wavelength of the incident light. Here again, a compensation could only be valid for a specific sample composition, as the wavelength distribution of the emitted light depends on the primary and secondary solutes and is also modified by other sample properties, such as colour. This left the gain of the electron multiplier as the only parameter for which temperature compensation could be achieved.

We considered, however, that this could still be valuable. All system parameters, with the exception of the electron multiplier gain, will affect both the coincidence sensitivity and the overall gain (see Fig. 1). A change in any of them, whether brought about by a change in sample composition or by a change in temperature, may be expected to cause

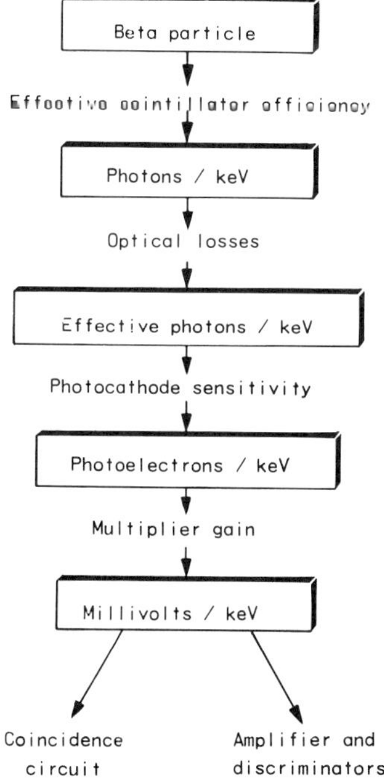

Fig. 1.

the point of a particular sample on the efficiency calibration curve to slide along that curve. A change in the electron multiplier gain, however, will affect the overall gain of the system but not the coincidence sensitivity, as the coincidence circuit is operated in the single electron plateau and, consequently, is insensitive to moderate changes in the overall gain. Due to this difference, a change in the electron multiplier gain, reacting to a temperature variation, will generally cause a point, characterising the sample on the efficiency calibration curve, to deviate from that curve. Compensation for the temperature coefficient of the electron multiplier gain will eliminate this effect and therefore seems to offer the possibility of making the efficiency calibration curve insensitive to temperature changes.

Regarding the magnitude of the various temperature effects mentioned above, the literature appears to be rather confusing. Certain large temperature coefficients that have been reported, however, seem to have been influenced by photomultiplier fatigue. Cathey[2] has shown that fatigue may cause appreciable changes in the gain of photomultipliers, at collector currents in excess of 0.1 μA. This makes any measurement results obtained at higher currents suspect. Another difficulty is created by the fact that many authors report overall temperature coefficients, which include the photocathode sensitivity as well as the electron multiplier gain, and sometimes even the scintillator efficiency. This makes it impossible to assess the temperature coefficient of each of them separately.

Murray and Manning[3] give, at least for two of the photomultipliers they studied, detailed graphs for normal operation as a photomultiplier and for operation as a diode, i.e. without using the electron multiplier section. From these graphs it can be deduced that the temperature coefficients of the electron multiplier gain of these tubes had values of approximately -0.2 and -0.4 % per degree Centigrade in the temperature range between 23°C and 0°C. Incidentally, the temperature coefficient of the photocathode efficiency for light with a wavelength of 4000 Å appears to be practically zero in this temperature range.

EXPERIMENTAL RESULTS

For our experiments we used a prototype of the Philips Model 4540 liquid scintillation analyser. A small temperature sensor was mounted in the detector and connected to a simple D.C. amplifier. The output of the amplifier could be varied from zero to several volts per degree Centigrade, and was connected in series with the high voltage for the photomultipliers. Using progressively quenched standards of ^{3}H and ^{14}C, we made measurements at different system temperatures and different settings of the D.C. amplifier gain.

Some typical results are shown in Figs. 2, 3 and 4. They show channels ratio curves for standard ^{3}H and ^{14}C measuring windows. Figure 2 shows the effect of a temperature change if no compensation is applied. Although the ^{14}C efficiencies increase considerably, if the temperature is lowered, they stay on the same curve. The ^{3}H calibration curve, however, is shifted to the right. By increasing the D.C. amplifier gain the ^{3}H calibration curves, taken at different temperatures, approach each other until a point is reached where they coincide (Fig. 3). In this situation, the 'cold' ^{3}H points lie slightly above the 'warm' ^{3}H points on the same curve. The ^{14}C points have come much closer to one another and are still on the same curve. This is what one would expect in a situation where the effect of temperature on the electron multiplier is compensated for. The remaining differences in the counting efficiencies can safely be attributed to a change in the scintillation efficiency.

% counting efficiency
no temperature
compensation
^{14}C
^{3}H
18°C
6°C
chanrels ratio
% counting efficiency
with compensation of
temperature effect
on P.M. dynodes
^{14}C
^{3}H
21°C
8°C
channels ratio

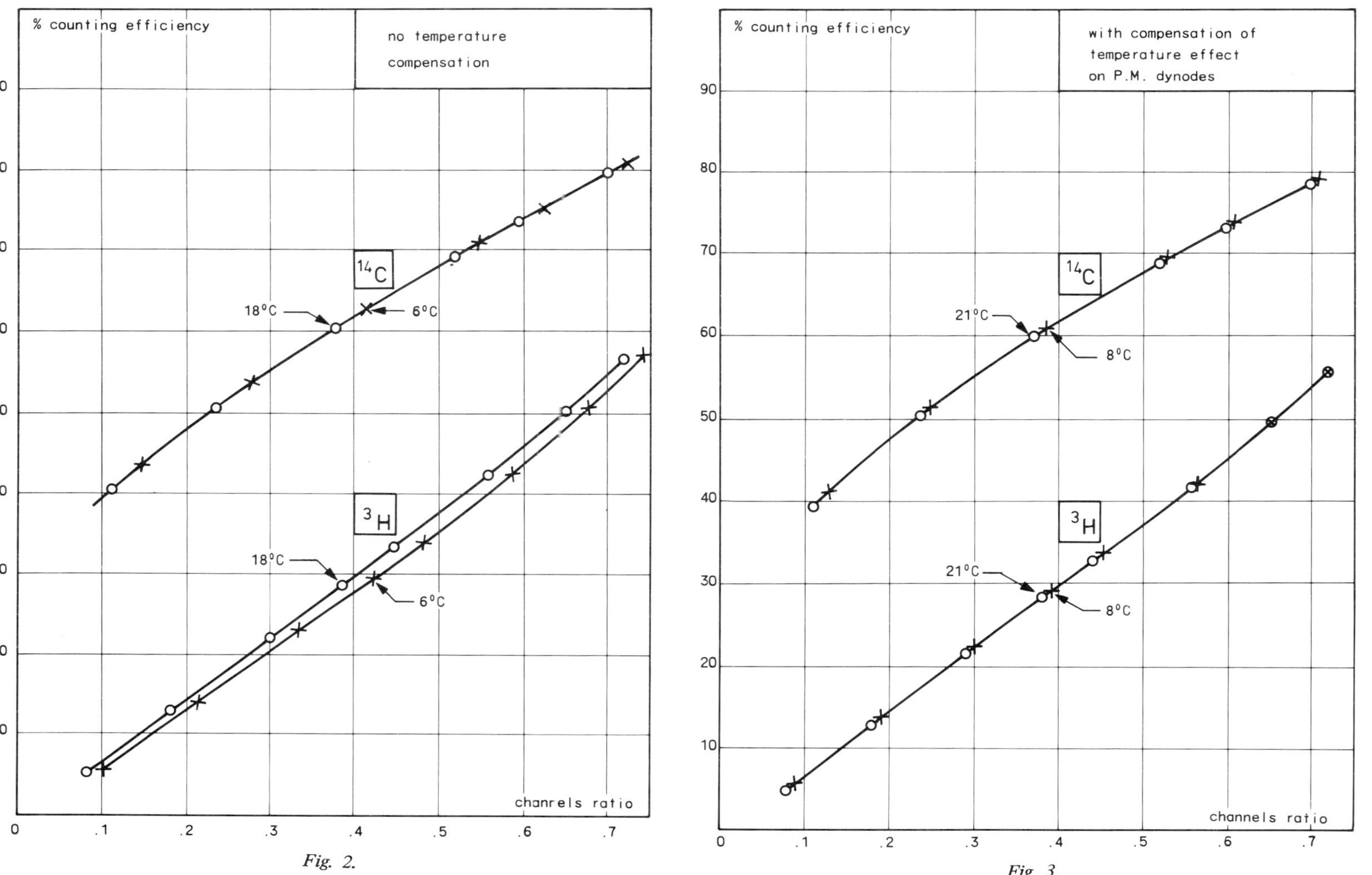

Fig. 2.

Fig. 3.

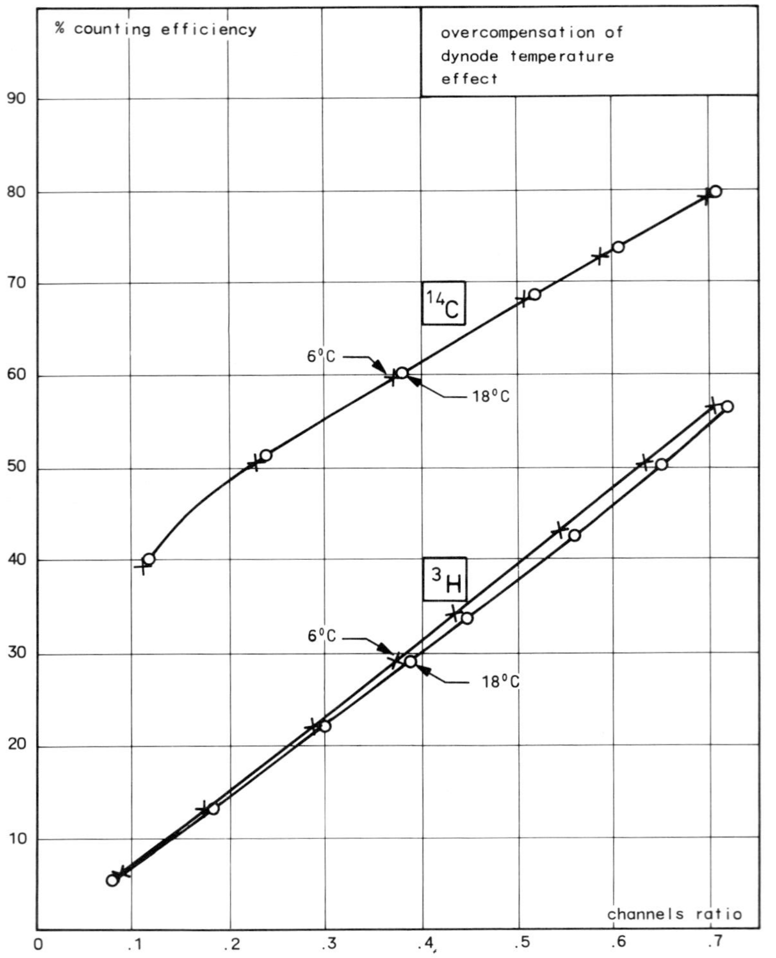

Fig. 4. (left).

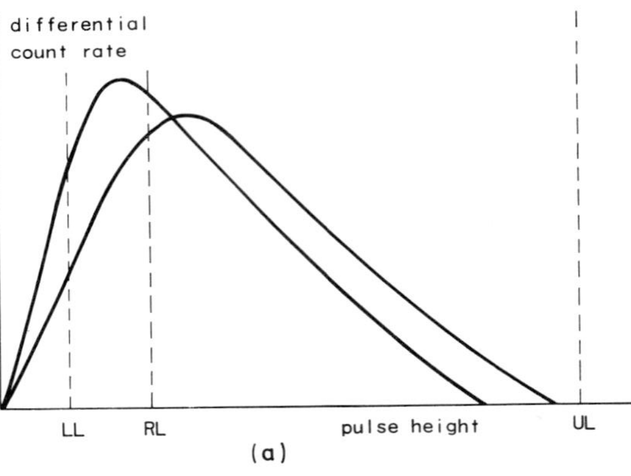

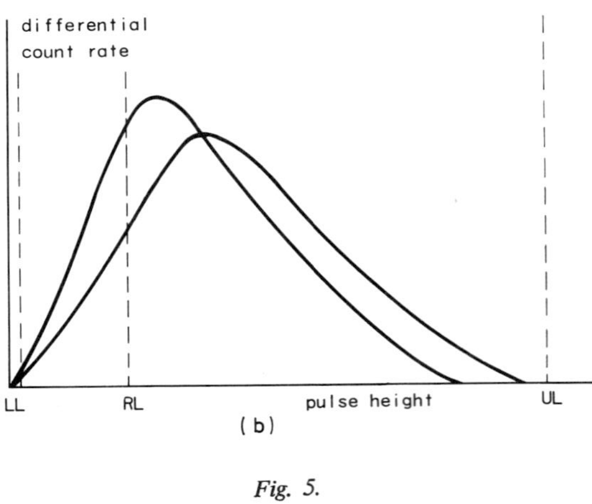

Fig. 5.

Figure 4 shows what happens if the D.C. amplifier gain is further increased. There are again two ^{3}H curves, but now the low temperature curve is to the left of the high temperature curve. The cold and the warm ^{14}C samples are still on the same curve, but the cold efficiencies are now slightly lower than the warm efficiencies. This clearly indicates an over-compensation. Similar results were obtained with external standardisation. Different pairs of photomultipliers generally required a different setting of the D.C. amplifier gain.

The fact that ^{3}H curves are sensitive to temperature changes (if no compensation is applied), whereas ^{14}C curves are not, can be explained as follows (see Fig. 5). In the absence of compensation, an increase in electron multiplier gain will shift the pulse height spectrum to the right. Consequently, both the counting efficiency and the channels ratio increase. It so happens that these two changes together keep the calibration points on the same curve in the case of ^{14}C.

For ^{3}H we obtain a different picture. Here an increase in electron multiplier gain will not alter the ^{3}H efficiency as the lower level of the ^{3}H window is practically at the coincidence threshold. The counts above the ratio level, however, increase considerably, causing the calibration curve to move to the right. It is this ratio change, which is not accompanied by a corresponding increase in efficiency, that is obviated by compensating for the gain temperature coefficient of the photomultiplier tubes.

CONCLUSION

Temperature changes affect the calibration of liquid scintillation counting systems, in particular for ^{3}H measurements. By compensating for the temperature coefficient of the gain of the photomultiplier tubes the validity of the calibration can be extended to a wide temperature range. Temperature compensation improves the accuracy of ambient temperature systems. For temperature-controlled counting systems it offers the advantage that calibration curves remain valid, if the temperature of the system is changed.

REFERENCES

1. J. B. Birks, *The Theory and Practice of Scintillation Counting*, Pergamon, London, 1964.
2. L. Cathey, *IRE Trans. Nucl. Sci.* **NS5** (3), 109 (1958).
3. R. B. Murray and J. J. Manning, *IRE Trans. Nucl. Sci.* **NS7** (2/3), 80 (1960).

DISCUSSION

A. Dyer: My question is not related to temperature effects, but to the way you select channels to obtain a channels ratio. Would it not be better to use counts in a lower ratio window divided by the counts in the full window rather than the counts in a high ratio window?

It has been argued that this gives better statistical accuracy when determining the efficiency of strongly quenched samples.

F. E. L. ten Haaf: I do not agree with that. The statistical accuracy with which the counting efficiency of a given sample is determined is in each case the same.

B. H. Laney: The statistical uncertainty in the disintegration rate (disintegrations min^{-1}) computed from the sample channels ratio (SCR) of quenched samples may be reduced by placing the middle (ratio) discriminator well below the level where it divides the unquenched spectrum into equal count rates.

In the following example, channel A is the ratio channel and channel B is the normal counting channel for ^{14}C. Channel A is completely contained within channel B.

The normalised statistical uncertainty in efficiency is computed from the SCR, its statistical uncertainty, the slope of the quench curve and the value of efficiency. It is normalised with respect to 10,000 counts in channel B. Therefore, an uncertainty value of 1 σ means that the percent statistical uncertainty in the efficiency computed from the SCR is equal to that for the count rate; 1%.

Even though channel A has less than 10% of the total counts from the least quenched sample, the statistical uncertainty in the efficiency determined from the SCR is less than one-quarter of that from the counts in channel B. Therefore, the efficiency determination contributes little uncertainty in the disintegrations min^{-1} computed from the SCR. The percent uncertainty in the efficiency is less than the percent uncertainty in the SCR from which it is derived, because the slope of the quench curve is small at that point.

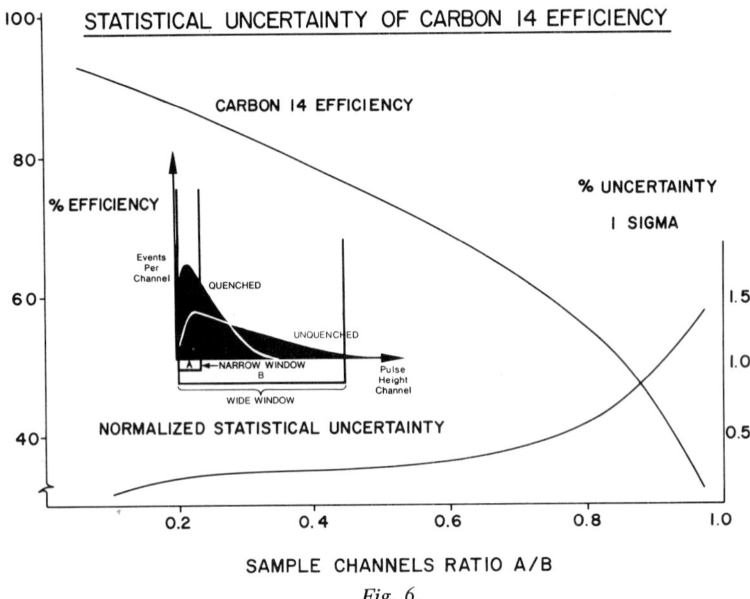

Fig. 6.

***F. E. L. ten Haaf:** There appears to be some confusion as to what should be understood by the terms 'high window' and 'low window'.

I quite agree with Dr. Laney's remark that a relatively low setting of the ratio level improves the statistical accuracy in the efficiency determination of quenched samples.

However, assuming that the ratio level is properly set (see Fig. 6), it divides the counting window B into a lower portion A (low window) and an upper portion, which we may call C (high window).

To obtain a ratio we may either divide the counts in A by the counts in B, or the counts in C by the counts in B. My point is, that it does not matter which one we take. Although the calibration curves are different, depending on whether we use one or the other ratio, the statistical accuracy in the efficiency determination of a given sample is in each case the same.

* Note added in proof.

Subject Index

Index to Contributors

Numbers in **bold type** are principal author pages, as opposed to discussion.